GEOMORPHOLOGY TEXTS
General Editor: K. M. CLAYTON, University of East Anglia

2

WEATHERING

This, the second study in the *Geomorphology Texts* series, considers the highly important process of weathering, the breakdown and alteration of material near the earth's surface into products that are more in equilibrium with the newly imposed physico-chemical conditions. 'Weathering' is thus not simply confined to the reduction of minerals to fundamental particles: the formation of new minerals in a weathering profile is accepted as part of the process.

The processes at work, the material operated upon (rocks, minerals and clay minerals) and the products of weathering, including soil profiles, weathering profiles and some landforms are all considered in this account together with the time factor which is treated on the geological scale.

An important feature of this study is that its scope is international rather than national, the illustrations being taken from the most appropriate example rather than from any specific country.

Not only geomorphologists and geographers but soil scientists, agriculturalists, civil engineers and architects will find this a valuable addition to their bookshelves.

C. D. Ollier is Head of the Department of Earth Sciences at the University of Papua and New Guinea.

OTHER TITLE IN THIS SERIES

1 *Geomorphology* F. MACHATSCHEK 1969

CLIFF OLLIER

Head of the Department of Earth Sciences
University of Papua and New Guinea

WEATHERING

Edited by K. M. Clayton

OLIVER & BOYD · EDINBURGH

OLIVER & BOYD LTD
Tweeddale Court Edinburgh 1

First published 1969

05 001795 0

Printed in Great Britain by
T. and A. Constable Ltd., Edinburgh

ACKNOWLEDGEMENTS

I AM most grateful to the many people who have helped in the preparation of this book by providing examples or illustrations or by discussing various aspects of weathering.

In particular I wish to thank D. T. Currey and E. B. Joyce, who read and helpfully criticised early drafts of most of the chapters; A. L. Walker and D. P. Drover who made helpful comments on chemical aspects of weathering; the Commonwealth Bureau of Soils, Rothamsted, and the Bureau Interafricain des Sols, Paris, who gave much assistance in bibliographic search; and the staff of the Geology Department, Melbourne University, who provided a great deal of help and encouragement. I would also like to thank Miss Joan Oldfield for redrawing many of the illustrations.

I would like to thank the following for permission to reproduce illustrations:

Dr M. M. Sweeting for Fig. 18 from *Essays in Geomorphology*, ed. G. H. Dury (Methuen and Co.), 1966. The Cave Research Group of Great Britain for Fig. 19 from their *Newsletter*, **100**, 1966. G. W. Moore for Fig. 43 from 'Guide Book to Carlsbad Caverns National Park'. *National Speleological Society, Guide Book Series*, **1**. J. N. Jennings for Fig. 44 from 'Geomorphology of Punchbowl and Signature Caves, Wee Jasper, N.S.W.', *Helictite*, **2**, 1966. L. Peltier for Figs 56-59, *Annals Association American Geographers*, **40**, 1950. Oliver and Boyd for Figs 60 and 61 from *Principles of Lithogenesis*, Volume 1, by N. M. Strakhov, 1967. L. Berry and B. P. Ruxton for Figs 64 and 65 from *Bulletin Geological Society America*, **68**, 1967. M. F. Thomas and the Council of the Institute of British Geographers for Figs 66 and 67 from *Trans. Institute of British Geographers*, **40**, 1966. M. F. Thomas and Gebrüder Borntraeger for Figs 118 and 122 from *Zeitschrift für Geomorphologie*, **9**, 1965. Brian T. Bunting for Fig. 73 from *Geography of Soil* (Hutchinson), 1965. Soil Conservation Service, United States Department of Agriculture for Fig. 75. B. W. Avery and A. J. Thomasson for Fig. 85 from *Trans. Hertfordshire Nat. Hist. Soc.*, **25**, 1963. D. Mackney and Clarendon Press, Oxford, for Fig. 86 from *Journal Soil Science*, **12**, 1961. S. A. Radwanski and Clarendon Press, Oxford, for Fig. 88 from *Journal Soil Science*, **10**, 1959. B. E. Butler and Australian National University Press for Figs 89 and 90 from *Landform Studies from Australia and New Guinea*, ed. J. N. Jennings and J. A. Mabbutt. T. M. Thomas and the Council of the Institute of British Geographers for Fig. 96 from *Trans. Inst. British Geographers*, **33**, 1963. D. L. Linton for Fig. 110 from *Geographical Journal*, **121**, 1955. J. D. McCraw for Figs 111 and 112 from *New Zealand Geographical Society, Misc. Ser.*, **5**, 1965. E. Ackermann and Gebrüder Borntraeger for Figs 113-115 from *Zeitschrift für Gemorphologie*, **6**, 1962. National Mapping Authority of Australia for Fig. 120 from mosaic photograph 792, zone 7. G. W. Moore and G. Nicholas for Fig. 125 from *Speleology: the study of caves* (D. C. Heath and Co.). B. Gèze for Fig. 130 from *La Spéléologie Scientifique* (Editions du Seuil, Paris) 1965. A. Wood and The Geologists' Association for Fig. 133 from *Proc. Geol. Assoc. London*, **53**, 1942. L. C. King for Fig. 134 from *Transactions Edinburgh Geological Society*, **17**, 1957. B. W. Sparks for Fig. 135 from *Geomorphology* (Longmans, Green and Co., London)

1960. C. R. Twidale and Australian National University Press for Fig. 136 from *Landform Studies from Australia and New Guinea*, ed. J. N. Jennings and J. A. Mabbutt. R. P. Moss and The Clarendon Press, Oxford, for Fig. 137 from *Journal Soil Science*, **16**, 1965. E. C. F. Bird and Australian National University Press for Fig. 145 from *Coastal landforms*, Canberra, 1964. E. P. Hodgkin and Gebrüder Borntraeger for Fig. 148 from *Zeitschrift für Gemorphologie*, **8**, 1964. M. Schwarzbach for Fig. 151 from *Climates of the past: an introduction to paleoclimatology*, Van Nostrand, London, 1963. E. Dorf for Fig. 152 from *Problems of palaeoclimatology*, ed. A. E. M. Nairn (Interscience, New York), 1964. L. R. Kittleman Jr. for Fig. 160 from *Journal Sedimentary Petrology*, **34**, 1964. I. Stephen and The Clarendon Press, Oxford, for Figs 160 and 165 from *Journal Soil Science*, **3**, 1962, **4**, 1953, respectively. R. V. Ruhe for Fig. 167 from *Soil Science*, **99**, 1659. P. Reiche for Figs 171 and 175 from *Journal Sedimentary Petrology*, **13**, 1943, and Figs 172 and 176 from *New Mexico University Publications in Geology*, **3**, 1950.

CONTENTS

I INTRODUCTION

WEATHERING is the breakdown and alteration of materials near the earth's surface to products that are more in equilibrium with newly imposed physico-chemical conditions. Many rocks were originally formed at high temperature, high pressure and in the absence of air and water, and a large part of weathering is, in fact, a response to low temperatures, low pressures, and the presence of air and water. Many kinds of alteration are possible, however, and even a material that is an end-product of weathering under one set of conditions may become the raw material of a weathering process if conditions change.

Before pursuing the subject further, it is worth looking at some proposed definitions of 'weathering'.

Perhaps the most widely accepted definition is that of Reiche (1950): 'Weathering is the response of materials which were in equilibrium within the lithosphere to conditions at or near its contact with the atmosphere, the hydrosphere, and perhaps still more importantly, the biosphere'. In my opinion the stress on the biosphere is not warranted, for biological weathering operates through chemical and physical effects and is not fundamentally different. There is no reason to assume that the starting products of weathering were ever in equilibrium with the lithosphere—some rocks may never have attained perfect equilibrium. For weathering to operate it is only necessary for the starting products to be less in equilibrium with surface conditions than potential weathering products. Keller (1957) is also of the opinion that Reiche's definition could be improved by deleting the words 'which were in equilibrium', for he says rocks are in equilibrium only momentarily, while the environment in which they were formed persists. The response of materials may lag behind changes in environment.

Weathering has been described by Polynov (1937) as 'the change of rocks from the massive to the clastic state'. This covers only a part of weathering according to the view taken in this book, and examples will be described of extensive chemical weathering with no breakdown into fragments.

The acceptance of chemical alteration *in situ* as a kind of weathering leads to another difficulty which is the distinction between diagenesis and weathering.

Diagenesis is the alteration of sediments by building up new minerals. As a sediment becomes more deeply buried, the pressure increases, the mineral grains are compacted, and water and air are excluded. In a way it is the reverse of weathering. However, weathering is not simply the reduction of minerals to fundamental particles: the formation of new minerals in a weathering profile is accepted as a part of weathering. Diagenesis is restricted to sedimentary environments.

Weathering makes a convenient place to enter the geological cycle. A rock weathers, and the weathering products may be removed either mechanically or in solution. The wearing away of rocks is termed erosion, and the movement of materials is called transport. Weathering and erosion together make up the process of denudation. Transported sediments are eventually deposited, usually in the sea. Here they accumulate, compact, experience mineral diagenesis, and eventually form sedimentary rocks. Earth movements may later lift such rocks above sea-level, and the cycle can go round again. There are a number of processes that can be regarded as weathering beneath the sea—what Reiche (1950) calls 'halmyrolysis'—but essentially weathering is a phenomenon of the land, and diagenesis is dominant in the sea.

In this study of weathering we shall consider the processes at work, the materials operated upon, and the products of weathering. The processes are various kinds of chemical and physical effect, sometimes acting through biological agents. Some environmental conditions—hydrology and climate—have a marked effect on these processes and will be considered in detail. The materials are rocks and minerals, including clay minerals which are also frequently a weathering product. The weathering products include weathering profiles, soil profiles and some landforms. The time factor in weathering will be examined; it is found that much weathering takes place so slowly that it must be considered on a geological time scale, and some aspects of weathering through geological time will be discussed.

At first sight weathering seems a hopelessly complicated subject, with a multitude of processes operating on an endless range of rocks and minerals under a great variety of climatic and hydrological conditions. Before embarking on a detailed study it may be as well to reduce this complicated picture to some conceptual simplicity.

The zone of weathering involves only half a dozen common rocks, made up of only half a dozen main mineral groups, and composed of only eight main chemical elements.

The earth's crust comprises 95 % igneous rock and only 5 % sedimentary and metamorphic rocks, but if the area of rock exposed at the surface (and thus to weathering) is considered, then 75 % of the earth's land surface is underlain by sedimentary rock. In order of area covered the most important rocks are shale, sandstone, granite, limestone and basalt. A more detailed table is given on p. 74.

The approximate proportions of mineral species exposed to weathering at the earth's surface are shown in Table 1.

Table 1 (from Leopold, Wolman and Miller, 1964)

Mineral	%
Feldspar	30%
Quartz	28%
Clay minerals and mica	18%
Calcite and dolomite	9%
Iron oxides	4%
Pyroxene and amphibole	1%
Others	10%

Table 2 shows the commoner elements of the earth's crust.

Table 2 (from Mason, 1966)

	Weight %	Volume %
O	46·60	93·77
Si	27·72	0·86
Al	8·13	0·47
Fe	5·00	0·43
Mg	2·09	0·29
Ca	3·63	1·03
Na	2·83	1·32
K	2·59	1·83

It is apparent that these eight elements account for almost the entire earth's crust on a weight basis, and on a volume basis oxygen is overwhelmingly dominant. The lithosphere is an 'oxysphere', and the earth's crust is made of oxygen anions packed together and bonded by interstitial silicon and metal cations.

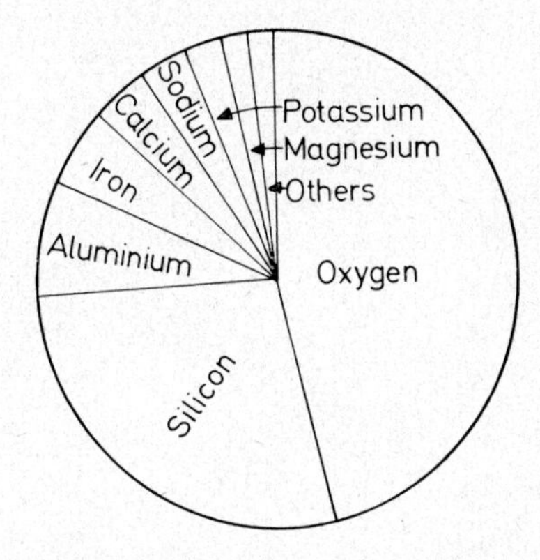

FIG. 1. Diagram showing the relative amounts of the most abundant elements in the Earth's crust.

Hydrogen and carbon have to be added to this list in assessing weathering products. According to Mackenzie and Garrels (1966), more than 99% of the dissolved material in the ocean, and more than 99% of the dissolved solids brought in by rivers, is accounted for by the constituents Na, Mg, Ca, K, Cl, SO_4, HCO_3, and SiO_2. Ocean water has a much higher concentration of all major constituents except silica, which is higher in river water.

Previous books on weathering, such as Reiche (1950), Keller (1957) and Marshall (1964), have been particularly strong on the chemistry of weathering, and I have not attempted their kind of treatment in this book, even though most advances in our knowledge of weathering in recent years have come from laboratory studies. This book is intended mainly for geomorphologists, and the emphasis is on weathering in the landscape. Nevertheless, a short treatment of chemical and physical aspects of weathering is presented here, and I have tried to make this part intelligible to those students of natural science who are not strong in physics and chemistry. Similarly mineralogy, petrology and hydrology are presented at a level which does not presuppose more than a slight knowledge of these subjects.

II PHYSICAL WEATHERING

Physical weathering is the breakdown of material by entirely mechanical methods brought about by a variety of causes. Some of the forces originate within the rock, while others are applied externally. The applied stresses lead to strain, and eventually to rupture.

SHEETING, UNLOADING AND SPALLING

Sheeting is the division of rock into 'sheets' or 'beds' by joint-like fractures that are generally parallel to the ground surface. The name 'topographic jointing' has been used for the phenomenon, and well describes the main feature. The sheets of rock are parallel or concentric, and have various relations with the ground surface in detail, but a near parallelism is very evident. This suggests that the fractures are directly related to the ground surface, and the generally accepted idea, first proposed by Gilbert (1904), is that expansion of rock masses when their confining pressures are reduced by uplift and erosion finds relief in the development of cracks. This pressure-release by erosion of a superincumbent load is seen mainly in granite, but also in massive sandstone (Bradley, 1963), massive arkose and conglomerate (Ollier and Tuddenham, 1962), bedded sandstone (Currey, 1968) and limestone (Kiersch and Asce, 1964).

Release from original stress may be mainly vertical, in which case the most apparent unloading fractures are sub-horizontal. At Mt Buffalo, Victoria, for example, the number of horizontal cracks increases towards the top of exposures, but the vertical cracks are not affected as much (Fig. 2). This unloading may be 'fossil', related as it is to the surface of the plateau top and not to the present overall slope shape.

Matthes (1930) studied exfoliation in the Yosemite valley and concluded that the formation of exfoliation sheets is a very slow process, and that many granite domes are of very great age.

Chapman and Rioux (1958) noticed that the well-developed sheeting in the Acadia National Park, Maine, showed numerous departures from parallelism with the present surface. The area is modified by valley glaciation, and the sheeting was in fact parallel to the pre-glacial land surface.

In contrast to the observations of Chapman and Rioux is the description by Lewis (1954) of pressure release cracks in gneiss bedrock behind a cirque glacier, parallel to the surface. Lewis thought the gneiss had been under the pressure of several thousand metres of overlying rock, which had been removed by glacial action. The strength of the gneiss was sufficient to resist 'bursting' until the final release of all the superincumbent load with the final retreat of the glacier.

The relation between sheeting and glacial landforms is displayed very clearly on Sermersoq, Greenland (Oen, 1965). In cirque walls sheeting occurs along steep planes characteristically concave to the sky. In peaks and horns it is nearly vertical. On cirque floors and in broad valleys the sheeting is sub-horizontal. On gently sloping, rounded hills and mountains the sheet structures show dome-shaped patterns. Where

FIG. 2. Unloading at Mount Buffalo, Victoria, Australia. There is an increase in the frequency of horizontal partings towards the surface.

these hills have been affected by later cirque erosion a typical intersecting of concave and convex sheeting is found.

Another example of intersecting unloading planes has been described from the Vaiont valley, scene of the world's greatest dam disaster (Kiersch and Asce, 1964). The topography is shown in Fig. 3. A broad, round-bottomed valley was exposed upon retreat of a valley glacier, not more than 18,000 years ago. The Vaiont river then cut a gorge-like valley, 650 to 1150 feet below the floor of the glacial valley. An old set of unloading planes (called 'rebound' joints by Kiersch and Asce) are parallel to the glacial valley floor, and a new set are parallel to the canyon wall. The rock is especially weak at the shoulder of the gorge where the two sets intersect. The rock in this instance is limestone, and during construction of the Vaiont dam there was a lot of 'slabbing' and 'rock burst'. The 'destressed zone' may be up to 500 ft thick according to Kiersch and Asce.

Bradley (1963) has described unloading fractures to a depth of 30 ft in the sandstones of the Colorado Plateau, U.S.A. River-cut valleys cross the plateau, and unloading sheets are parallel to the valley walls. Meandering of the valleys has sometimes resulted in new valleys oblique to the initial fractures, and new cracks develop parallel to the new valley walls. The fractures are discordant to older structural planes, but sub-parallel to topography, and they occur in all possible locations including permanent

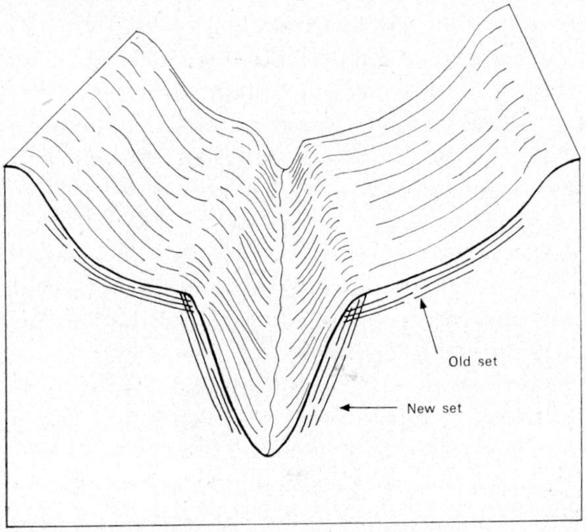

FIG. 3. Topography and unloading in limestone at Vaiont, Italy.

shade. Chemical weathering and scaling take place on the fracture surfaces, and have quite different characteristics from the major cracks. There is no possibility that they could be due to chemical weathering, and the evidence for unloading is very great, if not overwhelming.

A number of instances have been reported that illustrate the sudden appearance of unloading fractures after release of confining pressure.

A glacier burst occurred at the Franz Josef Glacier, New Zealand, in 1965, and Gage (1966) reported remarkable effects on the exposed glacier floor: 'Joints in the hard schist forming ridges across the uneven valley bottom near the snout were found to be newly opened into long cracks up to 1 inch wide. Some joint-bounded slabs and blocks of schist had sprung upwards, displacing the freshly ice-worn rock surface by as much as 2 ft. Bruises and gouged-out hollows the size of an armchair attest to the violent blows received from the large blocks in transit across the surface, and justify speculation that the rock having emerged only a few years previously from beneath hundreds of feet of ice had been still in decompression stress until relief was triggered by the hammering it experienced in the flood.'

Bain (1931) describes spontaneous mechanical rock expansion in limestone quarries in Vermont and Tennessee, U.S.A. As new faces are cut in the walls of the quarries the dense limestone expands, producing cracks parallel to the surface. In some instances quarries had to be closed down because of the danger of flying rock-sheets or spalls. The 'rockbursts' were most common on new faces. A more convincing demonstration of the reality of unloading would be hard to find.

Labasse (1965) has described how rockbursts occur with explosive violence in coal, potash and iron mines. Men and equipment are hit by fragments, projected at very high speed, and may be buried by debris. If the floor has little resistance, there may be sudden heaving which presses men and equipment against the roof.

Unloading or sheeting has been reported almost exclusively from massive, poorly jointed rock. One might expect, however, that all rocks would experience the same pressure release when a superincumbent load is removed. The lack of reported instances may be due to (*i*) a real absence of unloading fractures, the expansion being taken up by already existing planes of weakness, or (*ii*) a failure in observation, as it would be much harder to recognise unloading planes amidst a mass of other planes. Only very detailed investigation of the sort used for expensive construction and installation is likely to yield such evidence, and one example comes from a proposed tunnel site at Bellfield Dam, Victoria, Australia. The rock is well bedded and jointed Carboniferous sandstone. Nevertheless 'topographic jointing' was evident, and in fact proved to be so important that the proposed tunnel line had to be re-located and special measures taken in foundation treatment.

An experimental approach to the behaviour of rock on decompression supports the field evidence for the efficiency of unloading. Griggs (1936) subjected blocks of limestone to compression, and when the pressure was removed the blocks developed release fractures, crack planes at right angles to the direction of compression.

Unfortunately, deformation of limestone in laboratory conditions cannot be used directly as evidence for unloading, as limestone has high creep value, and a low ultimate tensile strength. This means that to some extent limestone can 'flow', as is shown by bending of limestone mantelpieces and tombstones under the action of gravity alone. Creep-flow at quite low stress is not known in other common rocks. There is a threshold value of stress in relation to flowage by creep. This is evident in glaciers, where the lower part of a glacier flows, but the upper part, with less load and less stress, is broken up by countless crevasses.

The cause of unloading may now be considered at a more fundamental level. Experiments show that as a rock is placed under increasingly great amounts of confining pressure, the differential stress required to produce the same amount of strain is greater. In other words, the ultimate strength of the rock is increased when confining pressure is increased. Thus the deeper a rock is buried beneath overlying material (rock or ice) the greater becomes its ultimate strength. Conversely, the ultimate strength of granite is decreased by lowering the confining pressure. Thus removal of surface material by erosion lowers the confining pressure, and hence the ultimate strength decreases. The internal stress then set up is relieved by expansion, namely elastic expansion. As the rock is confined laterally it can expand only upwards. Expansion of rock leads eventually to rupture, and a series of cracks is formed parallel to the surface.

Dapples (1959) has found the depth of unloading cracks, *d*, varies approximately as $t^{1 \cdot 5}$, where *t* is the thickness of the sheet. This relationship is approximately true to a limiting depth of 80 ft, below which *t* becomes so great that unloading can no longer be recognised.

The rock mechanics of unloading are further discussed by Labasse (1965).

SHEETING IN TECTONICALLY ACTIVE AREAS

Moye (1960) is of the opinion that many areas are much more tectonically active than previously thought. In the Snowy Mountains area of Australia there are high horizontal compressive stresses in the crust that cannot be ascribed to unloading of the surface.

It is possible that these forces could cause near horizontal fractures, and such tectonic tension cracks would give rise to sheeting similar to unloading. The process would be accentuated by stress concentrations at the surface due to changing topography.

However, it is apparent that tectonic sheeting would not closely parallel the surface as unloading does so often. One would also expect tension cracks due to tectonic forces to be developed in a regional scale and have regional trends, unlike the locally developed and directionally variable sheeting due to unloading.

Twidale (1964) believes that sheeting develops after cooling and solidification of granite. It has often been observed that sheeting is *not* parallel to structures in granite, but presumably the isotherms of cooling would not be parallel to structural or petrofabric planes. Twidale believes that granite was emplaced under radial compression, and when exposed the release of this compression gives rise to expansive arching, which he thinks offers the most comprehensive explanation for sheeting. This is open to question, however, for the mode of emplacement is very controversial and there is by no means any assurance that granite masses are emplaced under radial compression. 'Granitisation' granites appear to be formed by 'soaking' of other rocks until they are converted to granite, and the so-called 'permissive' granites appear to fill gaps rather than be forced into place against the forces of compression.

Yet another possibility is suggested by Oen (1965), who studied sheeting in post-tectonic granite massifs. Due to the mass deficiency represented by these massifs, they tend to rise, and the upper portions are under axial compression. When regional uplift ceases, a relative decompression enables rocks near the surface to exfoliate spontaneously by dilation in a direction normal to the free surface or topography.

The chief and obvious deficiency of this hypothesis is that it cannot account for identical sheeting on different rocks and in different tectonic settings.

SPALLING

Sheets, such as unloading sheets, are long in two dimensions in comparison to their thickness. There are some situations, as in cave or tunnels, where, although physical conditions favour cracking, spatial considerations prevent the formation of sheets. The process is called spalling, and produces spalls that are platy rock fragments, lozenge-shaped or irregular. They break from the wall by a combination of tension cracks parallel to the unloaded surface and shear fractures due to compression acting parallel to the wall.

The back surface of spalls in tunnels is made of a number of irregular or semi-radial cracks. These result from unloading producing stress differences causing a bi-axial bending stress on the rock (bending stress in two directions).

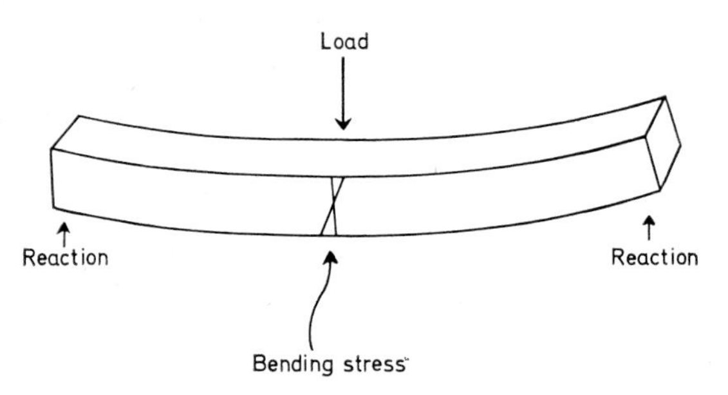

FIG. 4. Bending stress in a beam—uniaxial stress.

In the case of a beam, as shown in Fig. 4, bending stress is uniaxial, that is in one direction.

If a cross is loaded as shown (Fig. 5), bending stress at the centre is in two directions, that is biaxial.

By analogy of the cross with a loaded plate, it can be seen that biaxial bending stresses exist in a wall of rock bulging outwards into a cavity, or into the open air.

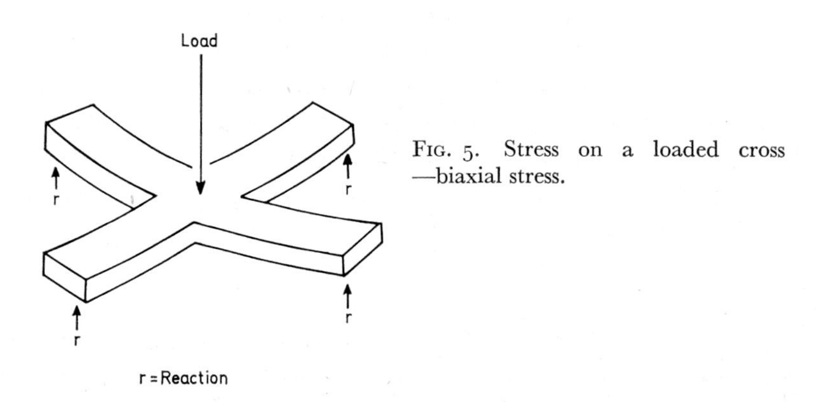

FIG. 5. Stress on a loaded cross —biaxial stress.

A system of radial fractures is produced as shown (Fig. 6). The irregular surface of spalls is different from the smooth face often developed on large unloading sheets, even in markedly heterogeneous rock such as the conglomerate described by Ollier and Tuddenham (1962).

CRYSTAL GROWTH

Volume changes due to crystal growth set up stresses within rocks that lead to breakdown, and thus constitute a part of physical weathering. The volume changes are due to freezing of water to form ice, to crystal growth from solution (salt weathering) or chemical alteration of pre-existing minerals.

(*a*) FROST WEATHERING

Water expands by about 9% on freezing at 0°C. The great change in volume has a potentially great disruptive effect, and frost shattering is one of the greatest, if not the greatest, mechanical agent in weathering. If water were to freeze in a confined space,

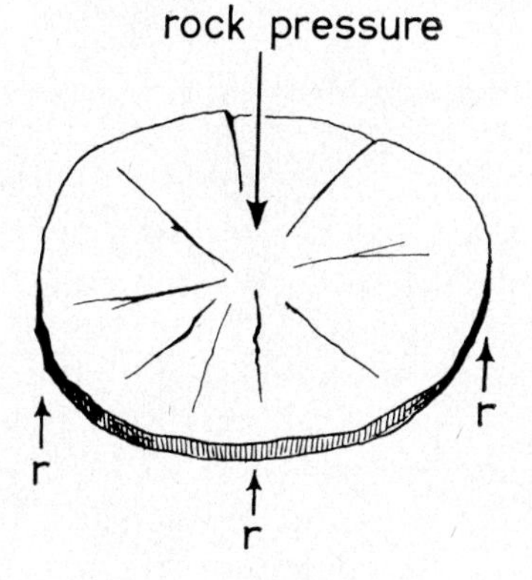

FIG. 6. Radial fractures on a rock spall.

very great pressures would be exerted on the enclosing rock. A theoretical pressure of several thousands of pounds per square inch is possible at lower temperatures (30,000 lb/in^2 at —22°C) which is far greater than the compressive strength of rock.

However, real freezing is never totally confined; for water to get into a crack in the first place it must be open. Then when the water freezes the ice can be extruded, and only part of the expansion is utilised in disrupting the enclosing rock. Nevertheless, at times it seems that ice behaves almost as if in a closed system. This may be due to super-cooling of water to as low as —5°C before freezing occurs very rapidly. The upper surface may suddenly freeze, thus closing the system.

Frost action acts directly and indirectly. The formation of ice can itself prise rock fragments apart. This works along planes of fissility in rocks, and produces angular rock debris. Another form of direct ice action is by heaving or wedging due to continued growth of ice masses. Frost wedges are commonly about 10 ft deep and 2 or 3 ft wide at the top, tapering to a point at the bottom. Frost wedging in north-western Canada in flat exposures of crystalline rocks, with foliation planes and joints, has been observed to lift blocks 25 ft across through a vertical distance of 12 ft (Yardley, 1951).

In unconsolidated sediments frost heaving takes place, which mixes material but does not break down small particles very much. However, a similar process may affect permeable rocks, in which case frost heaving would be a significant factor in the mechanical breakdown of the rock. Heaving is due to the movement of water through capillaries from unfrozen ground to ice nuclei, where it freezes to more ice, draws in more water, and so on. Thus irregular patches of ice grow in subsoils around original nuclei. The high tensile strength of water aids this process, as also the fact that it is apparently hard to freeze water in capillaries.

In soils subject to repeated freezing, the distribution of water in various horizons affects their physical weathering. St Arnaud and Whiteside (1963) examined soils whose particle size distribution and mineral distribution indicate changes within the profile due to physical breakdown of larger rock and mineral fragments, mainly by frost. Those soil horizons that held most water were attacked the most. Care is needed in such interpretations, for the horizons with the finer particles will tend to hold most water. With more severe freezing thick layers of ice may form giving rise to ground swelling. Muller (1945) has provided a table of swelling tendency of material showing this trend: ground swelling is absent in boulder and pebble deposits, minor in gravel and sand, and important in finer sediments and peat.

When there is alternate freezing and thawing, movement of rock particles occurs as an indirect effect of frost action. After a thaw the fragments may settle in new positions, and small particles may fall into cracks, preventing them from closing back to their original position. With repeated freezing and thawing it would be expected that there would be much more chance for rock disintegration than with steady low temperatures, but the actual relationship can be quite complex.

In coastal areas normal frost weathering can attack cliffs in cold regions, but there is also a special effect due to an ice-foot. This is a ledge of ice, attached to the base of a cliff, a foot or so above high-tide level, and which is not carried away at low tide. The ice is almost as hard as freshwater ice, having additions from snow and meltwater, and undergoes many freeze-thaw cycles due to temperature changes and also possibly by periodic splashing with salt water. This causes shattering of rock just above tide level, and the resultant debris is easily removed by storm waves. Nansen (1922) considered that ice-foot weathering was partly responsible for the formation of the Strandflat (see p. 239).

(*b*) SALT WEATHERING

The growth of salt crystals from solution can in some circumstances cause disaggregation of rocks. The disruptive effect is rather like that caused by frost, though crystallisation from solution is quite different from simple solidification of a liquid.

There is no problem about the formation of salt, which is due to evaporation, but it is rather hard to explain why crystal growth should continue against the pressure of the confining rock—the growing crystal actually exerts a force sufficient to prise apart grains. A possible mechanism has been described in some detail by Wellman and Wilson (1965).

In soils or porous rocks there is movement of solutions to areas of crystallisation. If there is sufficient salt, crusts or layers of salt may be formed at the ground surface, as in solonchak soils. Such efflorescence of salt does little, if any, actual weathering of mineral fragments.

In porous rocks, however, the crystallisation of salts within the pores at the surface of the rock may lead to granular disintegration, or possibly to some sort of exfoliation. There are no distinctive landforms that are definitely known to be caused by salt weathering, but various small-scale features have been attributed to this process. For example, the limestone of the Ma'aza Limestone Plateau in Upper Egypt contains sodium chloride, and it flakes and crumbles to such an extent that cliff faces are unclimbable. Wellman and Wilson (1965) attribute the scaling and cave formation at Ayers Rock, Central Australia, to salt weathering. The same authors even suggest that the actual shape of the inselberg may be due to salt weathering, though this seems to be an exaggerated claim.

The effects of salt weathering have been observed most carefully in building materials. Lucas (quoted in Fox, 1935) reported flaking produced in building materials of Cairo, and says: 'In the case of walls that have been plastered, the plaster is frequently forced bodily away from the wall, and in between the wall and the plaster a sheet of almost pure sodium chloride, sometimes one or two millimetres in thickness, was found. In other cases small cavities in the mortar or in the stone were filled with a powdery mass of crystals of almost pure sodium chloride.' The temples at Luxor and Karnak, Egypt, have been severely damaged by alternating solution and crystallisation of salt (Boute, 1963). In contrast, poorly soluble minerals such as manganese and iron oxides have formed a protective crust on granite blocks by the water's edge of the Nile. In the Houses of Parliament, London, built of Magnesian Limestone, it was reported that $MgSO_4$, which is readily soluble and generally easily removed, crystallised out in a dry spell with disruptive effect (Fox, 1935).

Though most salt weathering takes place in hot arid areas, it may occur even in Antarctica. Cavernous weathering of granite is widespread in the dry valley region of South Victoria Land, Antarctica. Almost all outcrops and large boulders are pitted by holes up to 6 ft in diameter, with floor and roof rising gradually upward. The holes are caused by granular disintegration due to the growth of salt crystals which prise minerals from the rock. At the base of the holes is a 'rock meal' made up of mineral fragments and salt (Wellman and Wilson, 1965). How the salt is dissolved in the first place in such a cold, dry area is not explained.

Salt weathering may also be important in coastal areas. For an area of periodically flooded, saline high tidal flats in northern Queensland, Coleman, Gagliano and Smith (1966) described and figured gravels of quartz and various rocks splintered into angular fragments. Disintegration is due to wedging by crystal growth, accompanied by some chemical weathering. The fracture surfaces commonly display a dull lustre suggesting rock alteration after or contemporary with fracture formation.

Tricart (1960) described experimental work on salt weathering of granite carried out because some rocks used in jetty construction were destroyed too rapidly. Rock samples were soaked in very salty water, dried, and then lightly brushed to remove weathered material, which was collected and weighed. The operation was repeated numerous times. Rates of weathering and the total amounts weathered were very variable, but the extreme case was a rock that lost 2% of its weight after only 12 cycles. From the brief report of this experiment it is not clear that salt weathering alone was responsible, or if solution or wetting-and-drying weathering could have played some part.

(*c*) CHEMICAL ALTERATION

Some chemical alteration of mineral or rock fragments may cause volume change, the weathered product having a greater volume than the original. The expansion consequent on this alteration gives rise to physical weathering, and in such cases it is difficult to separate the chemical from the physical weathering.

The volume increase usually gives rise to exfoliation, whereby thin shells of altered material peel from the parent rock, which may in turn alter chemically, peel off, expose a fresh surface and so on. Several parallel or concentric flakes may thus be formed.

In a rock made up of several minerals, further breakdown may occur because some minerals expand more than others, and the differential expansion gives rise to granular breakdown rather than flaking.

Hydration and oxidation seem to be the chief agents of chemical weathering associated with this mineral expansion weathering. In arid areas this phenomenon is often common, and possibly dew is the source of water for the hydrolysis. The flaking can take place equally well in shady and exposed sites, and even on overhangs and in shallow caves that are in permanent shade. It is found on many kinds of rocks, even on pure quartzites.

INSOLATION WEATHERING

Temperature changes cause expansion or shrinkage of rocks.

A rise in temperature causes a rock to expand, and a fall causes it to contract. Repeated temperature changes may cause a rock to break up, and if the heating agent is sunshine the weathering is called insolation weathering.

Rock is a poor conductor of heat, so a thermal gradient is set up between the surface and inside of a rock when it is heated. The surface of the rock therefore expands more than the inside, setting up a stress that might lead eventually to fracture.

Most rocks are made up of different minerals. These have different specific heats and coefficients of expansion, causing them to expand at different rates, and many are anisotropic and may expand more in one direction than another. Dark minerals will absorb heat faster than light ones, which may also give rise to differential expansion, leading to many small stresses within a rock, which may lead to the formation of minute cracks and possibly granular disintegration.

The diurnal range of air temperature may be very large, but that of a rock surface is often greater still. The minimum temperature of rock and air will be about the same, but the rock can have a higher maximum temperature. Hume (1924) reported that in Egypt an excess of 7°F was found on slate, 33°F on flint, and sands and gravels in sheltered places had temperatures exceeding air temperature by 36°F. Several factors affect the amount of excess, such as surface roughness, relation of the surface to wind direction, aspect, and possibly others.

Surface spalling of rocks by thermal stresses has been discussed from the rock mechanics point of view by Gray (1965).

It was once widely believed that thermal expansion and contraction of rocks could cause disintegration, and in arid regions especially the fragmentation of boulders was generally attributed to this process. However, experimental work by Blackwelder (1933) and Griggs (1936) appeared to demonstrate convincingly that the process was, to say the least, overrated.

Blackwelder subjected various rocks to large and sudden changes of temperature. Basalt and granite resisted sudden heating to 200 and 300°C before fracturing, and obsidian withstood sudden chilling of 200°C without fracture.

Griggs set up an apparatus so that he could heat a rock specimen with an electric heater and then cool it with a stream of cold, dry air. The temperature change produced was 110°C, and the rate of heating and cooling was rapid, so that it was possible to subject the specimens to as many heating and cooling cycles as would be achieved in 244 years of diurnal weathering. A polished granite surface had no detectable weathering after '244 years' of this treatment. However, when the rock was cooled with tap water instead of dry air it was found that only '2½ years' of weathering caused loss of polish, surface cracking and the beginnings of exfoliation. It was concluded that chemical weathering is much more potent than insolation weathering, and the impression was gained that insolation was almost completely ineffective.

It should be noted, however, that although Griggs established a temperature gradient in his specimens that was comparable with that due to insolation, his experiment has two important limitations. First, the specimens were unconfined, so that expansion could take place sideways, if not all round, whereas on the surface of a boulder an expanding patch of rock is confined and even compressed by neighbouring patches, and as it can only expand outwards the stress is much more concentrated and liable to lead to fracture. Secondly, the fatigue factor cannot be reproduced in rapid experiments. A rock heated for a few minutes may develop the same stress as a rock heated to the same temperature for half a day, but if the stress is maintained for a longer time it may lead to more permanent strain.

Nevertheless, the experimental work of Blackwelder and Griggs led to an almost complete abandonment of insolation for a number of years. In his review of weathering processes in 1950, Reiche thought the only possible effect of thermal expansion was the formation of 'Kernsprung' or heat-cracked boulders, and these he believed were restricted to dark, fine-grained, volcanic rocks.

There are, however, numerous reports of possible insolation weathering from a variety of rocks that are almost chemically inert, such as flint and quartzite, and from other rocks that are pale, coarse grained and non-volcanic.

Hume (1924) figured flints split by straight, parallel fractures, from Egypt. Bosworth (1922) illustrated quartzite pebbles from the Peruvian desert, some of which showed straight parallel fractures and others a very regular exfoliation. Also from

Peru, Brown (1924) reported pebbles that crack only on their exposed upper surfaces, and parts of the crust curl outwards (Fig. 7). Ollier (1963) described boulders of siliceous duricrust in the Central Australian desert that were cracked open into several

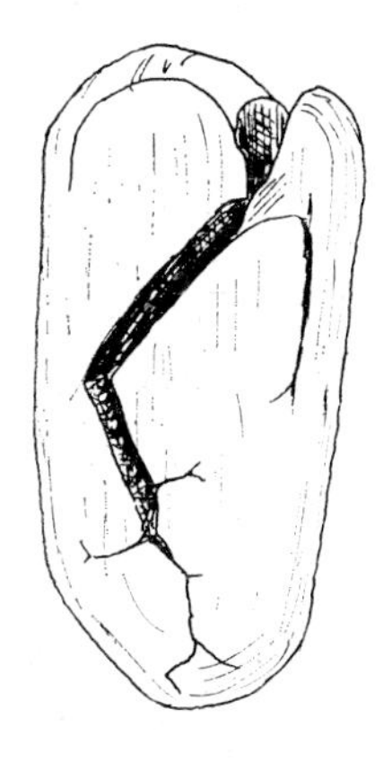

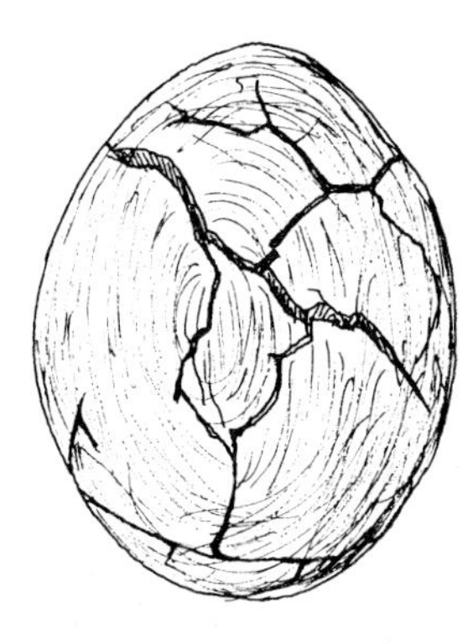

Fig. 7. Sun-cracked pebbles from the desert of Tumbez, Peru (after Brown, 1924).

angular fragments. Another type of weathering attributed to insolation is blocky disintegration, reported from quartzite in Central Australia. This is shown (Fig. 8), illustrating the way the boulder crazes and then breaks into blocks about 4 in across.

Fig. 8. Blocky disintegration of quartzite, Mount Conner, Central Australia.

It is thought that thermal expansion causes crazing of the surface, and at the same time a crack develops parallel to the boulder surface at a depth of a few inches. The two sets of cracks, one parallel to the boulder surface and the other normal to it, bound the individual blocks which eventually fall away. When a layer of blocks has been removed the process can be repeated on the newly exposed core, and so on.

The process takes place in very pure white quartzite, and in less pure quartzite in the same locality there is a tendency for exfoliation rather than blocky disintegration to occur. This suggests that chemical weathering gives rise to exfoliation but physical weathering gives blocky disintegration.

FIG. 9. Boulder cleaving. Mount Olga, Central Australia.

'Boulder cleaving' is the name given to another type of insolation weathering, reported from Mount Olga, Central Australia.

The Mount Olga group comprises a number of inselbergs over a thousand feet high made of a remarkable conglomerate. Boulders and pebbles of granite and basalt from a few inches to about two feet long are embedded in a matrix of fine sandstone that is largely epidotised and indurated and so is as tough as the boulders.

The hills are shaped to some extent at least by unloading (Ollier and Tuddenham, 1961), and the cracks formed by this process cut across most boulders. The surfaces of the mountains are generally smooth, as the boulders have all been cleaved flush with the general surface. A hillside surface presents a cross-section of a conglomerate, and the boulders look like so many circular tiles set in a fine-grained matrix. Nevertheless, occasional boulders project above the general surface, and these provide the evidence of immediate concern. Almost every boulder sticking up above the general surface has developed, or is developing, a crack that is flush with the general surface, so that very often the upper half of a boulder may be lifted up. Fig. 9 shows such a boulder. This process affects both the coarse-grained pink or grey granite and the dark and fine-grained basalt. In rare cases even complex boulders are affected.

A suggested mechanism for boulder cleavage is shown in Fig. 10, starting from the position where one boulder projects above a general surface, left perhaps after unloading.

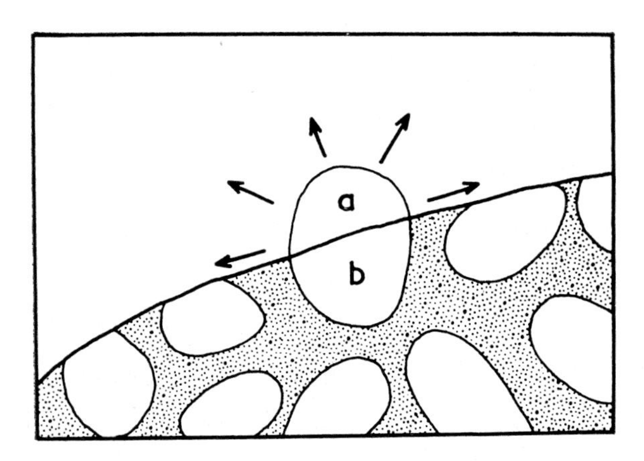

FIG. 10. The mechanism of boulder cleaving.

The upper, unconfined part of the boulder, *a*, is free to expand upon heating, but the confined lower part, *b*, is restricted. The unconfined part is likely to heat up quicker than the confined part, having more surface and being nearer to the top, and so will expand. This will put a stress between the upper and lower parts which will be released eventually by the formation of a crack between the two. The upper half is then liable to fall downhill as the slopes are generally steep, so it is not easy to find examples, and the process is probably more common than is apparent at first sight.

Expansion of the boulders for any reason could give the same sort of result, but the fact that granite, basalt and complex boulders all react in the same way suggests that thermal expansion is the most likely cause.

Another type of weathering phenomenon from the stony desert area of Coober Pedy, South Australia, is believed to be due to insolation weathering, acting rather indirectly, and is termed 'dirt cracking' (Ollier, 1965).

Evidence of dirt cracking is found in boulders with cracks that contain 'dirt' but do not completely split the rock, and also in completely broken boulders with dirt between the several portions. Fig. 11 shows the simplest sort of crack, with the fragments of the boulder moved apart several inches, and dirt of irregular grain size filling the gap. Sometimes the fragments are turned as they move apart, and the original arrangement is not always apparent. Figs 11 and 12 show before-and-after views of a boulder split into about five fragments (not all visible). The vertical viewpoint is rather deceptive and the fragments in Fig. 11 appear to be on the surface; in fact they are about two-thirds buried. After digging and re-assembly (Fig. 12) it is seen that the fragments fit into a simple form. Fragments had 'drifted' up to four inches apart. It is possible that they could move even further, but by this stage mass movement enters the picture, and dirt cracking is virtually inactive.

FIG. 11. Dirt-cracked boulder. Coober Pedy, Australia.

Dirt fragments fall into any open crack and prevent it closing as the rock cools. Further changes in temperature will force the cracks still further apart.

FIRE

Blackwelder pointed out that a good deal of thermal expansion and contraction of rock was due to forest fires. These were plentiful in the forested mountain areas of the semi-arid parts of the United States. Characteristic exfoliation results, and hillsides are sometimes littered with the spalls or flakes of rock that are produced. Many forest fires are started by lightning, and the process is largely natural. Blackwelder observed that spalling was absent in the higher mountains above the timber line, and only rarely seen on lower, brush-covered slopes.

The effects of brush fires were described by Emery (1944), who examined some parts of California soon after a fire had occurred. Spalling was found on rocks, especially quartz diorite, and from the blackening of inner and outer surfaces it appeared that the spalls were detached in a late stage of the fire. There was no evidence of any chemical weathering. The spalls of rock appear to be similar to those described by Blackwelder, though smaller. They are up to $2\frac{1}{2}$ inches at the centre, and taper to sharp edges. The rocks of the area were rounded by the fire spalling, and no rocks split clean across the centre were found.

FIG. 12. The same boulder after re-assembly.

MOISTURE SWELLING

It has been found that considerable changes in volume of rocks can be caused simply by a reversible absorption of moisture, and the volume changes may be sufficient to cause physical weathering.

Investigations of the shrinkage and swelling of a dense grey flint and a dense basalt were reported by Nepper-Christensen (1965). Specimens were first moisture saturated, and then brought to equilibrium in an atmosphere of 65% relative humidity at 20°C. The specimens were measured in both states, and the shrinkage expressed as percentage of the length. Shrinkage figures were:

Basalt 0·015% to 0·020% linear
Flint 0·006% linear

The mechanism of swelling is not clear, but it was found that the process is most pronounced in the region of 90 to 100% relative humidity.

Similar results were obtained by Nishioka and Harada (1958), who report linear expansion as a result of moisture absorption in 22 rocks, including sandstone, shale, limestone, pumice and granite.

The figures obtained for volume change by moisture swelling are of the same order of magnitude as those due to thermal expansion, and possibly even higher in some cases. This little known and little understood process may therefore be quite an important agent in physical weathering and some features attributed to other mechanisms, such as insolation weathering, may in fact be due to moisture swelling.

WETTING AND DRYING WEATHERING

Alternate wetting and drying of rocks can be a very important weathering process, though it has been little studied and is not even graced with a simple name as yet.

Some interesting experiments on this effect were carried out by M. A. Condon. These are reported in the files of the State Rivers and Water Supply Commission of Victoria, but are not published.

Samples of Silurian sandstone, siltstone, shale and mudstone were subjected to cycles of alternative wetting and drying, being first immersed in water for a day, and then left to dry in the open air (protected by a canvas fly sheet) for a day. In later experiments cycles of two days and even longer periods were used.

It was found that hard fresh sandstone showed no disintegration, and slightly oxidised sandstone showed very little. Fresh siltstone and oxidised massive mudstone were also little affected, but other fine-grained rocks showed very great disintegration.

The disintegration was of two types—a minor disintegration due to flaking at the surface of the sample, and a major type due to splitting of the sample into two or more large pieces of about equal size. Cracking tended to concentrate along the bedding planes and cleavage planes when these were present. It is noteworthy that disintegration was nearly always noticed on the second day of the drying period, and very seldom during the wetting period or on other days of the drying period. Because of this the optimum length of a cycle appears to be about four days. In the samples that disintegrated completely into small flakes, the first flaking occurred in less than five cycles, the first large crack appeared in less than 10 cycles, and the first major disintegration in anything from 15 to 76 cycles. Complete disintegration into small flakes took about 180 cycles.

A possible explanation of wetting and drying is provided by the mechanism of 'ordered-water' molecular pressure. Water, because two positively charged hydrogen atoms are arranged at one end of the molecule and the negatively charged oxygen atom at the other, is a polar liquid. The positively charged end of the water molecule is attracted to the negatively charged surface of a clay or other material, and other water molecules line up similarly, joining positive and negative ends like so many tiny bar magnets. This forms a layer of ordered water.

Possibly repeated wetting and drying allows the water molecules to become increasingly ordered, assuming a quasi-crystalline nature and exerting expansive force that thrusts against the confining walls.

In support of this hypothesis there are reports of experiments (Anon, 1966) in which rocks susceptible to this kind of weathering were treated with other liquids. Formaldehyde, a highly polar liquid, was very effective in causing disruption: carbon tetrachloride, a non-polar liquid, had no effect.

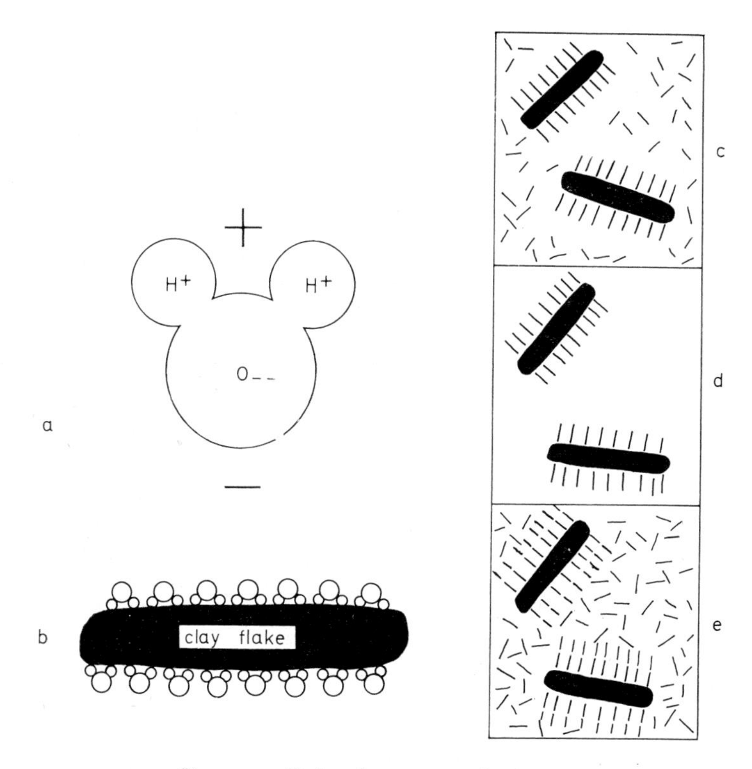

FIG. 13. Ordered water weathering.

a A molecule of water has the hydrogen ions at one end, and so is a polar molecule and can be regarded as a magnet. *b* Water molecules become oriented as they cluster around a clay fragment. *c* Water (represented as tiny bar magnets) is ordered around clay fragments, but remains disordered in bulk water. *d* When the clay material is drained the ordered water remains. *e* Renewed wetting builds up another layer of ordered water.

WATER LAYER WEATHERING

On shore platforms and flattened rocks near sea-level there are numerous small shallow pools, and it is often found that the edges of the pools are actively eroded at the water level, with the formation of small undercuts. This sort of erosion makes a pool wider, and eventually it will merge with others to make a more continuous, broad flat surface. Rocks affected by this process include basalt, limestones, sandstones, shales and weathered granite. Quartzite and hard granite seem resistant to the process, and limonite filled joints are reduced at a slower rate than neighbouring rock.

The process may be partly due to solution or to salt action, but Hills (1949) considers that the process of alternate wetting and drying is almost entirely responsible. In this coastal context the process is called water layer weathering. This term replaces the older 'water-level weathering' which might be confused with sea-level.

CAVITATION

Turbine vanes and other parts of hydraulic machinery can suffer a kind of corrosive erosion due to cavitation. This consists of the formation of bubbles in rapidly moving water, followed by their collapse. The bubbles are formed when the local pressure becomes less than the vapour pressure. The bubbles collapse violently (implode), and if in contact with a solid surface produce a shock like a hammer blow on a very small area.

Since cavitation can affect metal surfaces it might be thought that it is of significance in weathering and erosion, and indeed some geological phenomena have been attributed to cavitation (Barnes, 1956).

However, cavitation can only occur when water has a very high velocity, probably higher than occurs in nature. Even with an assumption of laminar flow (which would certainly not operate at high velocities) and allowing a generous factor for local effects, Barnes calculates a flow of about 25 to 30 feet/second as the minimum for cavitation. Hjulström (1935) states that the minimum velocity necessary for cavitation to appear in a stream is 12 metres, or 39 feet/second, but does not give the basis of his calculation. It is quite possible that even this figure is an underestimate. In any event, very high velocities are required, and if it occurs at all, cavitation is probably limited to exceptional sites such as the base of very high waterfalls.

ABRASION

Minerals and rocks may be worn away by simple mechanical abrasion. This may be due to friction, as when rocks slide over one another, or to impact, as between transported grains and bedrock.

The environment of weathering has a marked control on the nature and products of weathering.

In glaciers, mechanical grinding of bedrock and transported material is very marked, and is the dominant weathering agent.

In wind transport, cleavage minerals suffer fairly rapid attrition, and feldspars, hornblende and especially mica, are rare in wind-blown sand. When dunes do contain mica, like those of California described by Reed (1931), it is a sure indication that they are immature and have not been transported far.

The downcutting of river beds on hard bedrock is largely achieved by mechanical attrition, and is probably mainly due to the impact of boulders and cobbles. Transported sand grains do not break easily, and silt-sized particles are probably produced as chips by the collision of pebbles rather than the breakdown of sand grains.

As particles become smaller they become harder to break, for they can yield elastically. Possibly the limit for quartz is at about 0·02 mm, which may account for the comparatively large amount of quartz of about this size (silt size). However, Kuenen (1960) has suggested that much of the silt-size quartz (as in loess) is derived from glacial abrasion, from primary production from fine-grained rocks or from cracked quartz grains in igneous rocks. He does not regard particle collision as capable of reducing quartz grains below the size of fine sand.

The mechanical abrasion of minerals is discussed further in Chap. V, p. 62.

MECHANICAL COLLAPSE

Rock weathering may take the form of mechanical collapse following undercutting of various sorts. Bradley (1963) describes collapses in sandstone in Colorado where steep cliffs have been undercut by river meanders. Such collapse usually takes place along a curved, concave downwards, collapse face, and the fallen block commonly disintegrates partly due to the impact of its fall.

Various processes of undercutting can give rise to collapse besides river erosion. Wind undercutting, water seepage removal, or a clay band beneath a sandstone, solution caving, or exfoliation caving are examples. Fig. 14 shows a collapse due to exfoliation caving at Ayers Rock, Central Australia.

Many limestone caves grow by mechanical collapse, and karst landforms are frequently due to the interaction of solution and collapse.

COLLOID PLUCKING

A thin film of drying gelatine in a tumbler can pull flakes of glass from the sides. Reiche (1950) suggested that soil colloids might behave in a similar manner, and possibly pull flakes from mineral grains. This has not been seen, but Ollier and Tuddenham (1962) report a variety of colloid plucking whereby a film of clay, drying out and curling up from the surface of an arkose, pulled out entire grains from the rock. The process is probably very minor.

SOIL RIPENING

Soil ripening is the name given to the first stage in the weathering and evolution of soil from fresh alluvium, or other water-laid deposits that are exposed to the air for the first time. Practically all the work that has been done on this process comes from Holland, where there is great interest in reclaiming land. The processes are irreversible, and so, although changes are not great compared with many other types of alteration, soil ripening can be regarded as a kind of weathering.

The first stages of evolution of fresh alluvium have been described by Pons and Zonneveld (1965). Physical ripening is most important, and consists of loss of water by compaction, evaporation and transpiration.

When the soil colloids lose water their respective attraction forces increase, the consistency increases, the volume of the mineral parts is reduced and the relative proportions by volume of several soil compounds changes. Physical ripening is mostly an irreversible process. The smaller uptake of water in a wet period compared with the

FIG. 14. Collapse of rock mass due to undercutting. Ayers Rock, Central Australia.

original unripe soil and the consequent swelling of the particles prevent the soil from re-acquiring the original unripe status.

Physical ripening can be measured by the water content of the colloidal part of the soil, and a 'water factor', n, is introduced as a standard. It is the quantity of water in grammes which is absorbed by 1 gramme of the clay fraction. The n factor is defined in relation to illite clay (the commonest in western Europe), but Pons and Zonneveld provide formulae for standardising when other clays are present. The n value can be determined in the field approximately, as it is strongly reflected in the soil consistency, which in turn has a good correlation with penetrometer readings (see Table 3).

Table 3

n-value		Consistency
0·7	Ripe	Firm. Does not stick to hands much. Cannot be squeezed through fingers.
0·7–1·0	Nearly ripe	Fairly firm, rather sticky. Squeezed with difficulty.
1·0–1·4	Half ripe	Soft, sticky, easily squeezed.
1·4–2·0	Practically unripe	Soft, sticks fast to hands, easily squeezed through fingers.
2·0+	Unripe	Liquid mud, non-kneadable.

As physical ripening proceeds there is also chemical ripening, which involves mainly oxidation, cation exchange on the clays, and decalcification; and biological ripening which consists of microbiological alteration, and mixing by soil fauna.

III CHEMICAL WEATHERING

CHEMICAL EQUILIBRIUM

THE world of matter is made up of atoms, which may be combined as molecules, and which exist in gaseous, liquid or solid forms (phases). Any particular assemblage of atoms may be stable or undergoing change towards a new arrangement. The stable situation is called equilibrium. This is not necessarily static, but at equilibrium any changes in one direction are exactly compensated by changes in the opposite direction. Every physical or chemical process is the result of disturbed equilibrium, and is so directed as to tend to restore that equilibrium.

Physico-chemical equilibria demand that all substances should be present in the form of phases that are stable under the given pressure and temperature conditions. These phases must also be in equilibrium with one another. By 'phases' is meant not merely the gas and liquid phases, but also each individual crystalline species. If changes occur in pressure, temperature or chemical composition, certain phases or mineral assemblages will become unstable and may be converted into stable phases in mutual equilibrium, as happens for example in weathering.

The general principle of chemical reaction in weathering is Le Chatelier's principle, which states that any system in equilibrium will react to restore the equilibrium if any force is applied.

Equilibrium relations determine the maximum amount of a mineral that can be dissolved in any system. However, the actual amount that really is dissolved depends on rate factors. Rate factors have often been overlooked in weathering studies, as for instance in limestone solution which is discussed later. It is important to realise that simple chemical equations illustrate possible courses of chemical reactions but they do not tell us the rate of reaction, which really determines whether the reaction is important in weathering or not.

The greater the number of variables in a system, the harder it is to predict the course of a reaction, and the number of variables in weathering is often great. In the simple solution of calcite, $CaCO_3$, for example, there are seven variables, even at constant temperature and pressure (see p. 38). Furthermore, even a simple reaction may occur in a number of different stages; the solution of calcite, for instance, takes place in four distinct stages, and even more stages are involved in the alteration of complex silicate minerals to clays.

Removal of weathered products is of the utmost importance in determining the course of weathering reactions, as will be seen repeatedly. By removal of weathering products a reaction can continue in the same direction, whereas if they remain, a closed system will result, equilibrium will be reached and the reaction stopped at an early stage.

HYDROGEN ION CONCENTRATION

One type of ion that is always present in solution is the hydrogen ion, H^+. This is very important in controlling many reactions, so much so that the hydrogen ion concentration is often treated as a separate control on reactions. The hydrogen ion concentration is expressed as the pH, which is the log of the hydrogen ion concentration without the minus sign.

The pH of pure water is 7. A solution with a pH less than 7 is acid, with a greater value is alkaline. The table below shows the pH of some natural environments:

pH	Natural environments
10	
	alkali soils
9	
	sea water
8	
	calcareous soils
7	rain water
	river water
6	
	acid soils
5	
	peat water
4	
	mine water
3	
2	acid hot springs
1	

The solubility of many substances is very much affected by pH. Thus the solubility of iron is about 100,000 times greater at pH 6 than at pH 8·5. Weakly acid iron-bearing water will therefore precipitate iron when it reaches the alkaline sea. The solubilities of alumina and silica are very markedly affected by pH as shown in Fig. 15. At very low pH (<4) alumina becomes more soluble than silica. Environments at this pH are very rare, so the removal of alumina to leave a silica residue does not occur. On the other hand, between pH 5 and 9 alumina is virtually insoluble but silica becomes increasingly soluble, and such conditions could lead to differential leaching of silica and the formation of laterite or bauxite.

REDOX POTENTIAL (OXIDATION POTENTIAL, REDUCTION-OXIDATION POTENTIAL)

Some elements can exist in several oxidation states, e.g. iron, which can be native metal (oxidation state 0), ferrous compounds (oxidation state 2) or ferric compounds (oxidation state 3). The stability in any oxidation state depends on the energy change involved in adding or removing electrons. This can be measured quantitatively and expressed relative to the 'oxidation' of hydrogen to hydrogen ions. Oxidation potentials are symbolised E° or Eh.

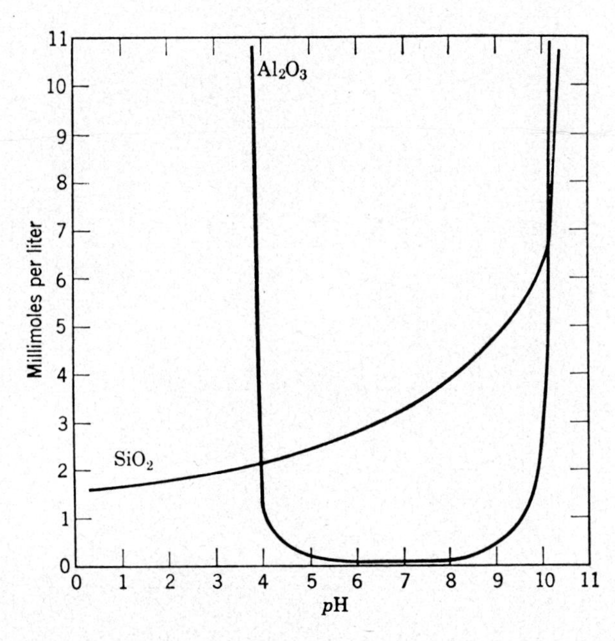

FIG. 15. The solubility of silica and alumina as a function of pH (from Mason, 1966).

The Eh varies with the concentration of the reacting substances, and if H or OH is involved, then the Eh varies with the pH of the solution. The Eh becomes lower as pH increases, and so oxidation proceeds more readily the more alkaline the solution.

In aqueous solutions in nature, only reactions between Eh 1·23 volt and Eh 0·00 can occur, for outside these limits water is decomposed. Environments can be plotted on a diagram of Eh against pH, as shown in Fig. 16.

IONIC POTENTIAL

Ions in solution attract water molecules to them and the hydration of an ion is proportional to its charge (symbolised Z) and its radius (r). The factor Z/r is known as the ionic potential, and provides a measure of the behaviour of ions towards water.

Elements with low ionic potential, such as Na, K and Mg, remain in solution during weathering; elements with intermediate ionic potential are precipitated by hydrolysis; elements with still higher ionic potentials form complex anions with oxygen that are once again soluble (Fig. 17).

The concept of ionic potential is useful in explaining which elements occur together in weathered rocks and which are removed during weathering. For instance, a number of fairly rare quadrivalent elements are concentrated in the range of ionic potential of hydrolysate sediments, and so bauxite may be enriched not only in beryllium and

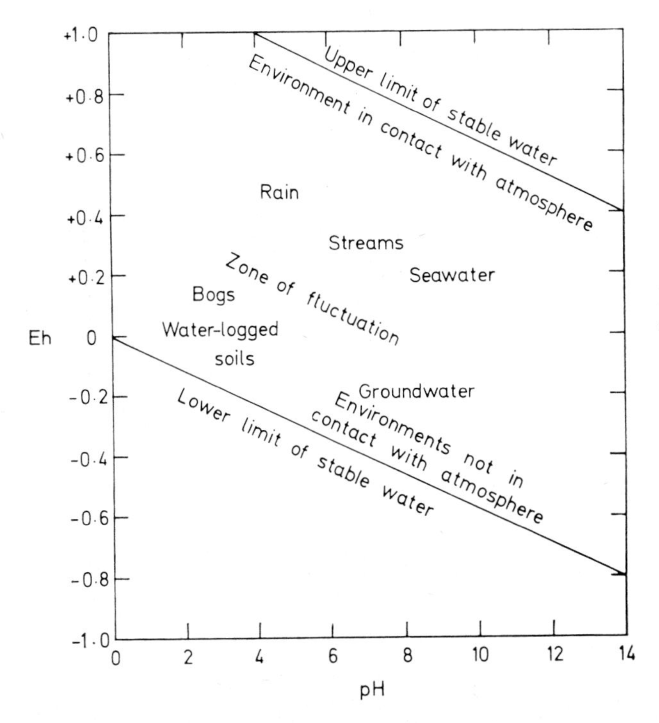

FIG. 16. Environments in relation to Eh and pH.

gallium, which are chemically similar to aluminium, but also in titanium, zirconium and niobium (Mason, 1966).

TYPES OF CHEMICAL REACTIONS

SOLUTION

Solution is usually the first stage of chemical weathering. It may take place in running water, or in a thin film of water around a solid particle. The amount of solution depends on the amount of water passing the surface of the particle, and the solubility of the solid being dissolved.

Common salt is very soluble, and only survives at the earth's crust in the most arid regions. Gypsum is less soluble, and carbonates less still. An idea of the weathering of rock can be gained from an analysis of run-off water in rivers or streams. They usually have carbonates in solution, and in warmer areas perhaps silica, indicating that different elements have different solubilities in different climatic regions.

Solutions may precipitate chemicals, which can lead to volume changes and enhance physical weathering. By carrying ions from one place to another solutions may lead to more combinations of reactions than if water did not move, or if there was no circulation.

Smyth (1913) and Polynov (1937) suggest that the direction of weathering as a whole would be indexed by the relative mobilities of the elements. To arrive at numeri-

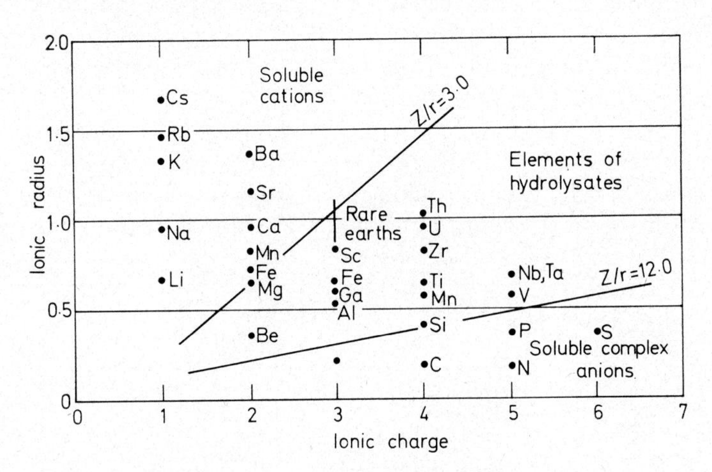

FIG. 17. Geochemical separation of selected elements by diagonal fences of constant ionic potential (charge divided by radius). (From Mason, 1966.)

cal values they compared percentage compositions of salts from river waters with estimated average compositions of the rocks composing or underlying their drainage basins. Smyth produced generalisation for all rocks and climates, and Polynov for all igneous rocks and climates. Their data proved to be comparable, but the order of mobilities may not hold in all conditions. The generalisation is that the group calcium, sodium, magnesium and potassium is more mobile than silica, which is more mobile than the sesquioxides. Polynov concluded that the individual mobility groups are removed successively during rock weathering, but in fact all three groups are removed simultaneously though at different rates.

Salt concentrations in rivers frequently reflect the nature of the rock (in a general sense) of the river basin. It is found in Victoria that the highest salinities are in areas of Tertiary laterite. At first sight it might appear that the laterites, being already thoroughly weathered, might have lost their salts. The high figures are probably due to the considerable degree of percolation of ground water in the old, deep soils and weathered rock so that they have enhanced opportunity to pick up salt, though the concentration is low. In fresh rocks, percolating water has a much shorter time to pick up salt.

A literature survey by Davis (1964) suggests silica is less variable in natural water than other dissolved constituents. This throws doubt on the idea that 'lateritisation' is confined to the tropics. The rock type is the most important factor in controlling the amount of silica in ground water.

The solution of limestone is further discussed on p. 38.

Sulphurous and sulphuric acids in industrial atmospheres can give rise to an artificial weathering and production of soluble salts. Raistrick and Gilbert (1963) described how weathering of Malham Tarn House was concentrated on the south-west face, which was exposed to the prevailing wind coming from an industrial region that provided the necessary sulphurous products. The Houses of Parliament in London, which are built of Magnesian Limestone, suffer in the same way. As Professor A. Beresford Pite wrote in *The Observer* of 23rd August 1925:' Ever since the magnesium limestone, of which the New Palace at Westminster is built, was first exposed to the chemical constituents of the London atmosphere, it has been slowly and surely transformed into some sort of a heap of Epsom salts'.

That solution can be quantitatively important in general denudation was strikingly illustrated by the work of Rapp (1960), who studied all the processes operating on slopes in a Lappland valley. He found that solution accomplished as much erosion as all the other processes put together, as shown in the table below.

Process	Downslope movement (m per year)	Tons-metres vertical
Rockfalls	225	20,000
Avalanches	200	22,000
Earthslides	600	96,000
Talus creep	0·01	2,700
Solifluction	0·02	5,300
Solution	700	136,500

OXIDATION AND REDUCTION

From the point of view of simple weathering, oxidation simply means a reaction with oxygen to form oxides, or if water is also incorporated, hydroxides. Since reaction with atmospheric oxygen is one of the commonest forms of natural weathering it is frequently useful to restrict the term to this simple meaning in this context.

Oxidation of minerals by gaseous oxygen probably always occurs through the intermediate action of water in which the oxygen is first dissolved. Water may be present in abundance or merely as a film, and most oxidation takes place in the aerated zone. Its effect is particularly marked by iron oxides and hydroxides, which give the characteristic red and yellow colours of many rocks and soils. It may be masked by black humus, but in the tropics humus may be colourless and red soils are characteristic of large areas. The red colour (and therefore oxidation) can penetrate to considerable depths. Reiche reports that in arid regions deep wells to the water table often reach green-grey rock that is beyond the oxidised zone.

Although inorganic oxidation is common and important, there is also a good deal of oxidation by bacterial action, the bacteria deriving energy from the oxidation of iron, manganese, sulphur and possibly other elements, including phosphorus, producing vivianite. The oxidation of sulphides produces sulphuric acid which is probably important in further weathering processes.

Chemists use the terms oxidation and reduction to mean the removal of electrons from and their addition to the atoms of some elements involved in the chemical change, and oxygen need not be involved at all. For example, iron may combine with sulphur to form FeS, in which case the iron is oxidised. It may further be oxidised to FeS_2, and yet further by combination with oxygen. Iron sulphide is thus an oxidised compound with respect to metallic iron, but a reduced compound with respect to haematite.

Reduction is the opposite of oxidation, and takes place usually in waterlogged anaerobic sites. The reduction of iron oxides is one of the most obvious changes and the red and yellow colours are changed in reducing environments to greens and greys. Much reduction is the work of bacteria; organic matter (humus), for instance, is reduced by fermenting bacteria, and sulphates are reduced by bacteria to sulphides.

Many silicate minerals contain cations that are easily converted into another oxidation state, such as Fe^{2+}/Fe^{3+}, Cr^{2+}/Cr^{3+}. A change of the redox potential in the environment of such minerals may cause oxidation or reduction of these cations. To maintain the electrostatic neutrality of the crystal structure, some other ion must enter or leave the lattice. This causes the lattice to become unstable and more susceptible to other weathering reactions such as hydrolysis, or to alteration to other mineral species.

In soil the oxidation-reduction potential is primarily determined by the oxygen and CO_2 dissolved in soil water and also by the soil organic matter. It therefore depends on the partial pressure of the gaseous oxygen and CO_2 in the soil air, and on the pH of the soil solution.

CARBONATION

Carbonation is the reaction of carbonate or bicarbonate ions with minerals. Carbonates are not common end products of *in situ* weathering, but the formation of carbonates is a step in the course of certain types of weathering, particularly the breakdown of feldspars. Carbon dioxide is abundant in soil atmospheres, and leads to much more rapid carbonation than might be suspected from the small amount of carbon dioxide in the atmosphere. H_2CO_3, a solution of carbon dioxide in water, is an acid whose presence greatly facilitates the base exchange process, and is vital in the solution of carbonates themselves. Numerous geological observations suggest that carbonic acid is a much more potent solvent than its weak acidity would suggest.

HYDRATION

Hydration is the addition of water to a mineral. Iron oxides, for instance, may absorb water and turn into hydrated iron oxides or iron hydroxides. Hydration is a very important process in clay mineral formation, and water may be actually incorporated as part of the crystal lattice. Hydration is an exothermic reaction, and involves a considerable volume change which may be important in physical weathering—exfoliation and granular disintegration. Hydration prepares mineral surfaces for further alteration by oxidation and carbonation, and enables the transfer of ions to take place with greater ease.

CHELATION

Chelation (or complexing) involves the holding of an ion, usually a metal, within a ring structure of organic origin. Chelating agents can extract ions from otherwise insoluble solids, and enable the transfer of ions in chemical environments where they would normally be precipitated.

Plants utilise chelating agents to extract ions (nutrients) from minerals, and thus enable mineral weathering to take place at a much greater rate than would be indicated by simple inorganic considerations. The actual mechanism is not known in any detail. Decomposing plant matter (humus) is rich in chelating agents, which may assist in the preparation of ions for absorption by living plants, but which also enhance differential leaching of ions in a weathering profile. Leaf leachates, for example, can give rise to rapid movement of iron down a profile, and can lead to very rapid podzolisation. Different species of trees have very different effects in this regard, and through chelation give rise to different 'phytogenetic' soils from the same parent material.

An instructive demonstration of chelation is described by Keller (1957). Powdered calcite is stirred into an aqueous solution of a sodium salt of EDTA (ethylene diamine tetra-acetic acid, a common chelating agent). No CO_2 is given off, the pH remains 10 to 11 (highly alkaline), and the Ca is taken up in solution by chelation and the carbonate ion remains in solution.

HYDROLYSIS

Hydrolysis is a chemical reaction between mineral and water, that is between the H or OH ions of water, and the ions of the mineral. This reaction takes place whenever a mineral is in contact with water—even pure distilled water or rain water—and it is a common misconception that water is neutral and inactive and only becomes important when acidified as by CO_2 solution.

A clear demonstration of hydrolysis can be given by measuring the pH of suspensions after a mineral is ground to a powder in pure water. Many minerals, including olivine, diopside, actinolite and nepheline, can give a pH of 11 when simply abraded in water. Stevens and Carron (1948) proposed identifying minerals by their 'abrasion pH', for the values obtained from minerals are sufficiently constant.

Table 4 shows the pH of some common minerals when abraded in water.

Table 4. Abrasion pH of selected minerals

Mineral	pH when abraded in water
Feldspars	
Albite	10
Oligoclase	9
Anorthite	8
Orthoclase	8
Microcline	8
Micas	
Biotite	8
Muscovite	7
Amphiboles	
Actinolite	11
Hornblende	10
Pyroxenes	
Augite	10
Hypersthene	8
Olivine	10
Nepheline	11
Carbonates	
Calcite	8
Dolomite	9
Clay minerals	
Kaolinite	6
Montmorillonite	7
Quartz	7

Data simplified from Keller, 1957.

The concentration of hydrogen ions is of fundamental importance in all weathering reactions, for the following reasons: (1) they are available from a large number of sources; increasing hydrogen ion concentration decreases the solubility of SiO_2 and Al_2O_3, which is conducive to the formation of clay minerals; (2) hydrogen ions combine with OH ions (which tend to increase during hydrolysis) thus removing them from the system and permitting continued hydrolysis; (3) they readily replace other cations (cation exchange).

Two important sources of hydrogen ions are acid hydrogen clays (a clay with a high proportion of H in its cation exchange sites), and living plants. The former is often in direct contact with a weathering mineral as a coating, and so attack by H^+ ions can be very effective. Experiments with suspensions show very effective attack, but in solid rocks and soil the process may be limited by the rate of removal of the weathered product.

Living plants provide a continuous source of H^+ ions, which create an acid environment and weather nearby minerals. The plants exchange H^+ for plant nutrients, and by removing the latter the reaction is kept in imbalance, so weathering is continuous as long as the plant lives, and weathering extends out from the contact zone at the root/clay interface.

It would be hard to exaggerate the importance of hydrolysis in weathering, and very extensive weathering can be achieved by nothing more than rock and water. Raggatt *et al.* (1945), for instance, show that only water is needed to convert basalt into bauxite, while Ca, Mg, Na and K are removed.

IMPORTANCE OF CRYSTAL STRUCTURE

Total chemical content is not a sufficient guide to reactions that may take place during weathering, and the crystal structure of minerals is important. Calcite and aragonite are both $CaCO_3$, but the atoms are arranged differently in the two minerals and aragonite is about 10 times more soluble than calcite. In a rather similar fashion, although sodium and potassium are both very mobile elements, in clay weathering sodium proves much more mobile because potassium is much more readily incorporated in clay mineral structures.

The oxidation of iron in a mineral can cause the collapse of the silicate lattice, so the entire mineral is affected.

SILICATE WEATHERING

Silicates are by far the most important group of rock-forming minerals, and will be considered in more detail, because a knowledge of silicate structure is a help in understanding the weathering of minerals.

The basic unit of silicate minerals is the silicate tetrahedron in which a small silicon atom (radius 0·39Å) lies at the centre of four oxygen atoms (radius 1·32 Å) in a tetrahedral arrangement. These tetrahedra are then arranged in discrete groups, chains, sheets, or three-dimensional structures. The Si-O-Si linkages are not very strong, and the stability of a given structure is maintained by the complex interplay of geometrical and electrical factors involving all the atoms in the final compound (unlike carbon chemistry where C-C linkage is associated with high energy which maintains structure).

Atoms of the same valency and similar size can proxy for each other in crystal structures. There can also be proxying of atoms of differing valencies but having similar sizes and identical co-ordination numbers (e.g. Al for Si, Mg or Fe (ferrous) for Al). The change in charge must be balanced by another in close proximity, which can happen in various ways. Packing of atoms is very important, but the radius to be ascribed to a given atom varies appreciably according to its co-ordination number. Pauling's law states that charges are balanced over the shortest possible distance. This brings free energy to a minimum, and expresses another important control in crystal structure. It appears to be a general rule in silicates containing hydroxyl that the OH group should not be linked to silicon.

The relationships of water molecules to silicate structures are largely controlled by the following facts:

1. Silicate surfaces show a strong adsorption for polar molecules.
2. In some cases simple geometrical relationships exist between silicate surfaces and regular associations of water molecules. This is important in clay minerals.
3. In some cases there is an association between water and metallic cations balancing framework charge.

Hydrogen thus has three different forms in silicate structures; that attributed to hydrogen bonds, that forming normal hydroxyl groups and that belonging to normal water molecules.

The breakdown of silicates by the action of water seems to involve three distinguishable but often contemporaneous processes—replacement of non-framework cations by the hydrogen ion from water; oxidation of ferrous ions to ferric; and hydration of fragments of the unstable lattice then exposed. The hydrated fragments with their numerous OH groups may pass into the aqueous phase entirely or they may remain attached to the surface forming a poorly organised layer. Such layers are very sensitive to changes in the outer solution which affect their hydroxyl groups. Hence their stability and subsequent fate depend markedly on the pH of the weathering solution.

The fundamental principles of silicate weathering have been summarised by Sticher and Bach (1966) as follows:

1. Stability of silicate minerals increases with increasing degree of condensation of the silicate framework.

Orthosilicates
Di and Ring silicates | degree of condensation increased
Chain silicates |
Sheet silicates ↓ stability increased
Framework silicates

2. Stability within a structural group decreases with increased isomorphous substitution of aluminium for silicon.

Quartz | increased isomorphous substitution
Orthoclase | stability decreased
Nepheline ↓

3. Stability within a structural group decreases with decreasing electronegativity of the metal ions (increasing electron-donating inductive effect).

Hypersthene | electronegativity of the metal ion decreased
Wollastonite ↓ stability decreased

4. Stability is also influenced by the type and structure of the metal ion-oxygen polyhedra linking the silicate units. Here, besides the size and charge of the ions, there are many other factors playing a directive role. Their influence may overshadow any difference arising from different structural framework. Thus zircon belongs to the most stable silicate minerals, although its structure is based on isolated SiO_4 tetrahedra.

SOLUTION OF LIMESTONE

Limestone is the only common rock in which solution plays a major part. Distinctive landforms and landscapes result from limestone solution, and limestone has been studied in considerable detail from the point of view of the origin of its scenery and the chemistry of its solution. In this section we will review the chemistry of limestone solution, and at the same time introduce chemical concepts that are valuable in the understanding of chemical weathering generally.

For any reversible reaction:

$$[A+B \rightleftharpoons C+D]$$

$$\text{the forward reaction rate} = k\,[A]\,[B]$$
$$\text{and the back reaction rate} = k'[C]\,[D]$$

where k and k' are the rate constants. The square brackets indicate concentrations, and [A], [B], etc., are the concentrations of the reactants and products.

At equilibrium these rates are equal and

$$\frac{[C]\,[D]}{[A]\,[B]} = \frac{k'}{k} = K, \text{ the equilibrium constant.}$$

This constant is derived assuming a closed system, or if the system is open and one of the reactants is being removed, that the reaction rates are fast enough and the rate of removal slow enough for the reactants and products to be at equilibrium concentrations.

The equilibria involved in the solution of limestone have been described by Garrels (1960). He points out that there are *seven* variables involved at constant temperature and pressure, which are the partial pressure of CO_2 and the concentrations of

$$[H_2CO_3]$$
$$[HCO_3^-]$$
$$[CO_3^{2-}]$$
$$[H^+]$$
$$[OH^-]$$
$$[Ca^{2+}]$$

He considers five cases:

1. The reaction involved in placing pure calcite in pure water with negligible gas phase present. This does not occur in nature, as a CO_2 phase is always present.

2. The reactions of calcite in pure water, but with the system open to CO_2. Here the pH of the system is entirely controlled by the carbonate equilibria. This applies roughly to streams and lakes open to the atmosphere, which have a fixed partial pressure of CO_2. However water is never pure in nature.

3. Equilibrium relations in a system with a fixed quantity of dissolved carbonate species, but with the pH arbitrarily fixed, that is controlled by other reactions in the system. Organic acids, for example, commonly control the pH.

4. Equilibrium in a system connected to an external reservoir of fixed partial pressure of CO_2, but with the pH arbitrarily fixed.

5. Equilibrium resulting from addition of $CaCO_3$ to a system originally open to a CO_2 reservoir, but closed to that reservoir before addition of $CaCO_3$.

By substituting the values of the equilibrium constants at different temperatures in the cases above, the effect of temperature on the solubility of $CaCO_3$ can be estimated, and from a consideration of the carbonate equilibria involved the causes of precipitation and dissolution can be predicted. More limestone can be dissolved by increasing the temperature, the CO_2 content of the water, or decreasing the pH of the solution.

The conditions under which solution took place must be known before the equations can be applied to find out whether or not a cave water is at equilibrium with the cave limestone. The full history of a cave water can never be known sufficiently well for this treatment to be applied accurately.

The situation is further complicated by reactions other than those of carbonates which may influence the pH of the solution. When the pH of a system is determined by factors other than carbonate equilibria, the amount of carbonate dissolved at equilibrium is estimated by using equations developed in cases 3 and 4 of Garrels. At present very little is known of the equilibrium relations of these miscellaneous reactions, or of their importance in limestone solution, but CO_2 may not always be so important a factor in limestone solution as is generally assumed. There may be reactions between dissolved components and clays or other cave sediments. Weathering products from igneous or other rocks may be washed into caves from outside and affect reactions, and organic acids are a further complication.

Holland, Kirsupu, Huebner and Oxburgh (1964) have discussed the evolution of cave waters, and the possible equilibrium relations that may be found in geological situations. Initially rain water is assumed to be in equilibrium with atmospheric CO_2 which is at a constant partial pressure. On percolating through soil the water comes into contact with soil air, which is likely to have a higher partial pressure of CO_2 than the atmosphere. The water may then react with limestone in the presence or absence of a CO_2 phase (cases 2 and 5 above). If the water reaches a cave, then a third stage is the equilibration of the cave waters with cave air, which usually has the same CO_2 concentration as the outside atmosphere. This brief account shows that even with a very detailed knowledge of equilibrium reactions it is still extremely difficult to follow the course of carbonate reactions in a real cave.

So far we have considered only equilibrium reactions. These determine the *maximum* solubility of limestone in any given case. The *actual* quantity dissolved depends on still more factors—those affecting the kinetics of reaction. It is of the utmost importance in understanding limestone solution to distinguish between equilibrium factors and kinetic factors.

The dissolving liquid can be either still standing, gently moving or turbulent. When the system does not have time to come to equilibrium the *rates* of carbonate solution are the determining factors of the *amount* of carbonate dissolved. The equilibrium relations will still determine the *maximum* amount that can be dissolved, but the *actual* amount is determined by the rate factors.

The kinetics of the dissolution of calcite have been investigated by Weyl (1958), who postulates that the dissolution occurs in the following four stages:

1. The dissociation of calcite at the solid-liquid interface into Ca^{2+} and $CO_3{}^{2-}$ ions, i.e. the diffusion of Ca^{2+} and $CO_3{}^{2-}$ ions from the solid-liquid interface into the solution.
2. Reaction between the $CO_3{}^{2-}$ ions and dissolved CO_2 to give $HCO_3{}^{-}$. This actually takes place in a number of steps.

$$CO_2\ (aq) + H_2O \rightleftharpoons HCO_3{}^{-} + H^{+}$$
$$H_2CO_3 \rightleftharpoons H^{+} + HCO_3{}^{-}$$
$$CO_3{}^{2-} + H^{+} \rightleftharpoons HCO_3{}^{-}$$

3. If the solution is in contact with a gas phase containing CO_2 there is an additional reaction CO_2 (gaseous) + $H_2O \rightleftharpoons CO_2$ (aqueous).
4. Transport of the various species through the solution by diffusion and fluid motion tending to equalise concentration gradients.

Results obtained by both Weyl and Kaye (1957) suggest that the step in the above scheme which determines the overall rate of dissolution is step 4.

In other words, the speed of water flow and its turbulence are far more important than the equilibrium reactions. The total amount of limestone dissolved will also be dependent on the area of the solid-liquid interface. The largest amount of $CaCO_3$ that can be dissolved in a given volume of water is predicted by the equilibria relations. However, since it is unlikely that equilibrium is attained in practice, the actual amount of $CaCO_3$ dissolved is determined by the kinetic factors outlined by Weyl.

In natural solution the total amount of limestone dissolved will also be dependent on the area of the solid-liquid interface. Hence the quantity of limestone dissolved will also depend on the lithology, porosity and geological relations of the rock. These factors enhance the importance of structure in the limestone. A rock with a lot of joints or other planes of weakness will be selectively dissolved. This gives a larger surface area compared with a uniformly dissolving limestone, and the greater area will cause even greater solution. Furthermore, the irregularities are likely to make water flow turbulent, which increases solution even further.

PRECIPITATION OF CARBONATE

A problem related to the dissolution of limestone is the re-precipitation of carbonates as travertine, stalactites and so on. Calcite can be precipitated from solution by a change in temperature, by loss of water, or by loss of CO_2. Stalactites are not likely to be due to evaporation, for air in caves with actively growing stalactites is always close to saturation. Loss of CO_2 to the atmosphere from a drip could cause precipitation. The CO_2 content of air is only 0·03 %, so the contribution of atmospheric CO_2 to stalactite formation is negligible. Cave atmospheres normally have the same percentage of CO_2, but soil atmospheres have up to 0·65 %. Water percolating through soil can therefore pick up CO_2 from the soil atmosphere, which enables it to dissolve more limestone, and then when the low CO_2 partial pressure of a cave atmosphere is reached, CO_2 can diffuse from water drops, causing precipitation of carbonate.

In flowing water there is likely to be more diffusion where the water flows rapidly as a thin film, and in such places, as on edges of rimstone pools, calcite is again likely to be precipitated. In the open air, as in surface rivers, precipitation is more often due to evaporation of water.

TEMPERATURE EFFECTS

Both the kinetics of dissolution and the constants for carbonate equilibria are affected by temperature changes. It is likely that temperature can affect rate constants rather more than other effects, for in general a 10°C rise in temperature gives a doubling of rate. However, since the most important rate controlling factor is transport of species through solution, which is not affected by temperature, the total effect of temperature is likely to be small. Corbel (1959) has suggested that the mean rates of limestone solution are dependent on variations in temperature, but this is too great a simplification, and Sweeting (1964) has stressed that non-climatic factors more than compensate for any variation due to temperature.

One alleged indicator of temperature effects is the absence of stalactites in Arctic regions. This is probably due to the lack of a high partial pressure of CO_2 in the soil air in cold regions, due in turn to lack of organisms. The direct effect of temperature is therefore not relevant, but the indirect effect, through organic activity, is important.

EFFECT OF OTHER ELEMENTS

The effects of small traces of other elements on the solution of limestone have been discussed by Picknett (1964) and Terjesen *et al.* (1961).

Traces of magnesium (less than 1 % Ca) increase the pH of a saturated carbonate solution by about 0·1 unit, whereas larger amounts have a smaller effect and may even reduce the pH. Thus it appears that these traces of magnesium in solution increase the amount of calcite which can be dissolved for a given carbon dioxide concentration. It is only when large amounts of magnesium are present that the expected happens, and the solubility of the calcite is reduced. This phenomenon is still under investigation.

Also known to have a marked effect are small traces of heavy metals such as lead, copper, manganese, etc. Terjesen *et al.* studied the rate at which calcite dissolved in water saturated with carbon dioxide at about 1 atmosphere pressure, and found that these metals had a marked inhibitory effect on the later stages of solution. In fact, the calcite apparently stopped dissolving when the solution should still have been unsaturated. This is illustrated by the following data on scandium, one of the more effective inhibitors:

Calcium content at 'saturation' gramme moles/litre	$9{\cdot}0 \times 10^{-3}$	$8{\cdot}6 \times 10^{-3}$	$7{\cdot}2 \times 10^{-3}$	$4{\cdot}0 \times 10^{-3}$
Scandium content gramme moles/litre	0	10^{-7}	10^{-6}	10^{-5}

Thus the effective solubility of calcite is reduced to a half by a scandium concentration only 0·2 % of the calcium concentration. The metals found to act in this way are listed below in order of effectiveness:

lead, lanthanum, yttrium, scandium, cadmium, copper, gold, zinc, germanium, manganese.

Several are commonly found in limestone.

TROMBÉ CURVES

Trombé (1952) has drawn up curves showing the relationship between saturation of $CaCO_3$, temperature and pH. Waters which are undersaturated with $CaCO_3$ at the

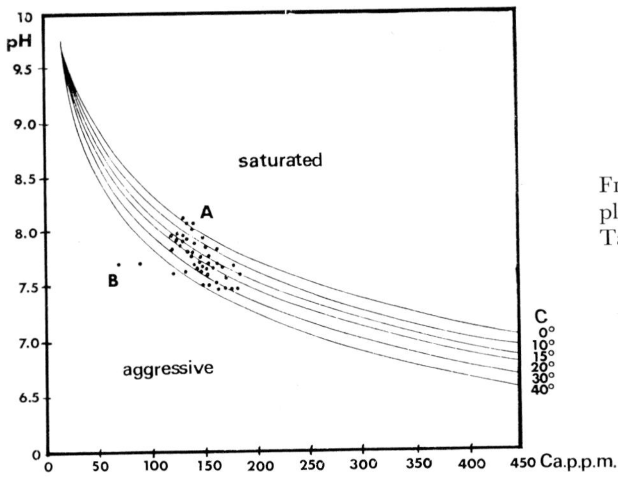

FIG. 18. Trombé curves. The dots plot samples of water from Malham Tarn (from Sweeting, 1966).

measured pH and temperature are termed 'aggressive' and the remainder are termed saturated. The curves are shown in Fig. 18; point A represents a saturated, passive sample, and B an aggressive sample. In the past decade a great deal of work on limestone solution has utilised Trombé curves, but it now appears that this approach is too simple.

When a water sample has been measured and plotted on a Trombé diagram, its place in relation to the curves probably does give an indication of the ability of the water to dissolve further limestone: however, the amount it is able to dissolve depends on the *rates* of the reactions as explained in the previous section (p. 39). It has frequently been found that small trickles of water have high carbonate content. This is because the water has had a relatively long time to approach equilibrium. Measures made on drips of water on stalactites give variable readings, but in general high carbonate contents are indicated. Again, a slowly formed drop may have had a long time to come into equilibrium with limestone, but as it may be actually diffusing CO_2 it is hard to evaluate the meaning of such measurements. Fast-flowing water generally appears "aggressive" when plotted on a Trombé curve, and in fact probably never does reach equilibrium, though it will carry away more carbonate than a saturated trickle.

Thus the usefulness of Trombé curves is extremely limited because in limestone solution the rate factors are much more important than equilibrium factors, and the curves only deal with the latter. It is of lesser importance that Picknett (1964) has shown that the Trombé curves are also inaccurate quantitatively.

MIXED WATER CORROSION

Mixed water corrosion (Mischungskorrosion) is a phenomenon postulated by Bögli (1964).

It is alleged that the relationship between CO_2 and $CaCO_3$ content in equilibrium solutions is not linear, but is curved as in Fig. 19.

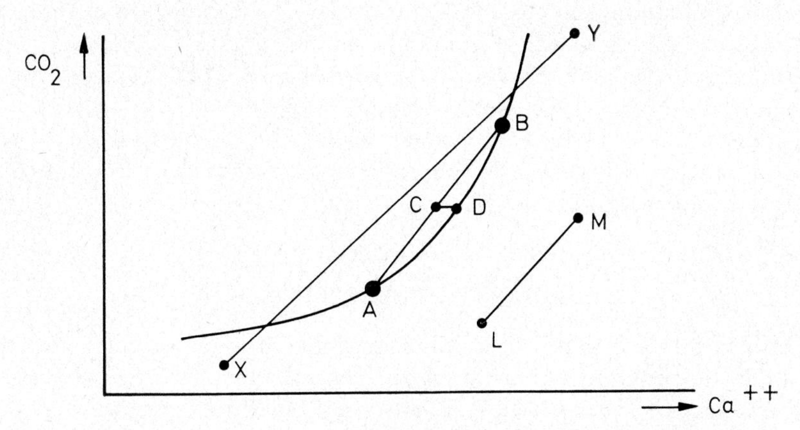

FIG. 19. Diagrammatic representation of mixed water corrosion (from Ashton, 1966).

If this is so, then in some circumstances corrosion occurs when two sample waters are mixed, though both were originally in equilibrium. Suppose two saturated water samples are represented by points A and B on Fig. 19. When they are mixed the resultant concentration would be somewhere on line AB, the position depending on the proportions of each sample in the mixture. Suppose the mixture plotted at point C. This is not in equilibrium and must give off CO_2 or absorb $CaCO_3$ to regain equilibrium. In this way solution of limestone is possible without external supply of CO_2, and Lehmann (1964) has suggested that this may be important in phreatic solution of limestone where no free CO_2 is available.

Samples such as L and M on Fig. 19, producing a mixture below the equilibrium line, would not give rise to mixed water corrosion, but in some circumstances exemplified by samples X and Y the mixture of two saturated solutions might yield an aggressive solution.

It will be noticed that once again rate factors are completely ignored and only equilibrium factors are considered. The feasibility and quantitative importance of mixed water corrosion remain to be determined.

HYDROTHERMAL ALTERATION

Rising waters, steam and other emanations from deep in the earth move upwards through enclosing country rock, and bring about some alteration. This is not weathering, but since hydrothermally altered rocks may come to look like weathered rocks it is necessary to find some way of distinguishing between the two.

Hydrothermal alteration causes argillation, and some major clay deposits such as the Cornwall kaolin deposits have been attributed to hydrothermal effects. Usually argillation is accompanied by mineralisation, and chemical criteria are the best for proving hydrothermal activity. Metal ions such as Au, Sn, Ag, Cu, Zn, Pb and others may be plentiful, and anions such as S, Cl, F, sulphates, arsenides and antimonides are typical, and may exceed carbonate compounds.

Hydrothermal argillation often occurs in an aureole around an ore body, but accompanying mineralisation can be completely absent.

The best non-chemical distinction between clay bodies formed by weathering and hydrothermal alteration is that weathering increases towards the surface, and hydrothermal alteration increases with depth.

COASTAL SOLUTION

The eroded appearance of many limestones on shore platforms, cliffs and boulders leaves little doubt that solution occurs along the coast. This is a little puzzling at first, because in warm seas at least the water is saturated or even supersaturated with carbonate at the surface.

Wentworth (1938) suggested that solution on shore platforms might be caused by fresh water from the land, and such a process could act down to the level permanently saturated by the sea, that is above low-tide level.

Guilcher (1958) put forward three objections to this idea:

1. It is doubtful whether fresh water could exceed the amount of sea water in the intertidal and spray zones, even in a relatively wet climate.
2. Corrosion forms are found at the foot of isolated rocks which are too small to provide appreciable run-off or seepage.
3. Corrosion forms are found in the Red Sea, an arid area where freshwater action can be discounted.

Changes in pH, CO_2 content, or biological activity on a daily basis may have important effects. If carbonate is dissolved by night and precipitated by day, for instance, there is always the possibility of its removal by moving water so that the total effect is of corrosion. It has been shown that some green algae living in pools raise the pH during the day, and that the CO_2 content in coastal pools in California is also raised due to the emission of CO_2 by algae and invertebrates at night. By day the algae absorb it in photosynthesis (Emery, 1956).

Nocturnal cooling may also be relevant, since CO_2 is more soluble in cold than in warm water. Finally surf, with its dissolved gases, may be more corrosive than still water.

The solubility of rocks in sea water is different from that in fresh water. Cigna *et al.* (1963) found that under laboratory conditions saline solutions dissolved up to 10% more calcium carbonate than did equivalent volumes of fresh water under the same conditions. Joly (1901) further showed that basalt, obsidian, hornblende and orthoclase are 3 to 14 times more soluble in salt than fresh water.

Weathering is a chemical problem, but a very complex one. Simple chemistry can hardly do more than describe the mere outline of the processes involved. Extremely complex chemistry is required to understand even quite simple alterations adequately, and very intricate experiments are required to discover mechanisms of weathering under even stringently controlled conditions. It is extremely hard to apply the most thorough laboratory results to real weathering, and although much work is being done on chemical aspects of weathering, mineral equilibria, and so on, it is not possible to treat real weathering with any chemical rigour.

IV BIOTIC WEATHERING

THE breakdown of rocks and minerals is very largely controlled by plants, animals and bacteria, and Polynov (1937) even believed that completely sterile weathering was impossible. This is going too far, for frost and unloading at least can happen unaided by organic forces, but the amount of weathering due to organic agencies, especially bacteria, is probably much greater than was previously thought. However, the amount actually known of the organic contribution to weathering is rather slight.

Physical break-up of rock or fragments may result from burrowing animals or even the passage through the gut of worms and other creatures. However, the main contribution of animals to weathering seems to be repeated mixing of soil materials, thus bringing fresh material into exposure to weathering agents, and allowing more ready access to the mineral particles for air and water. Organic matter may be carried down and assist weathering at lower depths than it might otherwise. Respiration of soil fauna, like that of vegetation, might increase the CO_2 content of the soil atmosphere, a significant feature in chemical weathering.

There is usually a certain depth below which soil fauna do not penetrate, even in soft rock or regolith that is freely drained. In tropical regions, for example, termites may be responsible for the sorting of the upper part of the solum. They can only carry particles up to about 1 mm across to build termitaria. Termitaria eventually collapse and new ones are built up, and in time a horizon of termite-mound material will cover the ground. At the base of termite activity, stones and coarser fragments will accumulate. The profiles in many tropical countries have rock (often weathered, but demonstrably in place) overlain by a stone line, overlain in turn by fairly uniform, fine grained 'soil'. It must be mentioned that this explanation of stone lines by termites is not universally acceptable. On Madagascar, for instance, stone lines are very common but termites are very rare or absent.

Cast-making earthworms burrow to about 1·5 m and pass 10 tons per acre per year as a mean and 20 tons per acre per year as a maximum (Ponomareva, 1950, quoted in Bunting, 1965), enriching the soil surface in exchangeable bases and water-retaining humus and fine mineral matter.

The burrows of rodents have several effects. Rabbits prefer to burrow in sandy material, when effects on weathering will be slight, but in other materials their mixing effect may be significant. Prairie dogs and other animals of the chernozem belt live in colonies occupying areas up to 40 acres. These tend to destroy soil structure, and by constantly re-mixing the soil they hinder leaching and other horizon forming processes.

Snails are common in lime-rich areas and can wear deep holes in limestone (Bunting, 1965). Bird droppings may provide organic matter for the start of soil formation and weathering, and in Antarctica penguins are the main source of organic matter. Ornithogenic soils are formed from guano and the keratin of feathers, with the assistance of blue-green algae. On certain islands bird droppings make enormous deposits of guano, which can weather the limestone and by reaction with carbonate make new materials. Bat guano in caves may erode pits in cave roofs, and possibly weather cave floors beneath guano, with the formation of new minerals. Most caves are in limestone, but they can occur in other rocks. Rare minerals have been formed in some lava caves by the interaction of bat guano and basalt.

Large animals compact soil, thus increasing run-off and soil erosion. The same effect may be achieved by destroying or reducing the vegetation by over-grazing. Erosion amounts to a removal of weathered product, which in turn affects the rate of weathering.

Soil bacteria may be grouped into heterotrophic bacteria, which obtain energy from organic sources, and two kinds of autotrophic bacteria: photosynthetic, which obtain energy from sunlight, and chemotrophic which oxidise mineral substances such as sulphur and iron for their metabolism. Chemotrophic bacteria are the ones of most importance in weathering.

They are extremely active in reducing conditions, and manufacture the sulphides, etc., that are typical of these environments. The iron, sulphide and manganese are the best known, but it has been suggested that they may be responsible for the removal of silica in tropical soils, and for some of the carbonate mineralisation in caves and possibly elsewhere. The nitrogen fixing bacteria (Azobacter) convert nitrogen to NH_4^+ compounds, and can also convert NO_3^- compounds to NH_4^+. They may affect soil pH, and certainly have a great effect on all other soil biota.

Bacterial weathering can be of economic significance, as in the Eliot Lake area in Canada, where water in old mines was found to become acid and to contain appreciable concentrations of uranium and ferric iron in solution which were economically recoverable. Tests showed that bacteria of the Ferrobacillus-Thiobacillus group promoted the production of an acid-oxidising solution effective in leaching, whereas in the absence of bacteria no leaching occurred (Harrison, Gow and Ivarson, 1966).

Soil algae are filamentous or single celled plants with chlorophyll. They are among the first colonisers of bare rock, utilising CO_2, N_2 (in the case of nitrogen fixing blue-green algae) and minute amounts of available nutrients from mineral matter.

The formation of desert varnish has been attributed to colonies of blue-green algae that mobilise Fe ions and produce a concentration of oxides on rock surfaces (Scheffer, 1963).

Varnish incrustations on rocks in the glacial region of the Central Tien Shan at an elevation of 4200 m were found to contain enormous amounts of micro-organisms (up to 1 million per gramme) including bacteria, algae and fungi (Parfenova and Yarilova, 1965).

David Ashton (personal communication) has found algae in cleavage planes of large feldspars and between quartz grains in granite at Wilsons Promontory and elsewhere in Victoria. Such algae are commonest in those places that remain moist for longest, such as crevices and the underside of boulders, and they have been found up to a quarter of an inch beneath the rock surface.

Lichens are symbiotic associations of algae and fungi, the former being the providers of carbohydrates manufactured by photosynthesis. Lichens can live on bare rock surfaces, holding a film of water and extracting nutrients from rock minerals by ion

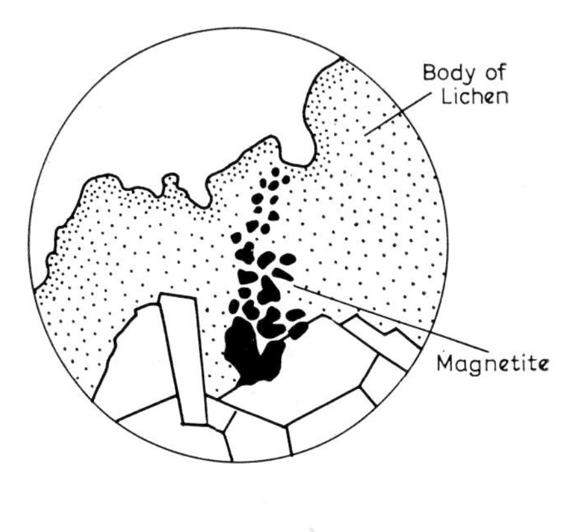

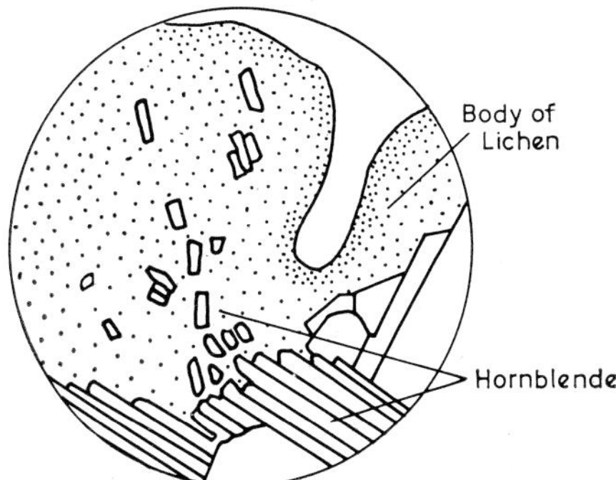

FIG. 20. Destruction of minerals by lichen (after Parfenova and Yarilova, 1965).

exchange. They effect the alteration of minerals both mechanically and chemically. They are swollen when moist and contract when drying, and may break off fragments of mineral which are absorbed into the lichen tissue. Fungal hyphae penetrate cleavage planes of micas, feldspars and other cleavable minerals, forming a dense network and breaking the mineral into minute parts.

Lichen die off to form a 'proto-litter' which provides a base for mosses and higher plants (though some tuft mosses come in with very little if any prior lichen stage). Eventually by mineral breakdown and bacterial humification a deposit is formed which might be called a 'soil' in the usual sense. Rock colonising lichens may be termed 'lithophyllic'.

Fungi are common in soils and may exert a physical effect, breaking the rock by expansion, and the fungal hyphae passing into fine pores of rock and soil aid in chemical effects by CO_2 production, acid secretion, etc. Webley *et al.* (1963) have described the microbiology of rocks and weathered stones. They found an increasing number of micro-organisms with increasing numbers of lichens. Bacteria, actinomycetes (a microfloral form between fungi and bacteria) and fungi were found in the interior of

FIG. 21. Roots prising apart joint blocks on a lava flow. Byaduk, Victoria.

weathered porous stones, but not in unweathered stones. A high proportion of these micro-organisms were able to dissolve silica in pure cultures in the laboratory.

Larger plants affect weathering in a number of ways. Cracks may be widened by root pressure, and the rock broken up a little faster than it otherwise might. Larger plants create a distinct microclimate at the ground surface, and affect the soil atmosphere considerably by root respiration; the increase in CO_2 content of the soil atmosphere is a significant feature in chemical weathering, especially in limestone (see Chap. 12).

The depth attained by roots may be very important. Tap roots of trees are commonly 10 ft deep, and fine roots 20 ft. The record for a traceable root is 175 ft for mesquite in U.S.A. At Wilsons Promontory, Victoria, under sclerophyll woodland a $\frac{1}{4}$ inch root was found at a depth of 17 ft. It was surrounded by $\frac{1}{4}$ inch of grey gritty clay, and outside that was gritty disintegrated granite for another inch. The surface soil in this area was only 2 ft deep, so the importance of roots in transferring weathering to greater depths is evident.

The accumulation of elements by plants and their return to the surface of the soil affect the nature of the soil and weathering profiles, and the course of weathering. Plants absorb certain elements, and on death return those elements to the surface of the soil so that there is a cyclic movement of elements. Cyclic movement of plant

Fig. 22. Root grooves on a joint face on limestone. Kuching, Sarawak. (Photograph by G. E. Wilford.)

nutrients is well known, but there is also movement of other elements, including silica and aluminium. Some plants are very marked silica accumulators, and opal is returned to the soil in the form of phytoliths—odd-shaped grains that for a long time were identified as sponge spicules. These form a significant part of the mineral fraction of some Australian soils. Chenery (1951) has described the aluminium cycle in soils and plants. Costin *et al.* (1952) describes an interesting example of the pedogenic influence of an aluminium accumulating plant, snowgrass (*Poa caespitosa*). This apparently gives rise to a thin layer of gibbsite in the soils, although the environment is cold and has alpine humus soils.

Vegetation litter and decaying vegetation—various stages of humus—are important in helping to conserve moisture, which in turn enhances weathering.

Another important effect of vegetation is the formation of 'leaf leachates' which are very important in cheluviation. The mobilisation of sesquioxides in soil has been ascribed to the action of colloidal humus or silica, but Bloomfield (1964) was unable to detect mobilisation by either. In contrast, aseptic soluble constituents of unhumified plant material were found to dissolve appreciable amounts of ferric and other oxides, and the effect was even greater under non-sterile conditions. Tree leaves are much more active in mobilising iron than are grasses. Polyphenols appear to be mainly responsible for these properties of leaf extracts.

Leaf leachates can speed up the process of podzolisation to an extraordinary degree, and different species of tree have different effects in this respect. Some trees that appear, from experiments on their leaves, to be quite capable of producing a podzol are in fact never found on podzolic soils. Because all the plants tested in the laboratory give extracts capable of mobilising iron, regardless of their behaviour in the field, factors such as the proneness of the phenolic compounds to suffer oxidation, and the palatability of the litter to soil fauna, may ultimately determine whether podzolisation will occur under a given species.

Ng and Bloomfield (1961) showed that elements other than iron are mobilised by decomposing plant remains. Mn, Zn, Pb, Ni, Co, Mo, V and Cu were dissolved, either when present as pure oxides, or in the forms in which they occurred in several soils. As with iron, the metals were in complex form, but unlike iron these elements were not precipitated from solution on aeration.

One most important role of vegetation is in controlling erosion, and the ease of removal of weathered products. Accumulation of weathering products acts as a brake on weathering; the more rapid removal speeds it up. Walker (1963) described the relative rates of erosion under grass and forest in Wyoming; it appears that erosion is at least twice as fast under grass, and it is most likely that weathering will therefore also be faster.

Butuzova (1962) has described the role of root systems of trees in the formation of micro-relief, in the case of butt hillocks formed under roots of pines up to 60 years old. It was found that soil is less leached under the roots due to protection from downward-percolating water.

SUMMARY OF BIOTIC WEATHERING

So far we have considered the action of different organisms in weathering; we can now summarise the total biological effects. Clearly biological weathering is a combination of chemical and physical weathering effects, of which the following are the most important.

1. Simple breaking of particles, as by the eating or burrowing of animals, or by the pressure exerted by growing roots.
2. Transfer and mixing of materials, mainly by animals, moving mineral materials into areas of different weathering effects.
3. Simple chemical effects, as when solution is enhanced by the CO_2 produced by respiration.
4. Complex chemical effects such as chelation, and the formation of complexes of organic-mineral substances.

5. Effects on soil moisture, and thus on weathering. These effects are partly due to the water-holding effects of root masses and humus, partly by the shade effects of plants. Shade may cut down evaporation, but transpiration may lose water at a greater rate.

6. Effect on ground temperature, by shade, by production of heat as in fermentation, and by moving material to and from the surface, that is into zones of different temperatures.

7. Effects on pH of surfaces. These are largely due to respiration, but also direct chemical effects. The exchange reactions by which plants absorb materials cause pH changes. Plants absorb CO_2 by day and not at night, so there can be a daily change in pH.

8. Protection from erosion, both water erosion and wind erosion, causes less exposure of new surfaces and therefore less total weathering under most circumstances. Complete exposure may render weathering slower due to less moisture retention.

BIOLOGICAL WEATHERING ON THE COAST

Some invertebrates such as *Pholas* and *Patella* bore holes into rock, either mechanically or by acid secretions. This primary attack also allows easier penetration of other weathering agents. Respiration of animals and plants produces CO_2 which aids solution of limestone. Some writers go so far as to ascribe all coastal corrosion of limestone to algae, but this is not so, for some corrosion, especially in the spray zone, is a purely chemical effect.

Animals also cause miscellaneous minor breakdown of rocks. Crabs may enlarge fissures by jamming themselves inside and thus levering the blocks apart. Some fish eat coral, and knock chunks from the living reef by butting. Many animals comminute and alter fine sediment as it passes through their bodies. On shore platforms the mats of algae and mussels protect the rock from abrasion and also aid in retaining water in small pools, preventing drying out. These effects are important in the genesis of such platforms, and Hills (1968) writes of the biological control of shore platform weathering as follows:

'Although it is easy to demonstrate that abrasion plays no effective role in the formation or modification of local platform surface, and that solution decreases in the wave notch nearer to the platform, no rational explanation based on a reasonable scientific hypothesis has been presented anywhere for the absolute preference of the platform level. Indeed, the only agents known to be so narrowly limited are the organisms that grow on the platform, and these certainly appear to call for further study by marine ecologists.'

V MINERAL WEATHERING

CHEMICAL composition alone is not a sure indication of weatherability, and the different rates of weathering of minerals of the same composition demonstrate the futility of a too 'chemical' approach to weathering. The densest mineral is the most stable isomorph. Kyanite (Al_2SiO_5) is denser and more stable than sillimanite and andalusite, which have the same composition. Rutile (TiO_2) is more stable than anatase or brookite, and quartz (SiO_2) is more stable than tridymite or crystobalite, which have the same formula.

The rate of weathering of a mineral depends on several other things besides its structure and composition. The main ones are:

1. *Crystal size.* Large minerals are harder to weather than small ones. This is because weathering can be regarded as a surface activity, and many small crystals have a much greater surface area than a single large crystal of the same volume. If a grain 1 mm across is broken into particles of 0·1 mm across, the surface area increases at least a thousand times.

2. *Crystal shape.* Platy crystals are more weatherable than chunky ones, for more of the crystal is near to a crystal face, that is near to the weathering surface.

3. *Crystal perfection.* Perfect crystals, that is those with a perfect geometrical lattice, are comparatively resistant to weathering, for each atom is securely in place. Where there are crystal defects—steps, faults, etc., and other blemishes, there are loose bonds and atoms are not well held, and the larger the number of defects the more rapidly can the mineral weather. However perfect the structure, there will be loose bonds on the surface, and especially at edges, and reaction is greatest at crystal edges. This is another reason why many small crystals weather faster than a single large one.

4. *Access of agent and removal of weathered product.* The more the weathering agent can get at a mineral, the more it will weather it. Thus, if a rock is porous and water can attack all grains, the weathering will be faster than if the rock is dense and compact and water can only penetrate from the rock surface, not from every mineral or grain surface. A porous limestone would thus dissolve faster than a massive one, other things being equal.

On a smaller scale, a mineral with good cleavage allows solutions to reach not only the mineral surface but also the cleavage planes and the greater ease of access allows more rapid weathering.

Cleavage may be picked out to give a saw-tooth or frayed appearance, as is sometimes seen in weathered grains of augite, hypersthene, hornblende and staurolite.

In any chemical reactions, the products must be removed for reaction to continue, or an equilibrium will be achieved. In weathering it is necessary for the products of weathering to be removed, or reaction will slow down or cease. Consider the case of a solution of calcite. If water is added to calcite and nothing further added or removed, a small amount of calcite will be dissolved until the water becomes a solution in equilibrium with the calcite, but unless the solution is replaced by fresh water (or some other change effected) no more calcite will dissolve. Running water will be much more effective in solution than stagnant water, for it will remove weathering products.

In more complex examples, removal of product is not always so easy. A grain of epidote, for example, may weather on the surface to clay and iron oxide, which is firmly attached to the crystal and immobile—inert chemically and hard to remove physically without moving the grain. This coating may be almost impenetrable to water and prevents water from attacking the rest of the mineral. It is largely because of clay films of this sort that real weathering series as determined in soil studies do not always correspond with theoretical or experimental series.

THE WEATHERING OF MINERALS

Rocks consist of minerals which are altered into secondary minerals during the course of weathering, often into clay minerals. Secondary minerals, including clay minerals, can alter into other minerals. A comparatively small number of minerals make up the bulk of all rocks, and are known as the rock-forming minerals. A slightly larger number of minerals are fairly common in rocks, but only in small quantities; these are the accessory minerals. A very large number of minerals are known to occur naturally, but are uncommon, and unimportant for a general review of mineral weathering.

ROCK-FORMING MINERALS

Quartz. Silicon dioxide (SiO_2) is the commonest mineral in sandstones. An essential mineral in granites and present in all acid and intermediate igneous rocks and many metamorphic rocks. It is found in small amounts in nearly all sediments. In igneous rocks it is often found to be the last mineral to crystallise from a magma and so often does not have the euhedral form of perfect crystals. The fact that it crystallises at lower temperatures suggests it might be fairly stable at normal temperatures of the surface, and this appears to be so.

Quartz is very resistant to chemical weathering, though it does dissolve in certain little-known conditions. It has no cleavage or fracture (although petrofabric studies of sandstones have indicated some preferential directions of fracture, not visible in individual crystals) and so is also resistant to physical weathering, or even attrition during transport by wind or water. Grains can eventually be rounded, as in the case of the so-called 'millet seed' grains of wind-blown sand. Quartz is almost sufficiently resistant to be a standard against which the weathering of other grains can be measured, and in many weathering profiles the proportion of quartz increases towards the surface. Polynov considered primary quartz such a stable substance that it may be considered to be quite passive in the zone of weathering, an opinion considered extreme by Reiche.

Feldspar. The feldspars are a group of alumino-silicate minerals. The main ones are orthoclase and microcline (both potash feldspars with the general formula (K,Na) $AlSi_3O_8$ and plagioclase feldspars with the general formula (Na, Ca) Al(Al, Si)Si_2O_8. The sodium-rich variety of plagioclase is albite, and there is a complete gradation through the intermediate species oligoclase, andesine, labradorite and bytownite to the calcium rich variety, anorthite.

Feldspars occur in most igneous rocks and metamorphic rocks. They are uncommon in most sediments, but abundant in arkoses and greywackes.

Feldspars are almost as hard as quartz, but cleavage is well marked, especially in the plagioclase feldspars, and this probably allows rapid attrition and penetration by water, enabling rapid alteration.

Microcline is the most resistant feldspar, followed by orthoclase. The plagioclases weather more rapidly, and the calcic members faster than the sodic.

Feldspars often alter to kaolin, but secondary mica (called sericite) and a number of other secondary minerals such as allophane can also be formed.

Feldspar is so weatherable that special conditions are needed to form feldspar-rich sediments, of the sort that might lithify into arkoses and greywackes. Weathering in arid regions, particularly by granular disintegration, may produce transportable debris with little chemical alteration. Nile muds contain fresh microcline and orthocase, indicating a derivation from an arid region. Glacial regions too have much physical weathering and little chemical, so can also provide feldspar-rich debris. In some instances very rapid erosion may produce sediments in which the feldspars simply do not have time to weather.

Mica. The mica group of minerals are basic aluminium silicates with a sheet structured crystal lattice, enabling the grains to break down easily into flakes. The two main micas are biotite, or black mica, with the formula $K(Mg, Fe)_3AlSi_3O_{10}(OH)_2$, and muscovite, or white mica, with the formula $KAl_3Si_3O_{10}(OH)_2$.

Micas occur abundantly in granites and many other igneous and metamorphic rocks, and are also found in smaller amounts in numerous sedimentary rocks. They are usually absent in wind-blown sediments or rocks derived from them, as mica is very prone to attrition during wind transport.

Mica is soft and is easily attacked by water because of the cleavage. Ion exchange is very common, and mica alters, often without great change in structure, into chlorite and other clay minerals. The structure of the micas is further described in the section on the clay minerals.

Olivine. A magnesium-iron silicate $(Mg, Fe)_2SiO_4$, found in basalt and other ultrabasic and basic igneous rocks. It has no cleavage but is often traversed by irregular cracks. It is a very weatherable mineral. Even on basalt shores olivine is rare, and never found 100 yards offshore. In fresh water and in soils weathering is also rapid. Alteration of olivine even starts during magma crystallisation (deuteric alteration) due to removal of magnesium from the rim of the olivine crystals and collapse of the crystal lattice into a rather chlorite-like structure and iron oxide. Similarly, 'iddingsite' is formed during weathering. This could easily be further altered to other clay minerals.

Pyroxene. The pyroxene group of minerals has a crystal structure in which silica tetrahedra are arranged in a chain-like manner. The crystals have good cleavage.

The commonest member of the group is augite, $Ca(Mg, Fe, Al)(Al, Si)_2O_6$. This is found in intermediate and basic igneous rocks such as dolerite and basalt, and in metamorphic rocks, but is rare in sediments except as an accessory mineral.

The cleavage assists in rapid weathering, and augite weathers by ion exchange and lattice alteration to clay minerals, but may first alter to hornblende, epidote, chlorite or other minerals.

Other pyroxene minerals are the magnesium-rich enstatite, iron-rich hypersthene and calcium-rich diopside. A variety of augite contains titanium, and is called titan-augite.

Titanaugite is probably the most resistant pyroxene to weathering, followed by augite, and then other varieties.

Amphibole. The amphibole group of minerals have a crystal structure made of silica tetrahedra linked in double chain formation, and are characterised by complicated chemical formulae. Tremolite, the simplest, is $Ca_2Mg_5Si_8O_{22}(OH)_2$.

In hornblende some Si is replaced by Al, and Na and K can be simultaneously incorporated into the structure. Mg can be replaced by Al, Fe_2, Fe_3 and Mn, and OH can be proxied by F.

Amphiboles occur in igneous and metamorphic rocks, but only as an accessory mineral in sediments. Some rocks (amphibolites) consist almost entirely of hornblende.

The amphiboles have good cleavage, and like pyroxenes alter by ion exchange and lattice alteration mainly to chlorite and other clay minerals, with release of bases to solution.

Amphiboles are usually more resistant to alteration than the pyroxenes, hornblende more resistant than other amphiboles, and so-called basaltic hornblende, which has a high proportion of Al_3 and Fe_3, is definitely more resistant in acid conditions than common hornblende.

Carbonates. Calcite, $CaCO_3$, and dolomite, $(Ca, Mg)CO_3$, are both trigonal minerals. They are the dominant minerals in limestones and are common in other sedimentary rocks such as calcareous sandstones and marls. Marble is metamorphosed limestone and consists very largely of carbonates. Carbonatite is a rare igneous form of carbonate rock.

Carbonates are the most soluble of the common minerals, and limestone gives rise to distinct landforms due to the prevalence of solution. The actual mechanism of solution is fairly complex and is discussed further in Chap. III. Dolomite is sometimes an end-product of weathering and diagenesis. It has a larger volume than calcite, and if calcite is converted to dolomite the expansion may cause physical weathering.

Two carbonates which are not rock-forming minerals may be mentioned here for completeness.

Siderite, $FeCO_3$, is much rarer than the other carbonates, and is readily altered and removed.

Aragonite is an orthorhombic form of calcium carbonate, and is more soluble than calcite. The relationships between the two have been summarised by Curl (1962).

ACCESSORY MINERALS

These are the minerals present in small amounts in rocks, and not essential constituents. Their total bulk is negligible in weathering, but they are extremely useful in certain studies of weathering, and their theoretical importance is much greater than their volumetric insignificance would suggest. Many accessory minerals are so very rare as to be beyond the scope of this book. Others are so easily altered that they are of no significance in real weathering and can be regarded as fugitive minerals or merely transition stages in alteration. Hence we shall be considering those accessory minerals that are in general resistant to weathering and are of significance in weathering studies.

1. *Zircon*, $ZrSiO_4$. Zircon occurs as orthorhombic crystals that are often euhedral (well shaped). Probably all zircons are originally formed in granites, but it is such a resistant mineral that it may survive through several cycles of weathering and erosion and even metamorphism and granitisation, so that one granite may contain zircons of different source and age. In weathering profiles it has been considered inert, and used as a standard against which changes in proportions of other minerals can be gauged. Zircon is rendered even more useful by its almost universal distribution and nearly all heavy mineral samples contain some zircon. Because of its usefulness it is often studied in greater detail than other minerals, as it may give clues to provenance (that is the source and environment from which it was derived). Thus proportions of euhedral to rounded grains may be counted, or zircons may be distinguished on colour. Most zircons are nearly colourless, but pink, purple, blue and other colours are also found. Inclusions, zoning, and radio-activity may also be utilised.

Some workers believe zircon to be less stable than is commonly supposed. Carroll (1953) distinguished between normal, hyacinth and malacon varieties of zircon, and believes that only the normal possesses the resistant character ascribed to zircon. Marshall (1965) writes that zircon should be more prone to attack in alkaline solutions than in acid.

2. *Tourmaline*. A complex silicate crystallising in the trigonal system, usually found as lath-shaped grains coloured, pink, green, brown or opaque, and markedly pleochroic. The grains are usually rounded, but are very resistant to weathering. It is derived from granites and metamorphic rocks, but can survive several cycles of weathering and erosion, and so is a very common accessory mineral in sediments.

3. *Rutile*, TiO_2. Tetragonal crystals, usually of dark red or brown colour, with very

high refractive index and birefringence. It is derived from granites, and survives several cycles of erosion.

4. *Garnet.* A group of minerals of which almandite $(Mg, Fe, Mn)_3Al_2(SiO_4)_3$ is the commonest. They occur as equidimensional grains, usually red in colour, isotropic and often displaying hackly fracture but no cleavage. They are derived from igneous and metamorphic rocks, especially schist and gneiss. In general, garnets are resistant to weathering, but the calcium-rich members weather more rapidly than the iron-magnesium and aluminium-magnesium members. In soil, under acid condition, they disappear at what is, in geological terms, a rapid rate (Marshall, 1965).

5. *Apatite.* This is a hard phosphate mineral lacking cleavage. It is often very stable but seems to be rather rapidly attacked and altered in acid environments. A cryptocrystalline form of apatite is collophane, derived from animal phosphates, teeth, bones, fish scales and the like. It may be a source of phosphates in soils. It is readily dissolved in acid, and is most commonly found in limestones or their insoluble residues. Brown and Ollier (1957) suggest that it may be commoner than generally believed, as normal laboratory treatment of samples tends to destroy it.

6. *Andalusite,* Al_2SiO_5. A metamorphic mineral found as tabular grains. It is moderately resistant to weathering and often has a distinctive appearance (pink to colourless pleochroism) and so is a useful indicator mineral in weathering studies.

7. *Anatase,* TiO_2. A yellow or blue mineral of distinctive appearance that may be derived from igneous or metamorphic rocks, but is very commonly an authigenic mineral, formed from decomposition of ilmenite or other titaniferous species.

8. *Epidote.* A mineral found mainly in metamorphic rocks rich in calcium, and in sediments subjected to extreme diagenesis. It is yellowish-green in colour, with one marked cleavage, and becomes cloudy and opaque with weathering.

9. *Kyanite,* Al_2SiO_5. A mineral derived from metamorphic rocks. It has good cleavage and gives distinctive tabular or lath-shaped grains which are fairly resistant to alteration and provide a good indicator in weathering studies.

10. *Sillimanite,* Al_2SiO_5. A metamorphic mineral, moderately resistant to weathering, found as fibrous or lath-shaped grains. Sillimanite is a useful indicator mineral.

11. *Staurolite.* A metamorphic mineral found as irregular or platy grains, and a useful indicator mineral. It sometimes displays 'concertina' edges, presumably due to solution, and superficial decomposition to chlorite is occasionally seen.

12. *Gypsum,* $CaSO_4 . 2H_2O$. Gypsum is found as well-formed colourless crystals or as fibrous grains. It is not uncommon in sedimentary rocks, but it is readily dissolved and re-formed, and much gypsum encountered in weathering studies is secondary or authigenic gypsum. In some soil profiles gypsum-rich horizons may be formed.

Opaque minerals

The opaque accessory minerals are conveniently grouped together because their study calls for different techniques from those used on transparent minerals.

1. *Magnetite*, Fe_3O_4. Occurs as equidimensional black grains in many igneous and metamorphic rocks. It is resistant to alteration in a reducing environment, but is fairly readily altered in oxidising conditions to limonite.

2. *Hematite*, Fe_2O_3. Found in sedimentary rocks mainly. Fairly easily altered to limonite.

3. *Limonite*. Limonite is a general name given to the red and yellow hydroxides of iron, which are partly amorphous and partly cryptocrystalline goethite and lepidocrocite. There are very complex relations between the many oxides and hydroxides of iron and determination of actual mineral species is not easy. Colour is not a certain guide to mineralogy or chemistry, but in general it seems that the yellow iron oxides are more hydrated than the red ones.

4. *Pyrite*, FeS_2. Derived mainly from igneous and sedimentary rocks, pyrite is generally stable in reducing conditions and may form authigenically in suitable environments. In oxidising conditions it breaks down readily, with the formation of sulphuric acid which affects the weathering of other minerals.

5. *Ilmenite*, $FeTiO_3$. Derived from basic and ultrabasic igneous rocks, ilmenite is found in many sediments and sedimentary rocks, and is far commoner than any other iron ore. It is usually stable, but may alter to leucoxene.

6. *Pyrolusite*, MnO_2. This is usually a secondary mineral in sediments and soils, occurring in many forms such as nodules, 'ferns', etc. It appears to be easily mobilised and reprecipitated.

WEATHERING SERIES

In the list of minerals given above, different species have been said to be 'readily weathered', 'resistant to weathering' and so on. There is no absolute scale of degree of weathering, and nor is there ever likely to be, for the weatherability depends on environment of the mineral as well as its innate structure and composition. However, the problem of putting degree into weathering may be approached by ranking minerals in a series—the weathering series—according to a general ease of weathering. The weathering series is based on many studies, as in any one rock or soil only a limited number of minerals are present.

Weathering profile studies (or soil profile studies) provide one sort of information. As an example the mineralogy of the Buwekula Shallow and Buwekula Red Soil profile data is shown below. The data shows, from the order of disappearance of minerals in Buwekula Shallow, that zircon, tourmaline and epidote are very stable (in this ennviroment), while muscovite is less stable, but more stable than biotite which is present only in the lower horizons. Comparisons with the more weathered Buwekula Red profile shows that epidote is the next mineral to disappear.

Table 5 Mineral Analysis of Buwekula Soils

Depth (inches)	% magnetite in heavy minerals	% feldspar in light minerals	Main non-opaque heavy minerals
Buwekula Shallow			
0–3	50	56	Z, T, E
3–10	50	45	Z, T, E, M
10–18	20	75	Z, T, E, M
18–36	50	72	Z, T, E, M, B
36–60	40	70	Z, T, E, M, B
60–72	60	75	Z, T, E, M, B
Buwekula Red			
0–3	5	0	Z, T
3–8	5	0	Z, T
8–18	10	1	Z, T
18–40	10	1	Z, T
40–60	20	1	Z, T
60–72	10	0	Z, T

Z, zircon; T, tourmaline; E, epidote; M, muscovite; B, biotite.
Data from Radwanski and Ollier (1959).

Because opaque minerals were studied separately the magnetite figures cannot be compared with the others, but the general and marked trend shows a fairly easy degree of weathering, perhaps comparable to that of biotite. Feldspar (mainly orthoclase in this case) shows a very marked degree of weathering, even greater than that of the micas. This is clearly revealed by comparison of Shallow and Red profiles.

The study of the Buwekula catena suggests a weathering sequence as follows:

zircon	quartz
tourmaline	
epidote	
muscovite	
biotite	magnetite
feldspar	

Many studies of mineral changes through soil profiles are available. For example, in a petrographic study of soils from Ghana, Stephens (1953) found that feldspars, pyroxene and hornblende were largely removed and epidote and garnet accumulated.

In Japan, Aomine and Wada (1962) found that in weathering volcanic ash the order of relative stability was

glass < feldspar (andesine/labradorite) $\leqq$ hypersthene, augite < magnetite

By using many such studies a more complete picture of the mineral weathering sequence can be obtained, and it is found that individual minerals do not always occupy the same position in the sequence. In the Buwekula soils, for example, epidote seems to be out of order, and in many cases is more weatherable than mica.

Another approach to weathering series suggested by Pettijohn (1941) is by means of geological studies, especially of sedimentary rocks. It is found that olivine is only present in recent sediments and never in pre-Quaternary rocks; it is incapable of resisting several cycles of weathering and is high on the list of weatherability.

Leucite, like olivine, is very unstable, and it is absent from beach sands along the Tyrrhenian coast although it is common in the volcanic rocks along the shore (Uzielli, 1875).

Thomas (1909) showed that though andalusite was well known in Pleistocene and Pliocene sediments it was rare in older rocks. It has since been found in earlier rocks, down to Permian and possibly Palaeozoic, but the observation and its implications regarding weathering rates is still valid. Edelman (1931) similarly showed that ferromagnesian minerals are rarer in older sediments than younger.

Sillimanite and topaz are not reported in sediments antedating the Mesozoic.

From many studies, the following brief weathering series may be constructed for the commoner minerals. Anatase, rutile and muscovite get negative numbers because they get more frequent in older sediments due to diagenesis.

Table 6

−3. Anatase
−2. Muscovite
−1. Rutile
1. Zircon
2. Tourmaline
3. Monazite
4. Garnet
5. Biotite
6. Apatite
7. Ilmenite
8. Magnetite
9. Staurolite
10. Kyanite
11. Epidote
12. Hornblende
13. Andalusite
14. Topaz
15. Sphene
16. Zoisite
17. Augite
18. Sillimanite
19. Hypersthene
20. Diopside
21. Actinolite
22. Olivine

The weathering sequence for very fine grained minerals is rather different. A sequence has been drawn up by Jackson *et al.* (1948) from studies of the relative abundance of residual minerals in the clay fractions of soils, but is confused by the presence of secondary minerals. Thus anatase, rutile and corundum are given as very stable minerals said to weather extremely slowly, but while rutile is probably a resistant mineral from parent material, anatase is frequently authigenic.

Table 7

Group	Mineral	Weathering
Primary minerals	1. Gypsum (halite)	soluble, unweathered or as secondary deposits
	2. Calcite (aragonite, dolomite)	
	3. Olivine, hornblende (diopside)	easily and rapidly weathered
	4. Biotite (chlorite, glauconite)	
	5. Albite (microcline, anorthite)	
Secondary minerals	6. Quartz	slowly weathered
	7. Illite (muscovite)	
	8. Hydrous mica intermediates	
	9. Montmorillonite	
	10. Kaolinite	weathered extremely slowly
	11. Gibbsite	
	12. Hematite (geothite, limonite)	
	13. Anatase (rutile, ilmenite, corundum)	

Another approach to weathering is by laboratory experiment. A simple method is to subject different minerals to identical weathering conditions and see what happens. In this way it can be shown that calcite is more weatherable (soluble) than dolomite. A more esoteric technique is to crush minerals and measure their reaction with water by the change in pH (see p. 35). Weathering has to be treated as a rate reaction, but it seems that the weathering series corresponds very closely to the series of simple reaction with water.

Many physical and chemical experiments have been performed on mineral weathering, but usually the more elaborate the experiment the harder it becomes to relate the results to natural weathering.

ABRASION WEATHERING

Minerals may be worn away by simple mechanical abrasion, and the resistance of a number of minerals has been determined by Freise (1931) and Thiel (1940). Mineral hardness is very variable, and is measured by comparison with a series of minerals in Mohs' scale. The differences in hardness on the scale are not equal, but for comparative purposes it is normally sufficient. A mineral such as calcite with a hardness of 3 would be scratched by quartz (7) but would be able to scratch gypsum (2). Freise constructed his abrasion series by comparison of minerals with crystalline haematite. It is not the same as hardness scale, because cleavage and 'tenacity' influence the abrasion characteristics.

The weathering environment affects the real rate of abrasion of minerals. Mica, for example, is fairly resistant to abrasion in water-borne alluvium, but in wind-blown deposits it is fairly rapidly destroyed by attrition.

Thiel (1940) performed abrasion experiments on minerals in a manner that he believed to approximate to that of minerals in running water. The results obtained were rather different from those of Freise, and no doubt the 'abrasion series' would vary from one environment to another. The lists obtained by Freise and Thiel are shown in Table 8.

Table 8

Freise		Thiel
Hematite	100	Apatite
Orthoclase	150	Hornblende
Quartz	245	Microcline
Apatite	275	Garnet
Garnet	420	Tourmaline
Tourmaline	850	Quartz

It is observed that there is a scarcity of certain particle sizes in sediments. Tanner (1958) quotes 0·03 to 0·12 mm and 1·0 to 8·0 mm as the particular values. These gaps divide sediments into gravel, sand and silt, which may be called blocks, grains and flakes. Rogers, Kreuger and Krog (1963) report experiments suggesting that there are two distinct modes of abrasion, one giving sand (grains) and the other giving silt (flakes). Smalley (1966), on the other hand, believes the size and shape of sand grains are largely determined at the moment the eutectic quartz solidifies in a cooling magma, and that so far as quartz is concerned abrasion plays no effective part in the formation of sand.

VI CLAY MINERALS

A CLAY material has fine particle size, and may contain tiny fragments of ordinary minerals, but in the main consists of distinctive clay minerals and some amorphous colloids.

The clay minerals have their atoms arranged in layer lattice structures, and knowledge of clay structure helps in understanding the formation, interrelationships and changes amongst clays during weathering.

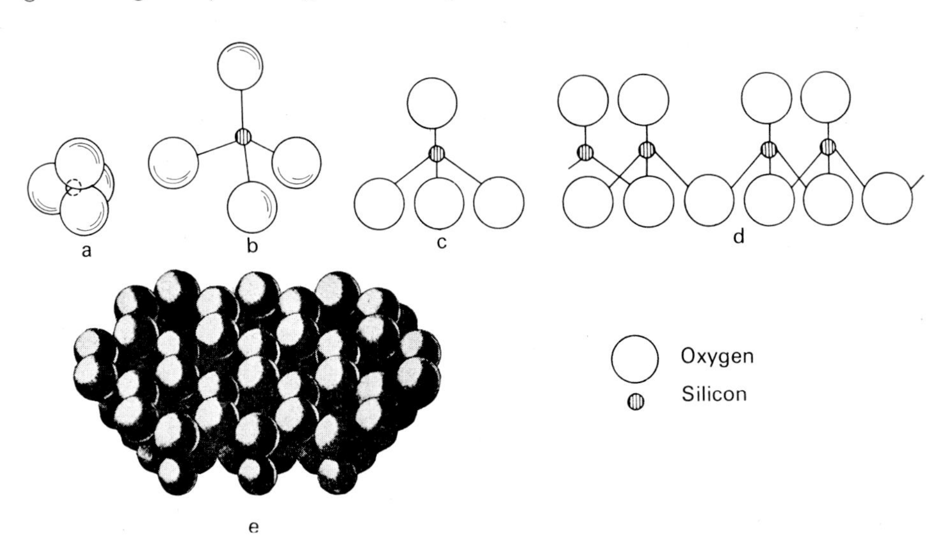

FIG. 23. *a-e* The silicon-oxygen tetrahedron and tetrahedral sheet.

a The atoms packed together. *b* The tetrahedron expanded. *c* Conventional representation of a tetrahedron. *d* Conventional representation of a tetrahedral sheet. *e* Model of a tetrahedral sheet.

A simple Si-O tetrahedral layer may be regarded as the basic structure, although it never occurs in isolation. Atoms of other elements may sometimes substitute for silicon in this structure. Another important layer, called octahedral, consists of aluminium and oxygen or hydroxyl (which is virtually the same size as oxygen) arranged as in Fig. 24. This layer can exist in isolation (or rather a stack of successive sheets can) and this structure is similar to that of gibbsite or brucite.

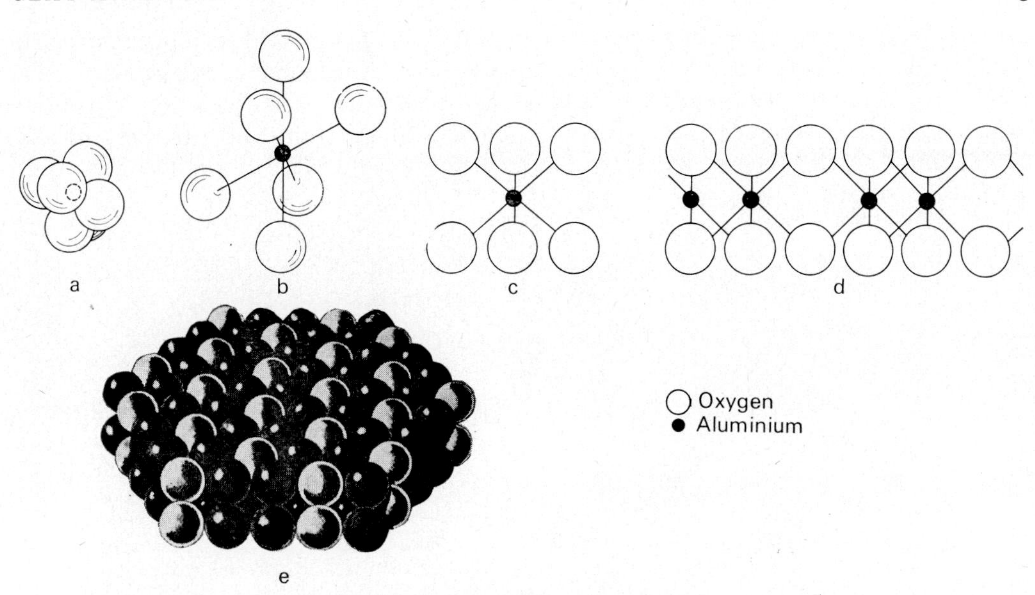

FIG. 24. *a-e* The aluminium-oxygen octahedron and the octahedral sheet.

a The atoms packed together. *b* The octahedron expanded. *c* Conventional representation of an octahedron. *d* Conventional representation of an octahedral sheet. *e* Model of an octahedral sheet.

A Si-O tetrahedral layer and a gibbsite-like octahedral layer can be arranged together as shown on Fig. 25. This is the basic unit of the kaolin minerals. As there is

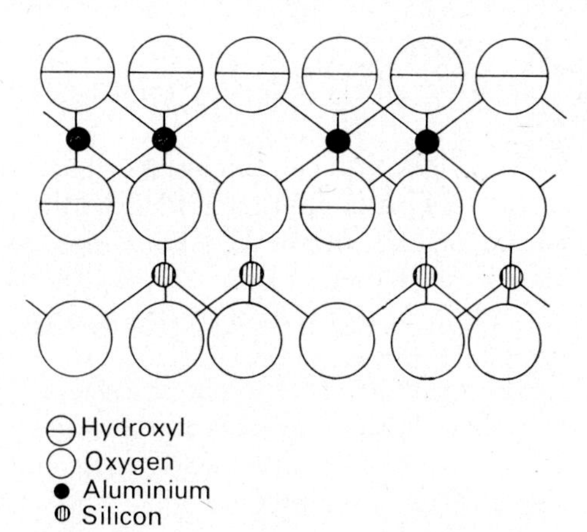

FIG. 25. One tetrahedral sheet combined with one octahedral sheet to give the kaolinite structure.

just one Si-O tetrahedral layer and one gibbsite-like layer this is called a 1 : 1 clay mineral. The sheets are of course repeatedly stacked on one another.

Another arrangement is to have two Si-O tetrahedral layers sandwiching one gibbsite-like layer symmetrically as in Fig. 26. These minerals are called 2 : 1 clay minerals. They are sometimes referred to as the mica group, and include montmorillonite and illite (hydrous mica). The 2 : 1 clays are also found in stacks, with other atoms as interlayer cations between silicate layers to balance the charges. In montmorillonite the interlayer cations involve sodium, magnesium and calcium; in illite potassium. An alternating sequence of a mica-type layer and a gibbsite-like layer gives a chlorite structure.

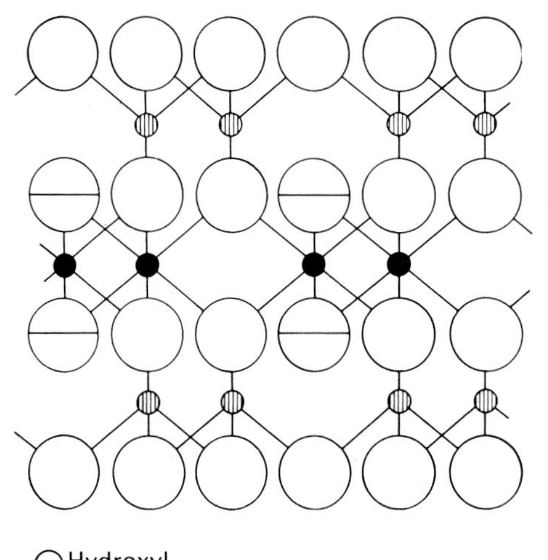

Fig. 26. One octahedral sheet sandwiched between two tetrahedral sheets to give the mica structure.

The properties of clays depend on their lattice structure to a large extent.

1. *Spacing.* The unit cell thickness or basal spacing is the distance between successive similar layers. This is characteristic for different minerals, and is used in the X-ray identification of clay minerals. Kaolinite, for example, has a basal spacing of 7·2 Ångstrom units, chlorite 14·1. Fig. 27 shows some more realistic diagrams of clay minerals, and illustrates the idea of basal spacing.

2. *Expanding lattice.* Some clays, especially montmorillonite, have the property of expanding their lattices by absorbing water (or in laboratory conditions, glycol or glycerol). A layer of oriented water molecules forms an extra sheet, the thickness of which depends on associated cations. For laboratory determination of montmorillonites it is normal to replace cations by Ca, expand the lattice with glycol and then X-ray. 1 : 1 clays do not expand except for hydrated halloysite.

The swelling and shrinking of clays upon wetting and drying is of some importance in rock weathering, and also in such practical matters as building foundations, self-mulching soils, and soil-moisture relationships.

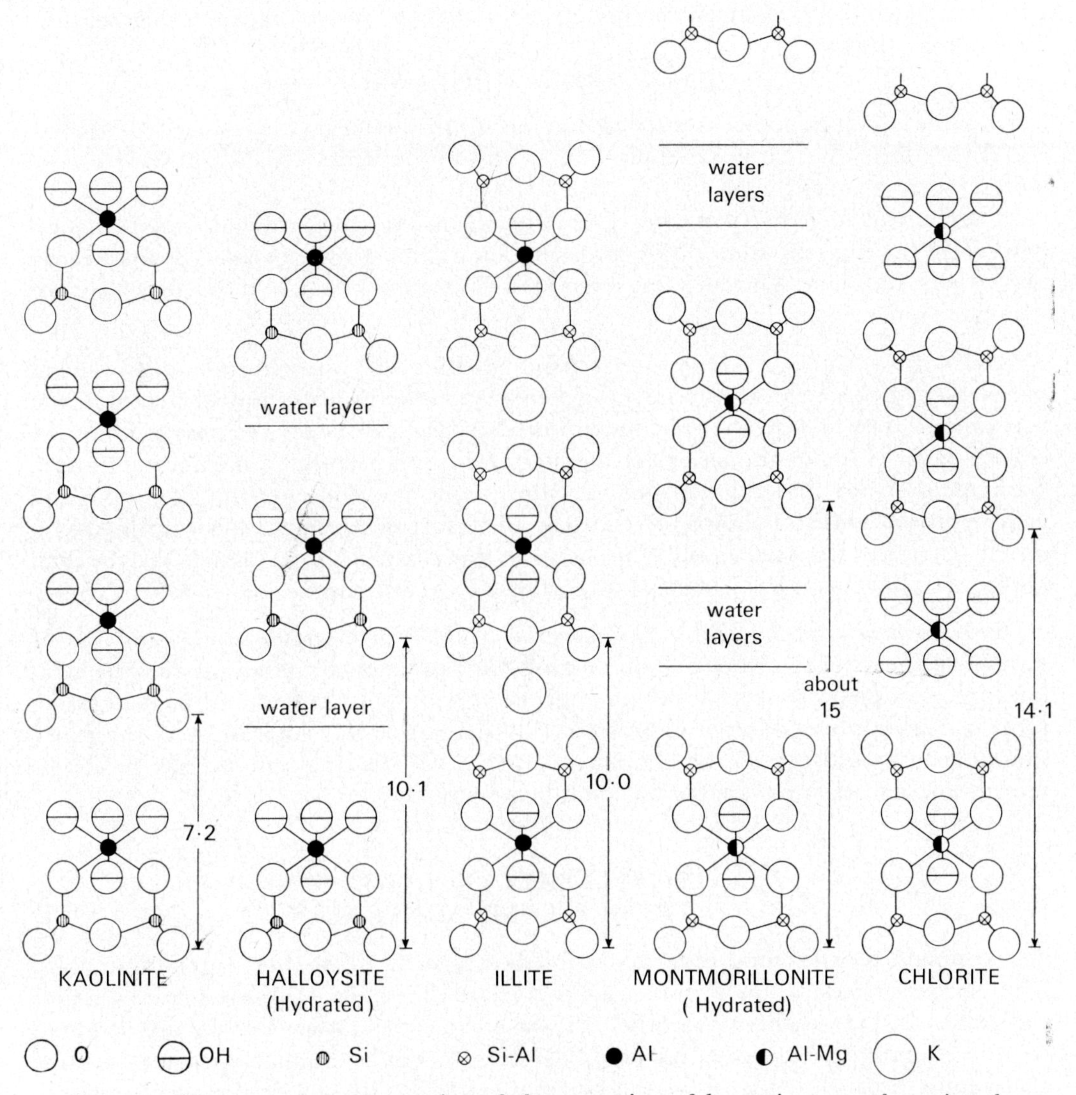

FIG. 27. Diagrammatic representation of the succession of layers in some clay mineral structures.

3. *Water absorption.* The more finely divided a clay is the more water it can hold, but the manner of attachment of water molecules is important. Water may be held by a clay mineral either adsorbed on the surface, as interlayer water, or as hydroxyl water attached to the crystal lattice.

Montmorillonite has a high absorption and both illite and kaolinite have a low one.

4. *Ion exchange.* At the edge of clay minerals there are unsatisfied bonds, on which

ions not related to the actual structure are held. These may be replaced if conditions are suitable, thus

$$\text{Na-clay} + \text{H}^+ \rightleftharpoons \text{H-clay} + \text{Na}^+$$

In kaolinite cation exchange is restricted to unsatisfied bonds on the edges of the lattice, and is therefore low. In micas there is also substitution within the actual structure, so ion exchange is much higher.

The ion exchange capacity therefore depends on the structure and chemical composition of the mineral and on the environment. Some ions are replaced more easily than others, but there is no constant series for any clay mineral. In most cases Ca is the dominant replacement ion.

5. *Dialysis of clays.* If a colloid is suspended within a semi-permeable membrane in pure water, then some ions will pass the membrane into the water until an equilibrium is attained. This is dialysis. The membrane has to allow the ions to pass, but not the colloid particles. Clay minerals act themselves as semi-permeable membranes, and if a fragment of clay mineral is placed in water it will lose some ions into solution, and gain hydrogen ions to balance the charges. The sort of equilibrium set up in such a system is called Donnan equilibrium, and is described in great detail by Marshall (1965) who believes it to be of great importance in weathering studies.

6 *Differential thermal effects.* Clay minerals usually have large endothermic and/or exothermic heats of reaction. Upon heating the clays, dehydration and dehydroxylation processes give rise to endothermic effects. Oxidation changes and crystallisation processes bring about exothermic effects. By heating a clay and an inert substance simultaneously and comparing their temperatures the thermal effects may be determined, and curves are available for identification of clay mineral species.

FORMATION OF CLAYS FROM OTHER MINERALS

The common rock-forming minerals other than quartz all weather to clay minerals. The change is easiest to follow in the case of a weathering mica, for the structure remains the same. In the weathering of other silicates the original lattice may be incorporated to some extent in the lattice of the new clay mineral, but as the spacing is not the same there must be some collapse of the original lattice. (In reality the mechanism of mineral weathering is very much more complicated than the simple outline given here.)

ALTERATION OF CLAYS

Clay minerals can change from one to another under suitable conditions. Desilicification and the converse, silicification, are the two main trends.

Desilicification commonly takes place in tropical regions, and gives rise to bauxite deposits (gibbsite) at the extreme. More commonly the reaction stops in the lateritic (ferrallitic) stage, with kaolin and iron oxide.

In flush sites, where cations are being added, and perhaps silica too, as for instance in tropical swamps, then there is a build-up to mortmorillonite or illite. Hence in tropical areas there are kaolin minerals on hillslopes, and montmorillonite in the valley (this is, of course, a simplification). In temperate regions there is less extreme variation and illite is more common and vermiculite for some reason is common in topsoils.

Parent rock, of course, affects the type of clay derived upon weathering. Basalt, and similar basic rocks, provide plenty of cations and tend to give rise to montmorillonite, though under certain conditions kaolin or even gibbsite can result. More siliceous rocks,

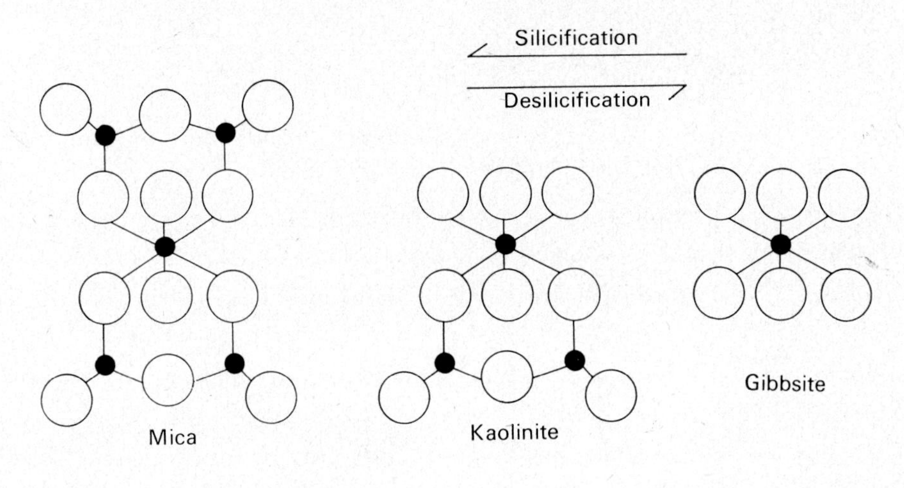

FIG. 28. Diagrammatic representation of the change of clay mineral structure with silicification or desilicification.

with sodium more common than calcium or magnesium, tend to give rise to kaolin minerals. Shales may produce original clay minerals by disintegration, but these will soon alter somewhat to accord with the local conditions.

Soil types are often associated with particular clays. Kaolin is dominant in lateritic (ferrallitic) soils; montmorillonite in chernozems, black earths, prairie soils and humic gleys; illite in podzols, brown forest soils and tundra soils.

Keller (1957) has given a summary of the chemical controls of clay mineral formation.

The formation of kaolin is favoured if Na, K, Ca, Mg and Fe ions are completely leached away and H ions introduced. It is especially important to remove the divalent ions (Ca, Mg, Fe) because:

(*i*) divalent ions are effective agents in flocculating otherwise soluble silica, and therefore hold up desilicification.

(*ii*) Ca and Mg ions would interfere with the formation of kaolin minerals, because the kaolinite lattice does not contain them. They only occur in clay minerals other than kaolinite.

Ferrous iron is not such a trouble because it can be oxidised, and the association of ferric iron and kaolin minerals is very common.

Monovalent ions, K and Na, do not flocculate silica like the divalent ions, but stabilise the silica sol. Silica and alkalis are therefore lost during weathering until the silica-alumina ratio of kaolin is attained, when kaolin will form. Alkali-rich rocks therefore are predisposed to weather to kaolin.

OTHER CLAY MINERALS

Halloysite. The chemical composition of this mineral shows about the same alumina: silica ratio as kaolinite, but water content is usually higher.

Vermiculite. Sheets of mica type separated by layers of water molecules occupying a definite space.

Palygorskite. Si-O tetrahedra arranged in chains instead of layers give rise to this clay mineral.

Allophane. This clay mineral may have some structural organisation, but is usually amorphous to X-rays.

Mixed layer minerals. Elementary layers of different clay minerals can be stacked together randomly or regularly to make 'mixed layer' or 'interstratified' clay minerals. The components may be either completely different clay minerals, such as kaolinite and montmorillonite, or may be one basic type of clay mineral with differing interlamellar material.

Further details of clay minerals and their relationships may be found in Dorothy Carroll's excellent chapter 'The Clay Minerals' in Milner's 'Sedimentary Petrography' (1962).

VII ROCK WEATHERING

Rocks are made up of minerals, but rock weathering is much more complicated than merely the sum of weathering of the component minerals. The style and rate of weathering is very much controlled by the porosity and permeability of the rock, which governs the ease with which water can enter and weathering products be removed. Some granular sediments have such a high porosity that practically every mineral grain is exposed to weathering; some massive igneous rocks have no intergranular porosity at all, and can only weather at the surface and along a few widely spaced joints.

Porosity and permeability are in turn controlled by grain packing and the amount of space between grains, as well as rock structures of various kinds that affect permeability. A fuller account of porosity and permeability is given in Chap. VIII, but for the present we shall concentrate on the rocks. These may be divided into four types:

1. Unconsolidated sediments
2. Sedimentary rocks
3. Igneous rocks
4. Metamorphic rocks

Unconsolidated sediments are found in various environments, such as alluvial plains, river beds, dune sheets, offshore marine deposits, glacial moraines and scree. They may be well sorted or poorly sorted, coarse or fine grained, have many or few minerals, and a wide range of sedimentary structures such as bedding, cross bedding, ripples, shrinkage cracks, etc.

Sedimentary rocks are formed by compaction and induration of sediments, and most of them are in fact derived from marine sediments. These have all the variations found in unconsolidated sediments, plus further features associated with cementation, induration and mineral alteration.

Sediments are usually laid down in successive layers, called strata or beds. Platy mineral grains tend to lie flat, and the preferred orientation of grains leads to planar anisotropy and preferential direction of water movement through the rock. Bedding planes are planes of fissility between beds which may be the contact plane of different lithologies, or may lie entirely within a rock of one composition. If these planes are closely spaced the rock is said to be 'thin-bedded'; if they are wide apart the beds are said to be thick or massive. The sedimentary structure may be further complicated by cross-bedding (Fig. 29), graded bedding (Fig. 30) and other features. A comprehensive account of sedimentary structures is given by Pettijohn and Potter (1964).

Sedimentary rocks are generally classified into *clastic* rocks such as sandstone, shale and conglomerate; rocks of *organic* origin such as coal and many limestones; and rocks of *chemical* origin such as travertine, chert and rock salt.

Igneous rocks may be extrusive, that is erupted at the earth's surface as lava flows or as volcanic ash showers; hypabyssal, that is emplaced as dykes, sills or other bodies in the upper layers of the earth's crust; or plutonic, that is formed at considerable depth in the earth's crust (Fig. 31).

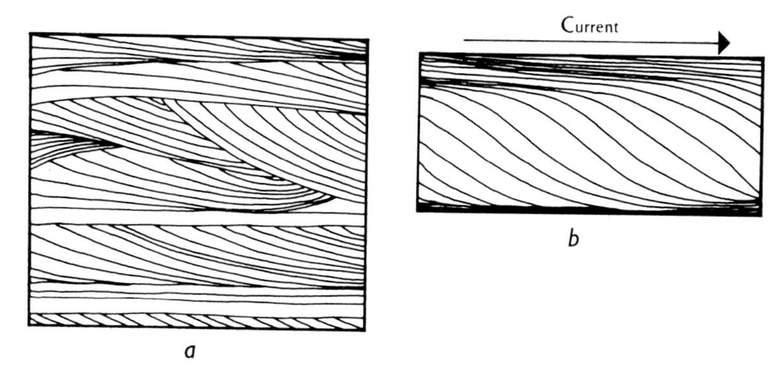

FIG. 29. *a* Cross bedding in a typical section. *b* A simple cross bedded unit, showing direction of current.

These three kinds of igneous rock—eruptive, hypabyssal and plutonic—are correlated with grain size of the rock minerals. Eruptive rocks, cooling rapidly, have small crystals as a rule, and may even be glassy due to extremely rapid chilling. A hand-lens

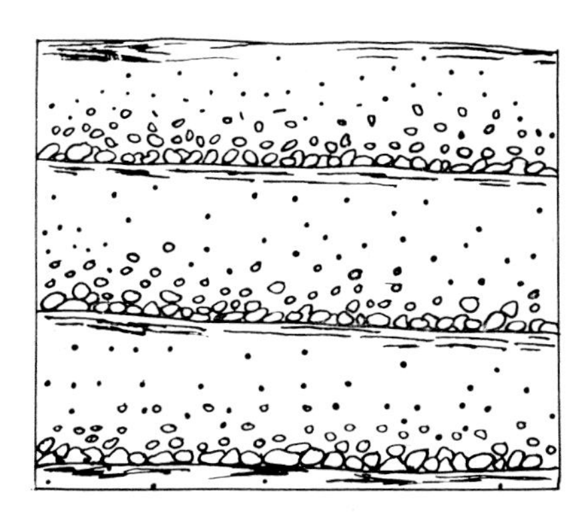

FIG. 30. Graded bedding. Each bed has coarse material at the base and grades upwards into fine sediment.

or microscope is usually required to see the crystals. Hypabyssal rocks have small to medium crystals that can be seen by the naked eye. Plutonic rocks, crystallising very slowly, grow large crystals that are plainly visible and may be inches or even feet across.

From a chemical point of view igneous rocks are classified as acid, intermediate or basic. 'Acid' rocks are those with a high proportion of silica, and such rocks contain the mineral quartz.

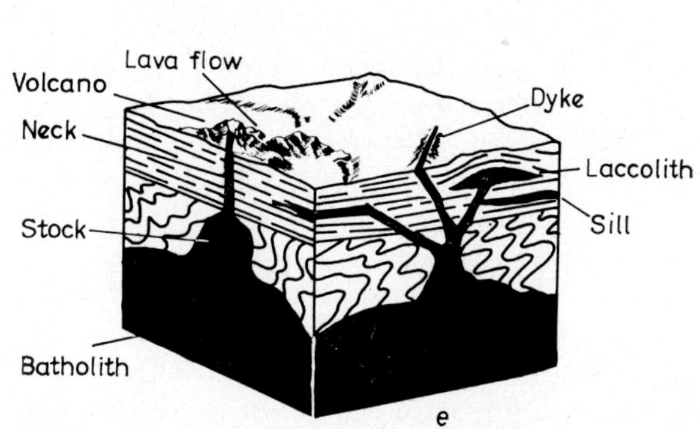

FIG. 31. Igneous rock bodies.

a Sill. *b* Dyke. *c* Plug. *d* Laccolith. *e* Diagrammatic representation of relationships between bodies of igneous rock.

A very rough classification of igneous rocks according to the features mentioned is shown in Table 9, and Fig. 32 summarises their mineral content. Further details of igneous rocks and their classification are contained in many textbooks, such as Williams, Turner and Gilbert (1954).

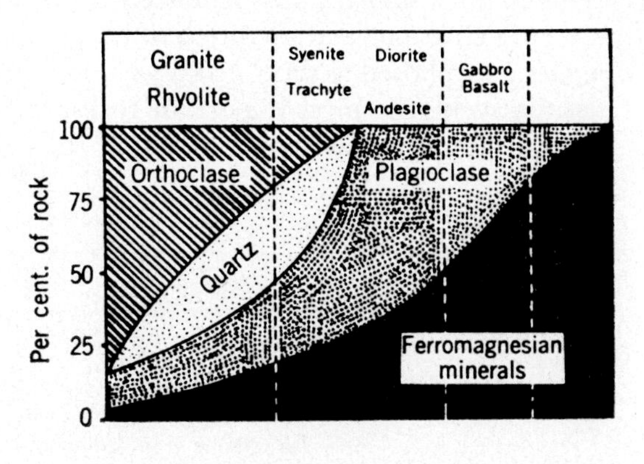

FIG. 32. Proportions of minerals in igneous rocks.

It happens that the vast majority of coarse-grained plutonic rocks are acid, that is granitic, whereas the great majority of fine-grained, eruptive igneous rocks are basaltic. Granite and basalt are thus the commonest igneous rocks in the earth's crust.

Table 9

	Acid	Intermediate	Basic
Coarse grained	Granite	Syenite	Gabbro
	Granodiorite	Diorite	
Medium grained	Microgranite	Porphyry	Dolerite
Fine grained	Rhyolite	Trachyte	Basalt
		Andesite	

Another form taken by igneous rock is the 'pyroclastic' deposits blown out of volcanoes as particles from dust size to huge blocks. These make beds or irregular deposits of scoria, lapilli or ash. Coarse pyroclastic rock is called agglomerate, or scoria, and fine pyroclastic rock is called tuff. Acid and intermediate lava gives rise sometimes to pumice, a very vesicular, porous rock, and sometimes to ignimbrite, a denser, layered rock which behaves in a similar manner to rhyolite.

Metamorphic rocks are due to the alteration of pre-existing rocks, whether igneous or sedimentary, by heat, pressure or both. Around the edges of a granite pluton there is usually a ring or 'aureole' of metamorphism that converts the country rock into hornfels, usually a fine grained, very tough rock.

Most metamorphic rocks are distinguished by very marked preferred orientation of minerals, the development of new minerals at the expense of original ones, and the formation of a very marked fissility in many parallel or near parallel partings as 'slatiness' or 'schistosity'. Coarse-grained metamorphic rocks are called gneiss, and finer grained ones phyllite, slate and schist. Dominant minerals give rise to distinct varieties, e.g. mica schist, andalusite schist, etc., and zones of different degree of metamorphism are characterised by different mineral assemblages. Very fine grained rocks may be metamorphosed to slate. Marble is metamorphosed limestone.

Quantitatively, the most important rocks, so far as area of land surface they cover gives an indication, are as shown in Table 10.

Table 10

	%
Shale	52
Sandstone	15
Granite (and granodiorite)	15
Limestone (and dolomite)	7
Basalt	3
Others	8

(Table from Leopold, Wolman and Miller, 1964)

It will be seen that sedimentary rocks cover about 75% of the earth's land surface, and that metamorphic rocks (included in 'others') cover only a small area.

Rock structures may be original sedimentary or igneous features, they may be due to alteration within the rock, or they may result from stresses applied from external sources.

Bedding planes are the most obvious of original structures in sedimentary rocks.

Joints are planes of fracture or potential fracture that are found in most rocks. They occur in *sets*, which have parallel planes, or *systems*, made up of several sets and usually having a pattern repeated over a considerable area. There is no displacement across the crack of a joint. Joints may be caused by shrinkage, as when a lava sheet cools, or a sedimentary layer loses volume by dehydration. Others are due to tectonic stresses. Joints commonly occur perpendicular to bedding planes, but many other attitudes are also possible. Joints are common and very important in the weathering of granite, basalt and limestone. Sandstones and other sedimentary rocks are usually jointed, but joints in these rocks are generally of less importance in weathering.

Plutonic rocks may have complex jointing related to the mode of emplacement, the structural setting, and the 'fabric' of the rock, and there are commonly complications due to unloading planes or joints. Joint spacing in granite is often very uneven, and plays a great part in controlling the rate and depth of weathering.

Cleavage is a direction of easy splitting in folded or metamorphosed rocks. It arises from the parallel orientation of platy minerals such as mica. This can happen by either rotation of the original mica or the growth of new micas under tectonic stresses on the new cleavage planes.

With increasing regional metamorphism, the minerals are further segregated until ultimately bands of micas alternate with bands of other minerals. The cleavage has now become a schistosity. Further change concentrates more and more minerals, and the bands become thicker and lenticular, with coarse minerals. The schist has now become a gneiss, and the partings are further apart, so the fissility and ease of weathering is decreased.

The many varieties of jointing and cleavage and numerous other rock structures are described in textbooks of structural geology such as Hills (1963).

The various features of rocks considered so far are important mainly in regard to chemical weathering. In considering some aspects of physical weathering, quite different data are required which will enable the explanation and possibly prediction of rock behaviour under varying stress conditions. Rock is an extremely complicated material, its physical properties varying not only with change of stress conditions but also with the rate of change of stress. For example, limestone may be deformed in either an elastic or plastic manner depending on the loading rate, and may

1. Not fail;
2. Fail, either
   (*a*) as a ductile material by plastic deformation, or
   (*b*) as a brittle substance failing by (*i*) brittle fracture, or (*ii*) shear fracture.

The type of behaviour of a rock also varies enormously with temperature and confining pressure. Brittle materials, such as rocks, tend to become ductile when subjected to compression from all sides. For example, limestone behaves as a brittle material under confining pressures of up to about 4000 atmospheres, but under confining pressures of 6000 atmospheres and over it is capable of considerable plastic deformation.

Rock behaviour is different again if stresses are repeatedly reversed. The branch of science that studies the physical properties of rock material is called Rock Mechanics, and this science has provided most of the data that is at present available for fundamental consideration of physical weathering.

Rock mechanics is a very complex subject, and it must be understood that the models provided by both theory and experiment provide only reasonable approximations to the behaviour of actual substances under very limited conditions. A textbook dealing with the physical principles involved is by Jaeger (1962).

The weathering of the commonest rock types will now be described. There are a great many other rocks; some can be regarded as variants or intermediates between those described here; others present unique features but are too rare, or their weathering features too little known, to be described here.

1. UNCONSOLIDATED DEPOSITS

Sand

Sand most commonly consists of the mineral quartz. This is chemically stable, but sands are always porous and hence easily leached, so that other minerals are easily attacked and altered, and weathering products removed. Well-drained sand thus tends to get even cleaner with weathering—that is, minerals other than quartz tend to be removed. In waterlogged sites the situation is different. Here removal of weathering products is hindered, but within the zone of saturation there is likely to be rapid movement of ions and redistribution of weathering products.

Sands are practically immune from physical weathering, and although it may be significant that quartz has a lower limit for mechanical breaking at 0·02 mm (below that size most stresses can be accommodated by elastic strain), it is unlikely that much breakdown of sand takes place in immobile, unconsolidated sand, for there is so much room for movement or expansion into the pore space.

Clay

Clay and mud consist of the clay minerals described in the last chapter together with varying amounts of water, and minor quantities of other minerals. Clay minerals have the ability of breaking up into colloidal fragments which can recombine into the same or different clay minerals in different conditions or at a different site. If new and different minerals are formed this can be regarded as weathering, that is one clay mineral weathers into another.

A water-laid deposit of clay is laminated, with the flaky clay minerals taking up a horizontal position. If the deposit is undisturbed, the clay slowly compacts under its own weight, and water is expelled. This compaction may eventually lead to diagenesis and the formation of new minerals. There may be a certain amount of ion-exchange and migration of salts if conditions permit.

Further alteration is likely to take place if a clay is occasionally dried out. Physical break-up of the clay sheet will occur due to shrinkage, deep cracks allow penetration of air, and there may be a certain amount of oxidation. Most changes, however, are likely to be reversed upon wetting, and the sum total of weathering of minerals is nil.

Bacterial activity is common in waterlogged clays and may lead to segregation of certain elements, the formation of secondary minerals, and consequent alteration of the clay minerals.

In soils the clay minerals are not generally oriented horizontally but are arranged in far more complex ways, making up crumbs, blocks and other sorts of soil structure. The units are called peds. The soil structure allows more rapid leaching or eluviation, and therefore more rapid weathering than in a simple clay deposit. Biotic activity is also great, again enhancing weathering due to the stirring activity and mixing besides its innate weathering. In soils the clays are so easily altered that different horizons in the soil may often be characterised by different clay mineral assemblages.

Waterlogged clays tend to be grey or green due to ferrous iron, while well-aerated clay soils tend to be brown. In transition zones there is usually colour mottling.

Alluvium

Sheets of detritus laid down by rivers, which although in any one spot are usually well sorted, can overall exhibit a wide range of grain sizes. Clays and sands behave as described in the sections above, as do mixed deposits. Waterlogged soils are common on floodplains, with all the weathering phenomena that are concomitant, but old alluvium of river terraces is usually well drained and oxidised.

Groundwater moves through alluvium, down the valley just like the surface water, although much more slowly. Moving groundwater is conducive to more rapid weathering than stagnant groundwater, because weathering products are carried away in solution.

The important thing about alluvium is that it is bedded and successive beds may be of different texture and mineral composition. Great care is required in studying alluvial weathering profiles to distinguish those layers due to deposition, that is what we might call geological layering, and those layers due to weathering or soil formation, which might be called pedological layers. For instance, a layer of non-micaceous silt over a layer of micaceous silt might be due to weathering of mica from the upper layer, or due to deposition of an upper alluvial layer which never contained mica. A mixture of sand and clay does not behave like either separately. Sand prevents 'panning' of clay, and the mutual relations of the sand and the clay—a part of the 'fabric'—deeply affects weathering and soil formation. Leaching and clay alteration and eluviation are enhanced by the presence of sand. In periglacial climates the process known as frost heaving is displayed at its best on mixed deposits. However, although this has a role in mixing or segregating fractions of the total deposit, the actual amount of weathering achieved is small. Colloid plucking (p. 24) may occur if a clay is adjacent to sand, although it is probably insignificant.

Changes in mineral content of alluvium in a downstream direction, unless due to addition from tributaries coming from a different source rock, are usually due to attrition, although other weathering phenomena such as solution may also be operative.

Loess

This is a peculiar deposit of uncertain origin. It is characterised by extremely well sorted material with a dominance in the silt size (·05 to ·002mm), which corresponds to the optimum particle size for wind transport and the minimum size of mechanical shattering of quartz. Loess shows no bedding, has widely spaced vertical joints, has carbonate disseminated throughout and also often in small globular concretions known as loess dolls. The dominant mineral is quartz. A rather curious feature is that much loess, at least in Europe, has the same suite of heavy minerals, including green hornblende and staurolite (Thomasson, 1965). Loess deposits may be found on any topographic site from valley bottoms to hilltops. Two features of loess weathering must be discussed; one is what happens when loess weathers, the other is whether loess itself is a deposit formed by weathering.

The weathering of loess is the simplest of these problems. Loess is very porous so there is scope for leaching and eluviation. Decalcification of the topsoil occurs and sometimes precipitation of a carbonate layer at depth. Clay may be reorganised at the surface or be eluviated down the profile and deposited in the B horizon. Other types of weathering and soil formation are found when certain soil-forming factors become dominant. Soils are formed very rapidly on loess and in Central Europe detailed chronologies have been built up on this basis.

The second problem is much more complex. Americans and most Europeans subscribe to the theory of aeólian origin for loess, with variations on the exact conditions under which it was deposited. A less commonly held idea is that it is an alluvial deposit although in many areas loess-like material has been redeposited by running water. Most of the deposits mapped by the British Geological Survey as 'brickearth' are of this type. The Russians, and particularly Berg (1964), have abandoned the aeolian hypothesis completely and believe loess is a weathering product (see also Thomasson, 1965). Kuenen's ideas are discussed on p. 24.

Till

This is the sediment deposited directly by glaciers. It is poorly sorted with a wide range of particle size, but commonly contains a large amount of clay which often makes till impermeable and so affects the course of weathering. Hydromorphic soil types such as surface water gleys are often dominant. Decalcification is important where the original till was calcareous, and in some conditions iron pans may be formed. In the United States the unctuous sticky soils that result from weathering of till are known as 'gumbotil'. The depth of weathering is dependent upon time, and it is thought that the

Great Interglacial gave rise to thick gumbotil and decalcification to a greater depth, while other, shorter interglacials gave thinner gumbotil and shallower decalcification.

Degree of weathering can be used to distinguish different tills, and conversely the weathering of tills tells us about weathering rates, although as the material is already physically divided it naturally weathers a good deal more rapidly than solid rock. In the northern Sierra Nevada, for instance, Birkeland (1964) has distinguished four glaciations, with tills of different weathering characteristics. Two glaciations, the Hobart and the Donner, are Pre-Wisconsin, and the oldest, the Hobart, is highly weathered. The Donner tills have a soil profile with A, B and C horizons, the whole profile being 6 to 8 ft thick, and contained boulders are completely weathered as in the Hobart till. The Tahoe is the first Post-Wisconsin glaciation, and has a till that contains fresh andesite boulders, though granite boulders are spheroidally weathered, and has only an AC soil profile. In the youngest till, the Tioga, all the boulders are fresh, and again an AC soil profile is present.

2. SEDIMENTARY ROCKS

Sandstone

Sandstone consists of sand grains, commonly quartz, more or less cemented by a matrix of other minerals. Sandstones often have widely spaced jointing, bedding and cross-bedding. Weathering of sandstone consists largely of attack on the cement and removal of support for the sand grains. The cementing minerals may be removed or altered, but if the sand grains are of quartz they are little affected.

Calcite-sandstones are largely affected by solution, and because of initial porosity solution goes to great depth. Calcarenite (sandstone with grains of calcite) weathers more like a limestone than a quartz sandstone.

Clay matrix sandstones weather by breakdown and eluviation of clay.

Iron-oxide cements tend to hydrate to hydroxides, and there is often migration of iron within the sandstone to form concretions or other accumulations.

Opal (or siliceous) cemented sandstones weather like quartzites (q.v.).

Sandstone generally has a high porosity so plenty of water is absorbed, and in cold regions frost weathering is important. The variations in water-holding ability due to joints, bedding and variations in porosity make lines of weakness for physical weathering.

Sandstone usually weathers to blocks, or crumbles, although in more impure types exfoliation is found. Only the most massive sandstones show exfoliation clearly. Irregular distribution of cement gives rise to differential weathering and removal to form boxworks, toadstones, tafoni etc.

With more clay or feldspars, sandstones become greywackes or arkoses. These have many properties in common with sandstones; most of the differences are attributable to greater ease of chemical weathering.

Shales, mudstones, marls

These rocks consist of clay minerals with mica and occasional other minerals, often with a high degree of orientation. Carbonate is present in marls. Bedding is usually well marked, and there is often further fissility by joints and cleavage. Weathering is often intense at the surface and decreases down the profile, but persists along cracks to a much greater depth. Thus the rock is broken into blocks which are fresh inside, but weathered on their faces. The surface weathering and the clay breakdown is very similar to that described in the section above for clays, although it takes a longer time to achieve if the rock is thoroughly indurated.

The attitude is also important. In well-bedded horizontal shale it is hard for water to penetrate, but if the strata are steeply inclined many shale edges and bedding planes are exposed which allow easy water penetration.

In the case of marls and calciferous shales, there is decalcification from the surface down and along fissility, with accumulation of insoluble residue at the surface. In cold regions frost weathering is most important, and large screes may be formed in this way with negligible chemical alteration.

In tropical regions there is often deep weathering to a rather irregular weathering front. There may be complete alteration of clay minerals and migration of ions to form lateritic profiles. Accumulation of iron can lead to iron-pan (ferricrete) formation, and in other areas a silica pan (silcrete) may develop.

Clays and shales erode faster than many other rocks, not only because they are soft and weather more easily, but because they are impermeable and so there is a lot of run-off. There is therefore a greater tendency for the weathered product to be removed. However, the difference is controlled by site, and many flat sites have retained a deep regolith. On slopes, however, it is safe to say that clays and shales will be eroded faster than most rocks.

Limestone

Limestone, including dolomite, is distinguished by its much greater solubility than other rocks. Except in extreme climatic environments, solution is the predominant weathering agent of limestone.

The actual process of weathering depends on the variety of limestone. Some is dense, massive and of very low porosity, and solution is concentrated along joints and bedding planes. This is the sort of limestone that gives rise to true 'karst' topography.

Other limestones are porous, such as chalk and dune limestone. Water penetrates the entire rock, giving rise to a more even weathering, and joints and bedding planes are of little or no importance.

Layers of secondary carbonate may form at or near the surface of soft limestones, forming a hard layer known as kunkur, caliche, calcrete, etc.

Limestone is so distinctive in its weathering, and has been studied so intensively, that it will be treated separately and at some length in Chap. XII.

3. IGNEOUS ROCKS

Granite

Typical granite is made up of quartz, feldspar and mica. It is coarse grained, well jointed, and without bedding. Granite can be regarded as a plutonic rock, formed from the cooling of a granite magma at depth. Some granites, however, merge through a more foliated rock such as granite-gneiss into metamorphic rocks and sediments. Such granites contain xenoliths and even 'ghost stratigraphy' indicating their transformation from pre-existing bedded rocks by the process of 'granitisation'.

In either case, granite is formed at a considerable depth in the earth's crust and under great pressure, so when exposed at the surface the release of pressure gives rise to marked unloading phenomena which are especially clear because of the homogeneity of the rock.

Frost action splits granite into blocks, and in cold regions granite felsenmeer are common.

Chemical weathering of granite is both marked and variable in its effect. Quartz remains unaltered, but feldspars are often converted to kaolinite and micas to various clay minerals. Granite frequently exhibits granular distintegration, now believed to be due to chemical depletion although the grains often remain rather fresh in appearance. Case hardening of granite boulders appears to be the corollary. Weathering often follows the joints, and isolated joint blocks weather spheroidally, leaving 'corestones' of unaltered granite in the centre. Other exfoliation takes places on exposed surfaces, and although rather different in detail from the spheroidal weathering at depth, is due to chemical alteration leading to volume change. Minor features of granite weathering are tafoni, gnamma holes and similar small hollows.

Deep weathering is common, and deep regoliths of varying kinds are known. Extreme weathering leads to china clay deposits consisting of almost pure kaolin and quartz grains, although china clay can also be formed by hydothermal action.

Other coarse-grained igneous rocks like granodiorite weather in a similar manner to granite so long as they have quartz. Gabbro, a basic coarse-grained igneous rock with no quartz, shows much more chemical alteration to clay, often montmorillonite, and fewer and smaller corestones. Case hardening, granular disintegration, etc., are absent, and simple clay cover is the usual resulting landform.

Basalt

A fine-grained, dark, effusive igneous rock, it has many minor depositional features, but is generally a dense rock with good, often columnar, jointing. Scoriaceous basalt and volcanic ash weather differently as they are extremely porous.

Basalt is attacked first along joint planes, leading eventually to spheroidal weathering. All the minerals are eventually converted to clay and iron oxides, with bases released in solution, and as there is no quartz in the original rock the ultimate weathering product is often a brown, base rich, heavy soil. The high proportion of clay, however, often prevents drainage and waterlogging results, so reducing conditions set in and complicate the weathering picture. Carbonate is commonly precipitated at various parts of the profile in suitable environments. 'Floaters' of basalt occasionally occur in the clay. They are irregular in shape, and do not have scales, but appear to be the equivalent of corestones. More normal, well-rounded corestones are also found, and it is possible that the regular ones are formed by deep weathering under a certain amount of compaction, and the irregular ones are formed near to the surface. In extreme conditions basalt can weather in strange ways. At Royal Park, Melbourne, a small amount of basalt has weathered to pure white clay, with magnesite in joints and small basalt corestones occasionally preserved in the centre of joint blocks. Elsewhere in Victoria basalt has been completely bauxitised (Raggat *et al.* 1945).

Dolerite

This has the same chemical and mineralogical composition as basalt, and weathers in a similar way. Jointing is well marked and very regular.

Rhyolite

This acid effusive igneous rock contains quartz and so can give rise to sand upon weathering. The rock is laminated (flow-banded) and jointed. It is not so weatherable as basalt, and gives rise to shallower and poorer soils.

Pumice

A frothy, glassy, igneous rock formed in violent eruptions of acid magmas, weathers slowly and with difficulty to form poor, sandy soils.

Andesite

This effusive igneous rock of intermediate composition has weathering properties intermediate between basalt and rhyolite.

4. METAMORPHIC ROCKS

Gneiss

This is a coarse-grained, foliated rock containing quartz, feldspar, amphiboles, pyroxenes or other minerals. It is intermediate in properties between granite and schist. It is rarely as well jointed and uniformly massive as granite, so unloading is not common, or at least harder to detect. Minerals are segregated into bands, and bands of the most weatherable mineral affect the total rock strength—a property that often proves troublesome in engineering. The distinct mineral orientation prevents good exfoliation weathering, and granular disintegration also seems to be rare.

Schists

These have marked fissility along the 'schistosity' and this is very important in weathering. They contain some very resistant minerals but weathering is moderately easy. Frost weathering can rapidly break up schist.

Marble

This is metamorphosed limestone, and it weathers in all major respects like a dense limestone.

Quartzite

Metamorphosed sandstone. This has lost the porosity of sandstone, so is even less weatherable. It is almost inert chemically, especially if pure, and physical weathering is dominant.

Amphibolites

These present a special case of metamorphic rocks, consisting almost entirely of hornblende. These rocks weather rather like basalt, often very deeply, giving rise to a clay with abundant bases and iron. Cleavage enables easy water penetration and deep weathering.

MUTUAL EFFECTS OF ROCKS

The weathering of a rock may be affected by its structural and topographical relationship to other rocks. Thus a cap rock may protect a lower stratum from wetting, and hence weathering, or may concentrate water along certain zones whose disposition follows features in the cap rock rather than the lower stratum. In addition to water-concentrating effects, a lower rock may also be affected by leachates derived from overlying rock. Thus both structural and lithological features may be relevant in the mutual effect of rocks in weathering. It is often found that rocks are more weathered at their contacts with other rocks than in the mass, so lithological boundaries are picked out by weathering—a fact commonly utilised in geological mapping.

On a small scale, numerous examples of weathering out of the edges of pebbles in conglomerate have been reported by Schwarzbach (1966), see Fig. 33.

On a larger scale Currey (personal communication) has investigated the effect of a basalt flow on the weathering of underlying Ordovician sedimentary rocks exposed at two dam sites in Victoria, Australia. It might be expected that a basalt flow would protect underlying material from weathering and that neighbouring sediments with no basalt cover, exposed to atmosphere, would weather faster. In fact the reverse happens, and weathering is much more intense below the flow. Tullaroop and Eppalock dam foundations are composed of rocks of identical geological age, and both dam sites are in valleys which have been eroded along the junction between Pleistocene basalt flows and Ordovician sedimentary rock. In each case the rivers have cut across a minor tongue of basalt, forming a basalt outlier on one side of the valley (Fig. 34).

The foundation excavations in the Ordovician sediments revealed that the weathered rock changed abruptly from soft crumbly clay to hard rock that rang dully under a hammer blow. The junction of the soft and hard weathered zone crosses the strike of the beds. The spatial distribution of the soft-weathered rock is shown in Fig. 34.

Two possibilities might be envisaged to account for the distribution of weathered rock:

(*i*) The soft-weathered rock was originally of uniform thickness and has been eroded away where not protected by basalt.

(*ii*) There has been an increase in the weathering of Ordovician rocks beneath the basalt.

FIG. 33. Conglomerate weathering along the junction of boulders and matrix. (Photograph by M. Schwarzbach.)

Since deep, soft-weathered rock is found under basalt, both of the main flow and of outliers, and on the slopes between outliers and main flow, and yet is absent from all valley sides which never had a basalt cover, possibility (*ii*) seems much more probable. Furthermore, if the soft-weathered zone were prebasaltic, it would be expected that the weathered zone could persist along the old depression beyond the outlier of basalt, but it does not. In excavations it was found that the base of the soft-weathered zone dipped steeply precisely at the presumed position of the basalt edge, at a gradient which could not have been parallel to an earlier valley side. In brief, the distribution of soft-weathered rock, both in plan and in section, indicates that weathering of Ordovician sedimentary rock was greatly increased below a basalt cover.

It is also possible for an underlying rock to affect the weathering above, mainly by an effect on drainage. Thus Hutcheson and Bailey (1965) found that weathering of loess was greater where it was underlain by coarse-textured material.

The effect of one rock on the weathering of a neighbouring stone of different lithology is well known to builders and architects, and building stones which weather adversely when used together are said to be 'incompatible'.

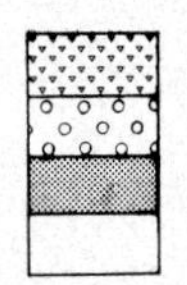

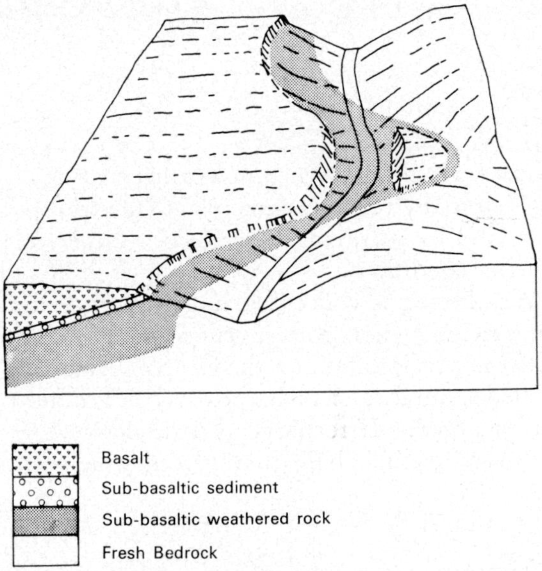

FIG. 34. Diagrammatic representation of the sub-basaltic weathering at Tullaroop dam site.

In the Geelong area one of the most durable building stones is the Barrabool Hills sandstone, which is unfortunately of a dull and drab colour. Light-coloured limestone was occasionally used to relieve the dark appearance of the buildings, but unfortunately caused excessive weathering of the sandstone in contact with it. In this instance the comparatively rapid water absorption of the limestone appeared to be mainly responsible, but rocks can be incompatible for a variety of reasons (Finch, 1955).

VIII HYDROLOGY AND WEATHERING

Water is vitally concerned in weathering. It is one of the major reactants, the solvent in which many reactions take place, the transporting agent for many weathering products, and by its presence or absence largely controls the separation of oxidising and reducing conditions.

Weathering is concerned largely with water in the ground, called groundwater. This originates not from juvenile water in the rocks, but from infiltration of water gained in precipitation on the earth's surface. The addition of water to groundwater is called groundwater recharge. When groundwater is lost, as at a spring, it is said to be discharged. If recharge and discharge are in equilibrium, the water table will remain stable, though fluctuating somewhat with seasonal and other climatic variations.

ROCKS AS RECEPTACLES FOR WATER

There are many cavities in rocks, such as open joints and interstices between grains which can become filled with water, and the size and distribution of these spaces has a profound effect on weathering. The properties of porosity and permeability of different rocks were treated briefly in Chap. 5; this will now be recapitulated with emphasis on groundwater hydrology.

The water-holding spaces in a rock will generally form an interconnecting meshwork of channels through which water can move. The percentage of total space in a rock is a measure of its porosity. The rate at which water can move through the rock is its permeability.

Some approximate porosity and permeability values are shown in Table 11.

Table 11

Porosity %		Relative permeability	
Granite	1	Igneous	1
Basalt	1		
Shale	18	Shale	5
Limestone	10	Limestone	30
Sandstone	18	Sandstone	500
Gravel	25	Gravel	10,000
Sand	35	Sand	1,100
Silt	40		
Clay	45	Clay	10

Some water is held by molecular attraction on to individual mineral surfaces, and such water would not flow freely. If the rock porosity is made up of many very small pores, much of the water is held in this manner and permeability will be very low. If the pores are large, the greater part of the water will be unaffected by molecular attraction and can flow freely.

Hydrologists distinguish between the water that can move freely and that which is strongly held by the rock. In groundwater hydrology the water that is retained by the rock is called 'dead storage'. It is measured in the laboratory by filling a column of dry rock with water, draining away the free-flowing water, and measuring the quantity held by the rock.

Workers in soil water have a similar measure called 'field capacity', which is a measure of the water retained in a column of soil when freely flowing water has drained away.

Sand, or weakly cemented sandstone, will have a fairly high porosity. Little water will be retained by molecular attraction, and so there will be little dead storage in the case of a rock, and a low field capacity in the case of a sandy soil. Gravel will exhibit the same trend to a greater extent.

Clay may have a very high porosity, but because of the great molecular attraction there will be low permeability, high dead storage and, in the case of a soil, high field capacity.

A cylinder of fresh, dense granite, basalt or limestone may exhibit virtually no porosity, and the only water retained is what wets the surface of the block. Small specimens of such rock therefore indicate no water-holding capacity at all. In the field, however, such rocks have fissility due to bedding, jointing, cleavage or other structural features and can in fact contain a lot of water. Laboratory measurements therefore do not reveal all the hydrological properties of a rock, and must be supported by field evidence.

THE WATER TABLE

Fig. 35 shows a simple basin on a geological scale. The rock with pore space that holds water is called an aquifer, and the impermeable rock that holds up the water is called an aquiclude.

Fig. 36 shows a sloping ground surface intersecting an aquifer overlying impermeable rock. The water table will reach the surface near the lithological contact and will provide a spring.

Fig. 37 shows a cross-section of permeable rock such as chalk or sandstone. When the water table reaches the surface, water flows freely as a surface stream. If the water table is below the surface, valleys are dry. It will be seen in the diagram that the water table in this kind of situation has a profile similar to the topographic profile but with a lower amplitude.

It is possible for a rock succession to provide a series of aquicludes and aquifers, and water may not be at the same level in separate aquifers. Excess water from a higher aquifer can flow over an aquiclude and into a lower aquifer, which in turn could fill up

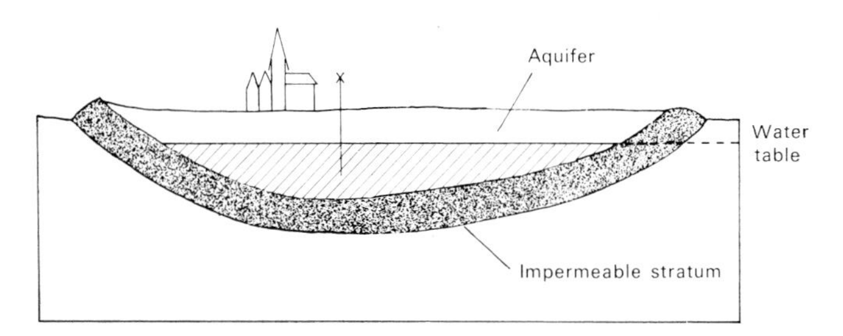

FIG. 35. Water in permeable rocks within a basin of impermeable rock.

and overflow from a spring (Fig. 38). The upper aquifer is said to have a perched water table.

In rocks that have no intergranular porosity but only cracks of various sorts, the continuity of groundwater is imperfect. Some cracks may be dry when nearby ones are full of water, and the depth to water or to dry rock is hard to predict. In these situations the water table becomes more of a theoretical concept than a physical reality,

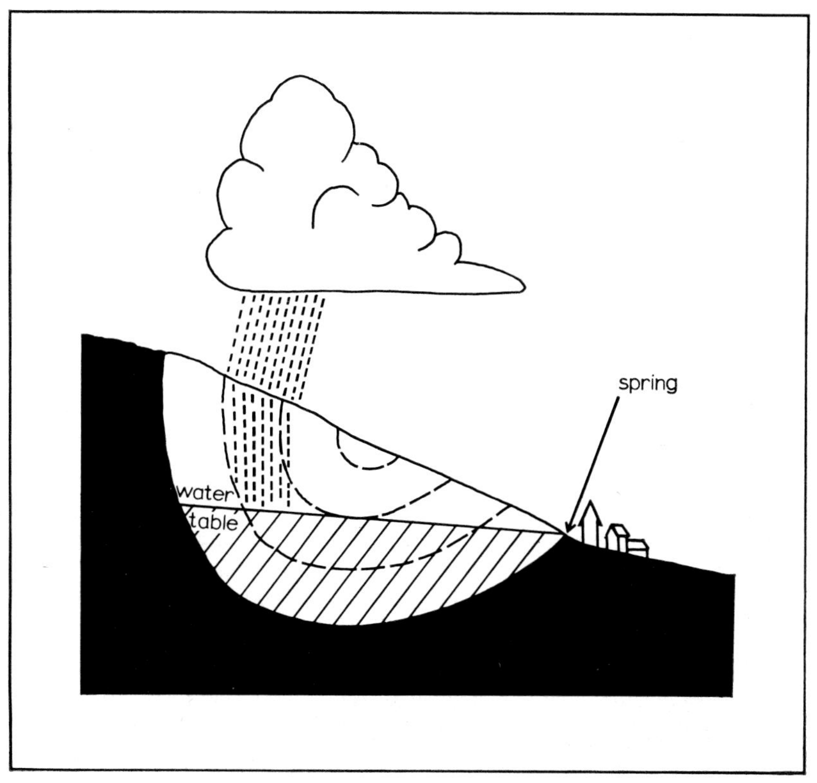

FIG. 36. A geological example of an overflowing basin of groundwater.

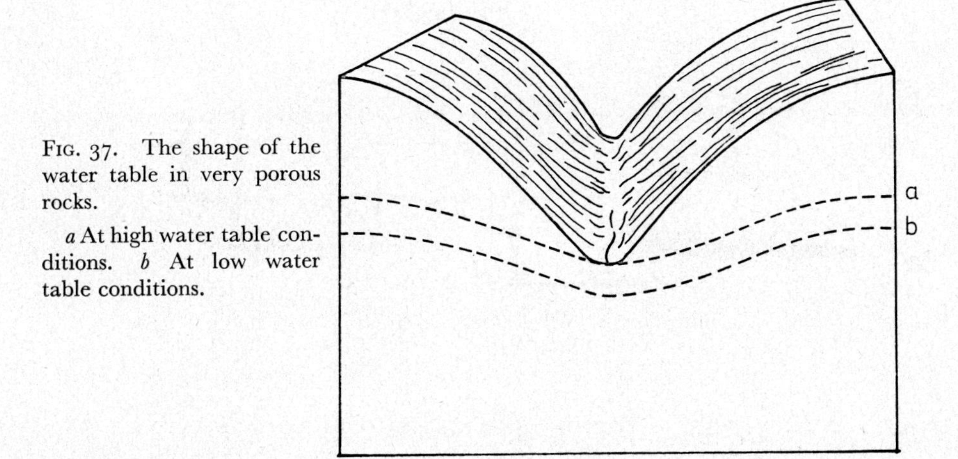

FIG. 37. The shape of the water table in very porous rocks.

a At high water table conditions. *b* At low water table conditions.

but is nevertheless a useful and valuable concept. In general it can be thought of as the level to which water will rise in a bore which strikes water, though there may be 'islands' of rock with no groundwater at all. Fig. 39 shows a typical groundwater arrangement of this type. A bore at X would strike water which would rise to the level of the water table at *p*, but a bore at Y remains dry. Another way to conceive the water table in such a situation is as an imaginary surface across the highest parts of all water-filled fissures.

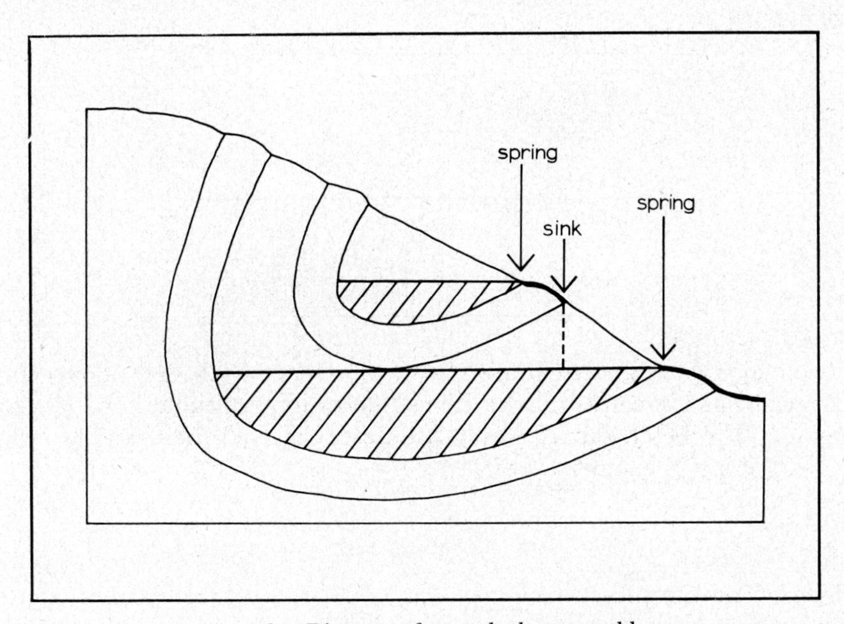

FIG. 38. Diagram of a perched water table.

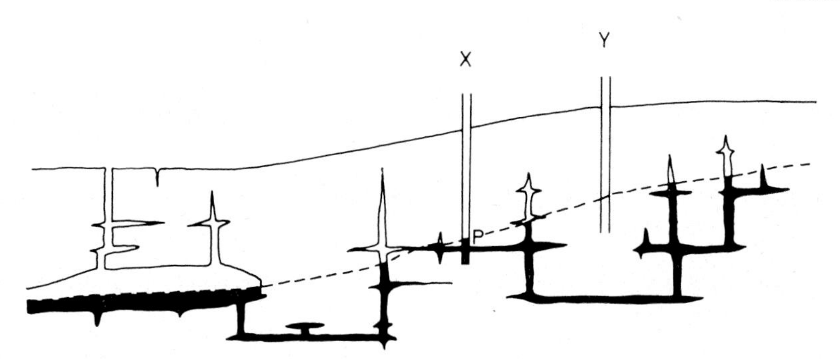

FIG. 39. Groundwater in rock permeable along cracks, but not porous. This example is in karstic limestone.

ZONES AND BELTS OF GROUNDWATER

Groundwater occurs in a number of zones and belts shown diagrammatically in Fig. 40. From the surface down, these are:

1. Belt of soil moisture.
2. Intermediate belt.
3. Capillary fringe and belt of fluctuation.
4. Belt of discharge.
5. Belt of stagnation.

Quite commonly these five belts are grouped into two main zones, which have various names, shown below:

Belts				
1, 2, 3	Aeration	Vadose	Oxidation	Oxidation
4, 5	Saturation	Phreatic	Reduction	Supergene Enrichment

The two major zones are essentially based on presence or absence of oxygen, though the soil water belt may cause complications due to impermeable horizons, and local waterlogging. The behaviour of water is different in each belt, and is reflected in different weathering phenomena.

THE BELT OF SOIL MOISTURE

Rain falling on the ground surface may evaporate, run off or infiltrate, and even on any one site the proportion of rain going in these different ways is not fixed. The prior state of the ground is the first relevant variable; if dry, the first rain may be entirely absorbed, but as the soil becomes wetter so penetration of water becomes more difficult for clays have absorbed water and swelled, pore spaces have filled, and at a certain stage clays may start to puddle, lose their structure and behave as an impermeable layer. Infiltration and run-off are complementary, and as infiltration decreases, run-

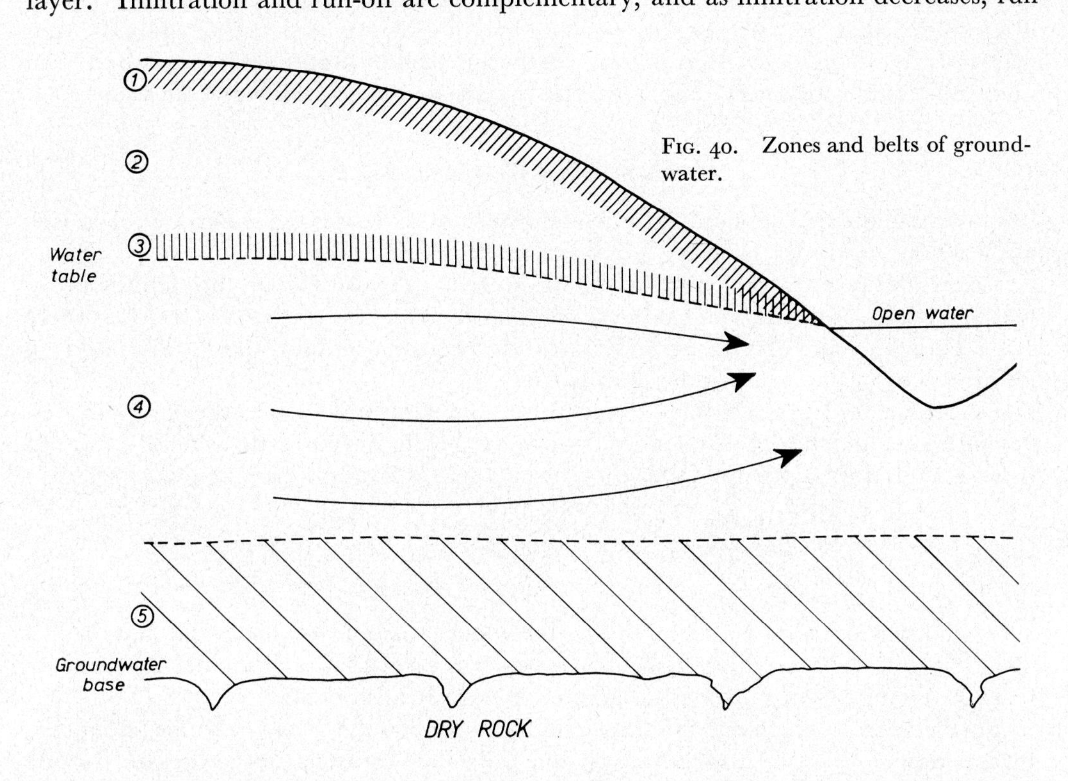

FIG. 40. Zones and belts of groundwater.

off increases. A saturated soil gives rise to run-off, but at the same time water may percolate further down the profile to the lower zones, thus replenishing the groundwater. It is significant that the soil-water zone acts as a great retainer of water that must be filled before the lower zones are recharged, and the soil zone acts as a barrier to recharge.

In very porous materials the surface of soil water makes a distinct water table. If it is very close to the ground surface, there is liable to be a very thin oxidising zone and ground water gleys are the probable soil formation.

On very impermeable material the water table concept is scarcely applicable to soil water. The upper parts of soils will be aerated and rendered porous by soil structure, but would soon be temporarily waterlogged by storms. In these localities surface water gleys are found, and below them the ground gets drier. There is no recharge to underlying rocks, however porous, and such rocks have to derive their groundwater laterally from more permeable localities. Changes in soil water conditions may be progressive. For example, a sandy shale may originally be moderately porous but eventually may have clay eluviation from the A layer and deposition (illuviation) in the B horizon, or at the top of the stagnant zone. This may eventually cause a clay pan that prevents further percolation, and then a new system forms in what was originally just the upper zone. This gives rise to a kind of perched water table.

In general, weathering is intense and rapid in the belt of soil moisture, for there are frequent changes of conditions, generally easy access of air and water, and the presence of organisms that produce organic acids, chelating agents and carbon dioxide, and mechanically stir the soil. In this zone weathering cannot profitably be separated from other soil-forming processes, and this zone is discussed at greater length in Chap. XI.

THE INTERMEDIATE BELT

The intermediate belt is mainly a zone of downward movement of water, but occasionally dries out, and may even have upward movement of water.

Water that penetrates the belt of soil moisture then trickles down through the intermediate belt towards the water table. This belt therefore is occasionally wet, but there is also abundant air. Leaching and eluviation are considerable, and various kinds of chemical weathering are possible in this zone.

Occasional drying out of this belt may lead to precipitation, especially of iron oxides. Discontinuous patches of water may be formed during drying, with generally convex surfaces, and, if they precipitate as they shrink, colour banding may be produced.

THE CAPILLARY FRINGE AND BELT OF FLUCTUATION

In very porous rock with wide pore spaces the water table may be nearly flat, but there is always likely to be a capillary fringe where water is drawn into the smaller pore spaces. Thus the change from saturated to aerated zones is transitional.

Furthermore, with changes in level of the water table, the capillary fringe also moves up and down. This is especially well marked where seasonal variations in climate prevail, as in savanna lands. If such a zone of fluctuation is present it suffers alternation between oxidising the reducing conditions, frequently revealed by grey and rusty mottling of iron oxides in the zone. It can be regarded as usual for the water table to fluctuate, but occasionally it remains stationary long enough to leave distinct evidence of its level. In general, oxidation will take place above the water table, and reduction and hydrolysis below.

In limestone country there is evidence that weathering is most intense at and just below the level of the water table, that is in the zone of fluctuation. Here movement of water is most rapid, and so also removal of dissolved material. Organic and carbonic acids are also most likely to be concentrated at this level. Caves are frequently found in the 'shallow phreatic' zone (see p. 224) and the results of rates of solution given on p. 248 show more solution underground than on the surface.

In regions where evaporation is for some time greater than precipitation, there will be evaporation of water, and possible movement of ions upwards in solution. This seems to be significant in laterite profiles, and may be part of the reason for the abundant iron in the true laterite or ironstone horizon and the lack of iron in the pallid zone below.

When the water table is high there is considerable evaporation, but when the water table has been lowered to a certain point evaporation is greatly reduced. In the Murrumbidgee irrigation area, where evaporation control is important to prevent salinity increase, Talsma (1963) reported the following figures relating depth of water table to evaporation rate:

Table 12

Evaporation rate (cm/day)	Depth of water table (cm)
0·7	112
0·4	134
0·2	164
0·1	201

Further reduction of evaporation is achieved only by lowering the table to a much greater depth.

THE ZONE OF SATURATION

Here all pore spaces are permanently filled with water. This waterlogging leads to anaerobic conditions and anaerobic bacteria may weather some minerals. Solutions from above are added to this zone and may be precipitated, and a clay pan may also form from particles mechanically eluviated from above.

Discharge belt

The upper part of the saturated zone is the zone of discharge, where water moves towards spring or seepage sites. Groundwater movement is generally fairly slow, and is usually greatest towards the water table and least at depth. The flow of water can, of course, transport the products of weathering in solution, thus enhancing further weathering. However, it is likely that conditions are always near equilibrium, and chemical reactions will be slow.

There is lateral movement of solution downslope as well as vertical movement through the profile, and the maximum lateral movement is likely to take place at the top of the saturated zone.

Stagnant belt

The lowest part of the groundwater may be quite immobile. This is called the belt of stagnation, and is particularly recognised in thick sedimentary basins. There is usually an underground barrier to flow, and the confined water is frequently highly saline. A stagnant belt may also be present in other situations where the saturated zone is fairly thick, such as in closed pockets of regolith in deeply weathered granite, or thick sandstone beds.

Weathering will be slower in the stagnant belt than in the discharge belt, as flowing water cannot carry away weathering products. However, this does not mean that weathering ceases entirely. Hydrolysis can still take place, though slowly, and weathering products are removed by ionic diffusion through the stagnant water rather than by flow within the water. Ionic migration is quite rapid enough for thorough weathering to occur, given sufficient time. As areas of deep weathering may have formed over a length of time measured in geological periods, the time available is quite sufficient.

THE GROUNDWATER BASE

There is a level beyond which groundwater does not penetrate. This is usually irregular, and water will penetrate deeply along joints or other lines of weakness while the bulk of the rock remains dry. Nevertheless, the actual contact between saturated and dry rock is frequently sharp. The groundwater base is equivalent to the weathering front (see p. 121).

EVIDENCE FROM METALLIFEROUS MINES

In Chap. X the formation of oxidised and supergene sulphide ores is explained. The position of the water table controls oxidation; above it, oxidation can take place readily, but at the water table oxidised minerals give way to sulphides. Mines provide some of the best evidence on the lower belts of groundwater and associated mineral alteration. The sulphide-enrichment zone is very markedly affected by groundwater conditions. A high, nearly stationary water table, such as might occur in a humid region of low relief, gives rise to a thin but well-enriched supergene sulphide zone, as at Ducktown, Tennessee (Fig. 70, p. 132). A deep water table, especially if slowly falling, favours a thick supergene sulphide zone. The ideal condition for sulphide enrichment is active erosion with progressive depression of the water table at a rate such that oxidation and enrichment can keep pace with it (Bateman, 1950). The contact between oxide and sulphide zones is frequently smooth, and sometimes abrupt. However, in many cases the contact is somewhat indistinct due to oscillation of the water table, and there may be large penetrations of oxidised zone minerals below the general water level in special circumstances, as at Cananea (see Fig. 69, p. 132).

More complex geological structures may cause even greater complications in the distribution of groundwater and hence of weathering zones. Thus inclined permeable beds in a generally impermeable sequence may conduct oxidation downwards, and faster downward circulation of water near faults may have the same effect, as at Copper Mountain, Morenci, Arizona (Fig. 41).

CHANGES OF GROUNDWATER LEVEL

Besides the seasonal or other short-term fluctuations of the water table, there are also long-term changes if the weathering profile is traced through long periods of time. The normal change is downward, owing to progressive downcutting of valleys.

The position of the water table can also be altered by climatic change and by geological 'accidents' such as faulting, or drainage modifications following volcanic eruptions or alluvial burial.

If the changes are sufficiently slow, all the zones of weathering will move down the profile, which remains adjusted to the new hydrological regime. However, the changes may be sufficiently rapid for the weathering zones to be stranded above the water table or drowned below it. Many examples are provided by economic mineral mines of oxidised zone minerals below the water table, and sulphide minerals above the water table (Bateman, 1950).

Examples of submerged oxidised zones include Globe, Arizona, where the water table was raised by Recent sedimentation 2000 ft thick; East Butte mine, Butte, Montana, where the oxidised zone was downfaulted below the water table; and the Zambia

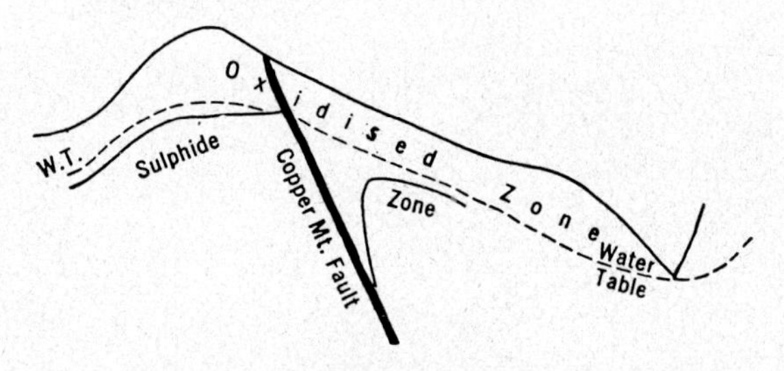

FIG. 41. Sketch of oxidised sulphide zones at Copper Mountain, Morenci, Arizona, showing deep penetration of oxidation along the hanging wall of an impervious fault, due to circulation of aerated groundwater.

copper belt where a change of climate from arid to humid raised the water level and submerged oxidised ores to depths of 2000 ft.

Examples of stranded sulphide zones include Bingham, Utah, where deep valley cutting caused the fall of the water table; similar downcutting by rivers left stranded sulphide ores at Rio Tinto, Spain, at Morenci, Arizona, and many other places; in Kenya former sinking of the surface of Lake Victoria has lowered the surrounding groundwater level, and left sulphides stranded above the water table.

VERTICAL AND LATERAL MOVEMENT OF WATER

So far in this chapter the vertical downward movement of water has received a lot of attention, but it must be realised that lateral movement is also very important.

Lines of flow within the belt of discharge are generally curved towards the spring or other water outlet, rather as shown in Fig. 42, the flow lines connecting points of equal pressure. The most rapid flow will normally be near the top of the discharge belt just below the water table.

In the soil moisture belt also, lateral movement of water is very important and is frequently responsible for changes in soil morphology on hillslopes. At the top of a hill, soils are freely or even excessively drained, the mid slopes are areas of through drainage, and lower slopes tend to be wetter, less perfectly drained, and enriched by leachates and solid matter derived from upslope. Soils with consistent hydrological and topographical relationships of this sort form the basis of the *soil association*, a grouping used by the Soil Survey of Scotland. Soils formed from the same parent material and differing mainly due to hydrological factors are called hydromorphic variants.

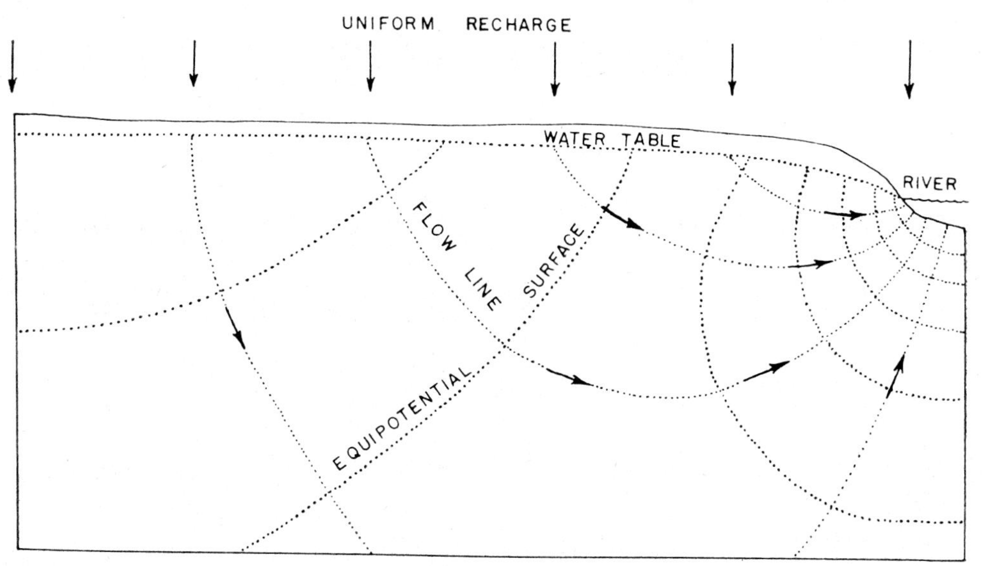

FIG. 42. Flow lines in the discharge zone in uniform rock. The spacing between equipotential surfaces is inversely proportional to average water velocity.

One of the more extreme effects of lateral seepage is postulated by Maignien (1959). Iron is believed to be leached from upper slopes and deposited as a lateritic ironstone sheet on lower valley sides. This eventually becomes impermeable, and new valleys are cut around the laterite so that it forms small terraces or plateaus. These in turn are then leached and a new ironstone layer begins to form on the younger valley sides, and so on. In this manner, it is postulated, lateral drainage and iron movement cause periodic terraced land surfaces each with a laterite cap. The higher flat surfaces are the older, with very leached lateritic ironstones, and the lower ones are younger, with incipient laterite.

On a large scale, lateral migration of solution affects geographic zoning of soils. Thus in areas of internal drainage the central part has chlorides and possibly other very soluble salts, which of course will have a profound effect on weathering. Around the chloride zone is a zone with sulphates, usually gypsum, in the soil profile, and beyond this a zone of soils with carbonates. In well-flushed areas not even carbonate is present in the soil profiles and pedalfers are formed rather than pedocals (see p. 147).

A similar zonation of soils has been suggested by Stephens (1966) to account for the distribution of silcrete and ferricrete in lateritic profiles of Australia. In general, it is suggested, silcrete is fixed in the drier, precipitating zones, and ferricrete in the wetter, leached zones.

Vertical movement is commonly downwards, but there may be upward movement of either solutions, or of ions through solution. Upward movement of solution is only an important factor in the soil moisture belt. Here evaporation at the surface and within the surface soil, perhaps aided by capillary rise, causes upward movement of soil water.

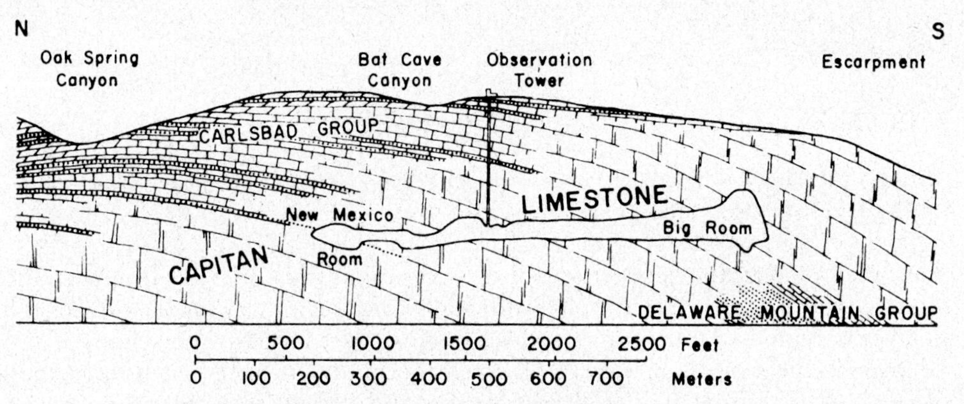

FIG. 43. Partly hypothetical section through Carlsbad Caverns, New Mexico (from Moore, 1960).

At lower levels evaporation has no effect, but ionic movement through solution can still be effective.

In laterite profiles there is commonly a bleached, iron poor, 'pallid' zone below the iron-enriched zones of the lateritic ironstone and the mottled zone. The pallid zone appears to have been originally in the saturated zone, and iron has been released and moved upward, and precipitated in the oxidised zone. Laterite profiles can be extremely thick, and it is doubtful if evapotranspiration could cause sufficient upward migration of solution. However, when iron is removed by precipitation a chemical gradient could be set up allowing migration of Fe ions through solution, thus depleting the pallid zone.

HYDROLOGY AND LIMESTONE WEATHERING

The study of limestone caves provides an excellent opportunity to study the relationships between groundwater and weathering.

It is found that many caves are nearly horizontal, even if formed in steeply dipping limestone (see Figs. 43 and 44).

This relationship immediately suggests that the position of the caves is related to a water table. The 'water table' and 'shallow phreatic' theories of cave formation suggest that caves are formed at the water table, and the many caves with active stream passages flowing through them and enlarging them support this hypothesis. The caves of County Clare in Ireland seem to be of this type. Water derived from shale caprock enters the limestone through sinkholes and then flows along the surface of the water table through almost horizontal passages either towards the sea or to springs (Ollier and Tratman, 1956). A fall in water table is required to convert the submerged passages into air-filled caves.

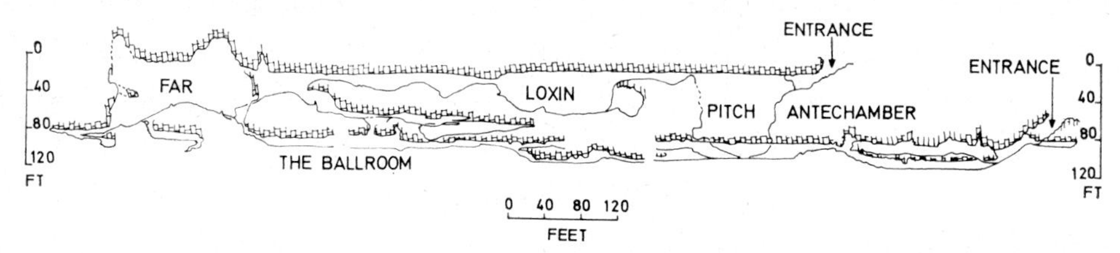

FIG. 44. Longitudinal section of Punchbowl and Signature Caves, Wee Jasper, New South Wales (from Jennings, 1964).

The 'shallow phreatic zone', the belt of fluctuation, may offer ideal opportunities for solution, having rapid flow, coupled with ready access of air, carbon dioxide and other agents, both organic and inorganic, and a free air surface for gaseous exchange.

Changes in the level of the water table, such as occur following downcutting of major rivers, lead to the formation of galleries of caves at different levels but all nearly horizontal. Near springs the water table tends to remain at a constant level, while in places

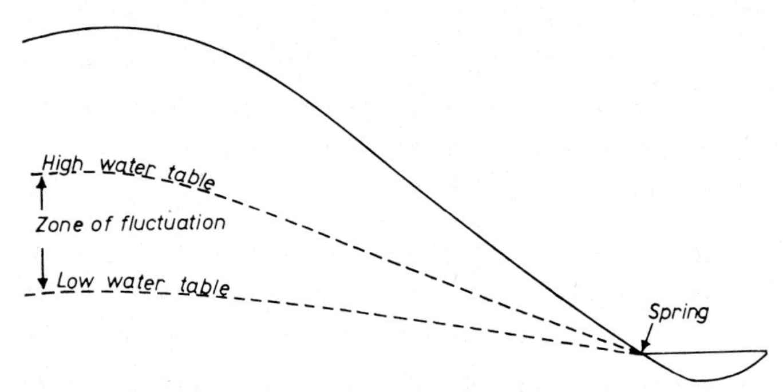

FIG. 45. The water table fluctuates through a much wider zone under interfluves than at streams.

far removed from springs there may be large fluctuations of the water table if there are large variations in groundwater recharge (Fig. 45). Deike (1961) reported just such a situation from Virginia; near the spring, caves were shallow and horizontal, but away from the springs the caves have less horizontal control and more features of 'phreatic' caves.

A 'two-cycle' theory of cave formation was proposed by Davis in 1930, in which it was suggested that most caves were formed deep in the phreatic zone, where water would be circulating very slowly, and only following a cycle of erosion was the water table lowered, exposing the caves already formed. This theory accounts for those cave systems with an irregular three-dimensional network and no marked near-horizontal control, but it seems that more and more caves are turning out, on closer investigation, to be shallow phreatic or water table caves. In the deep phreatic zone, with very slow removal of weathering products, a very long time would be needed for cave formation. This is quite possible, but only in specially favoured localities.

Above the water table in limestone, both solution and precipitation take place. The solution is largely in the belt of soil moisture and the upper part of the intermediate belt, and precipitation of calcite occurs in the lower part of the intermediate belt, forming stalactites and other dripstone and flowstone decorations.

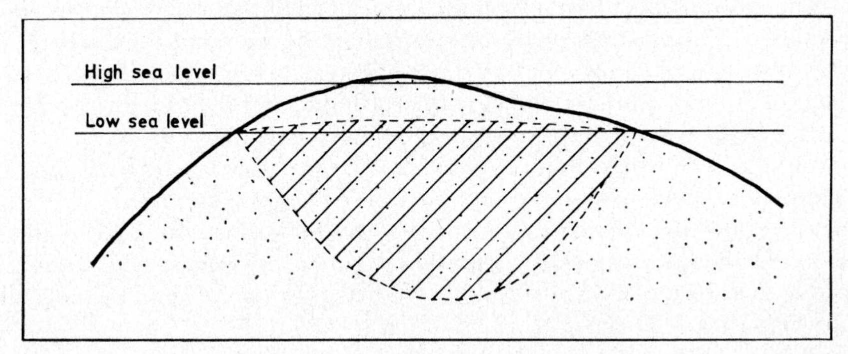

FIG. 46. At low sea level a coral island is large and holds a lens of fresh water. At high sea level the island is small and cannot hold fresh water.

A rather special condition occurs in coral reefs that affects groundwater conditions and weathering. A lens of fresh water may be held in the coral, and this projects much further downwards than upwards (Fig. 46) floating rather like an iceberg on the salt water. Within the lens of fresh water weathering conditions will be different from those in sea water, and in fact caves may be formed.

There is a certain minimum size for an island that can retain a lens of fresh water. In Bermuda there are many caves, now full of sea water, and the island's fresh water has to be collected from rain or delivered by tanker. Bretz (1961) has suggested that the caves were formed in glacial times when sea level was lower than at present, the islands were bigger, and a lens of fresh water could be retained.

In massive, but jointed, limestone of low porosity, water is contained in widened joints above a groundwater base that extends downwards irregularly until joints are quite tight. With continued solution the joints become still wider, and so can hold more water. If the quantity of water available remains constant, then the water table must fall. At the same time the groundwater base will probably be lowered as the extremities of open joints are dissolved, and this will tend to lower the water table further. However, the evidence from caves suggests that the water table is lowered by periodic, rapid falls, rather than by a steady continuous decline.

HYDROLOGY AND GRANITE WEATHERING

Solid granite has virtually no porosity. The initiation of weathering on bare granite surfaces would be quite slow, but, as can be seen from weathering of granite monuments, even the freshest surfaces are attacked. The weathering crust holds water, and aids further alteration, and the more the rock is altered, the more water can be held. Weathering proceeds fastest along joints, and corestones may be isolated from the main bedrock mass by weathering along the joints. In the early stages of weathering the residues may be dry or thoroughly wet, as if all the rock were in the soil moisture belt.

When weathering residues are deep, it is possible for all groundwater zones to be present. The groundwater base is commonly very irregular, controlled by joints and other structures. The groundwater base corresponds to the weathering front. Because of the marked structural control and presence of corestones, the boundaries between the various belts of groundwater are also of complex shape, and the distribution of different degrees and type of weathering are correspondingly irregular.

The saturated zone would roughly correspond to the grey 'weathered' zone (p. 124) and the oxidised zone to the red and pink weathered zones above.

In many exposures in quarries and at the ground surface, as around the base of tors, it is found that the oxidation is present right to the weathering front. This is generally due to lowering of the water table following downcutting in the normal course of erosion (Fig. 47).

Only in deep engineering works is the saturated weathering zone encountered. Such was the case in the excavation of the Tarago tunnel. It was found that the dry regolith could be blasted and sealed without difficulty if done quickly, but if it was left for a few days it became wet and started to run, even in narrow seams less than a foot wide. One extensive zone of saturated material, in which the contractors advanced too far without sealing, flowed extensively and eventually filled the tunnel for a distance of 120 ft.

Granite undergoes constant volume weathering in all belts below the soil moisture belt.

SOME MISCONCEPTIONS

Geomorphologists have long been influenced by two erroneous concepts which must be corrected in the light of the mass of evidence provided by groundwater exploration and metalliferous mining investigations.

The first major error is the supposition that weathering is more or less confined to the oxidised zone, and great stress is frequently placed not only on oxygen but also on carbon dioxide, in the belief that acidity is essential to weathering. This belief naturally leads on to another erroneous idea, namely that no weathering takes place below the water table.

Two statements by Cotton (1942, p. 21) exemplify these errors: 'The depth to which weathering proceeds depends upon the depth to which rain-water can sink vertically before reaching a continuous body of water (the *ground water*)'. 'The deepest level reached by the water surface sets a limit to the depth at which ordinary rock-decay can take place, for below this the continuous body of groundwater preserves rocks from attack by atmospheric oxygen.' Another example is provided by Ruxton and Berry (1957), who wrote, 'The agents of chemical weathering are gas (carbon dioxide and oxygen) and water (often with organic and other acids), and they act effectively only

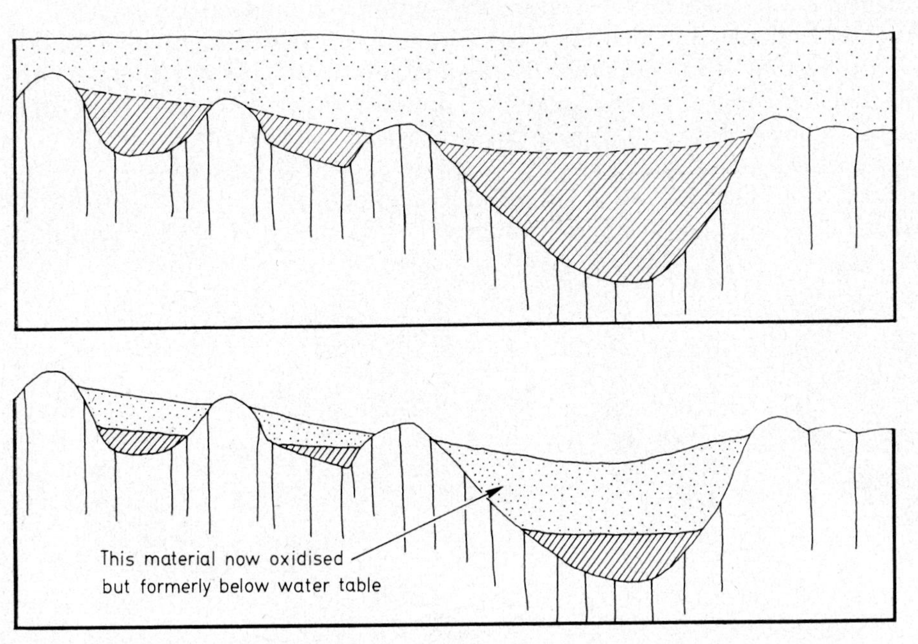

FIG. 47. Diagram to show how erosion can lead to oxidation of material formerly below the water table.

together', and 'The lowest level of the water table acts as a base level in the normal process of weathering'. Similarly, Linton (1955) wrote of the water table, 'Below its level the restricted circulation will inhibit chemical reaction since equilibrium concentrations of the soluble reaction products will soon be built up in the nearly stagnant water'.

These ideas equate a zone of weathering with the zone of oxidation, and ignore the weathering that occurs in the zone of saturation, especially reduction, hydrolysis and ionic substitution. The many supergene ore bodies are sufficient evidence, if any is needed, of the alterations that can occur below the water table.

The second misconception is less common, and can be exemplified by a statement by Linton (1955) concerning deep weathering on Dartmoor. He had a figure showing a deep layer of saprolite overlying fresh rock, the contact between them being sharp, and in places nearly flat (Fig. 110, p. 200). The flat surface was labelled BB, and Linton wrote '. . . the surface BB . . . has acted in some way as a downward limit to rock decay

. . .', and in seeking a mechanism to control the position of surface BB he wrote '. . . the surface most likely to do so is the water table'.

Surface BB is the 'basal platform', coincident with what is now termed the weathering front. It marks the base of weathering, and below it is hard, solid granite. Hard solid granite does not hold water, and its top is *not* a water table. Weathered granite can hold water, and the water table is situated at some height within the regolith. In Linton's statement the water table has been confused with the groundwater base.

Another example, in which the error is stated very clearly, is provided by Browne's (1964) description of tor formation, 'It is therefore most probable that, as envisaged by Blackwelder (1925), Davis (1938) and Linton (1955), tor topography originated through prolonged differential chemical weathering extending from the surface down to the *water-table where weathering ceases to operate*, and that only when the rotted rock is removed by erosion, following, for example, slight or moderate uplift, are the residual blocks of undecomposed rock brought together and exposed in outcrop' (my italics).

IX CLIMATE AND WEATHERING

One would expect from the very word that the group of processes known as weathering would be fairly strongly related to weather, and in fact climate, which can be regarded as average weather, is a very important factor.

The chief climatic controls are related to water and temperature. Water involves the total amount of precipitation, intensity of rain, proportion of precipitation that forms run-off, and the precipitation-evaporation ratio, amongst others. Temperature involves mean temperature, temperature range, and fluctuations about freezing point amongst others. Many other factors such as cloud cover, relative humidity, drying winds and climatic changeability may also be important locally.

Several climatic conditions affect the rapidity and depth of mechanical weathering; others affect leaching, groundwater recharge and salt movement, biotic weathering, and erosion (removal of weathered product). Some climatic effects are direct, others operate through their effects upon vegetation and soil.

PRECIPITATION

Water is the most important reactant in almost all forms of weathering, and clearly its supply is a great factor in the amount and style of weathering. Leopold, Wolman and Miller (1964) have suggested that there might be some threshold value of minimum total precipitation below which weathering does not occur, although it seems that very small amounts of water may be adequate for certain types of weathering.

Total rainfall is an inadequate measure of the effectiveness of precipitation in weathering. With three inches of rain a year Arizona is arid, while northern Alaska with much the same is a land of swamps, and weathering processes in the two areas are quite different. The extent and style of weathering is therefore only partly controlled by the total water supply, and is greatly affected by other local conditions, some climatic and others hydrological.

Because so many other factors are involved only a few generalisations can be drawn about the control of weathering by total water supply. Leaching of some cations may be sensitive to total precipitation. Soils in humid regions are strongly leached of sodium, potassium, magnesium and calcium compared with soils of arid regions. Sand and silt fractions of soils contain more quartz in humid regions, indicating increased weathering of less stable minerals with more total precipitation. In the United States, dissolved load carried in rivers increases with increasing annual run-off up to about 10 in, but in wetter areas the dissolved load depends on the availability of salt, which is in turn related to the composition and weathering rate of local rocks.

Some reactions may be favoured by equable distribution of rain, and others by seasonal changes. The weathering in the humid tropics is different from that of savanna countries with marked dry seasons. In the latter there is alternate solution and precipitation, irreversible chemical changes on drying, upward movement of

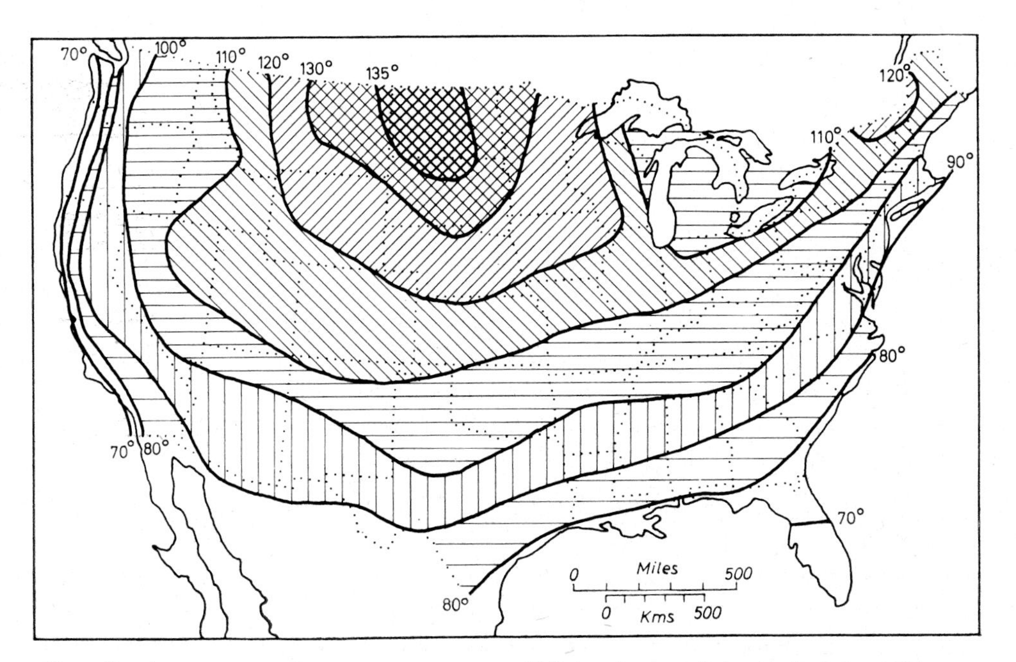

FIG. 48. Average annual temperature range, U.S.A.; degrees Fahrenheit (after Visher, 1945).

solutions in the dry season, and so on, while in the humid tropics weathering is continuous and unvarying.

Precipitation-evaporation ratio is also very important. If precipitation exceeds evaporation there is largely a through movement of solutions, and continued removal of weathered products. If evaporation is dominant there is periodic upward movement of water, drying out of soil, crystallisation of salts, and lack of removal of weathered products. The precipitation-evaporation ratio is therefore expressed through kind of soil, type of clay mineral, salt accumulation, salt weathering, and other related features. Alkaline soils and saline lakes are found only where evaporation normally exceeds precipitation.

Weinert (1961) has reported an instance of weathering apparently controlled largely by the precipitation-evaporation ratio. In the areas where Karoo dolerite is used for road making it is found that badly weathered dolerite only occurs in the east where moisture aids its decomposition. In the west it suffers more physical weathering, and the boundary between good and badly weathered rock appears to follow the line of 3·0 evaporation/precipitation ratio.

Aridity may also affect the course of mineral weathering, and Hay (1963) has recorded that in the Olduvai Gorge, Tanzania, which is hot and dry for most of the year, weathered tuffs give rise to zeolites rather than clay minerals in the alkaline soils of the area.

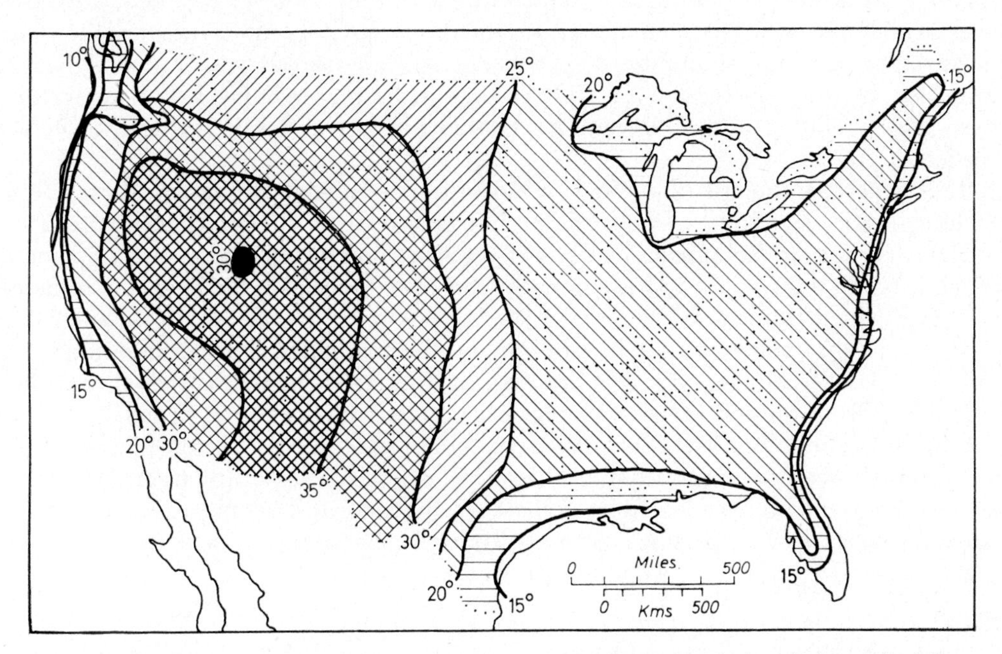

FIG. 49. Average daily temperature range in the shade, U.S.A.; degrees Fahrenheit (after Visher, 1945).

TEMPERATURE

Average annual temperature range may be significant in mechanical weathering. Visher (1945), who has provided one of the most detailed accounts of the geological effects of climate, believes that a range of even 90°F subjects exposed rocks to considerable expansion and contraction. As the temperatures presented on Fig. 48 are those recorded in the shade of official instrument shelters at an elevation of about 5 ft above the ground they are less extreme than those which occur on exposed rocks.

The average daily temperature range, in the shade, is shown in Fig. 49. In clear weather the daily range normally is fully 50 % greater at ground level, even in the shade, than the averages shown, which are based on both clear and cloudy conditions.

There is a tendency for the rate of chemical reaction to increase as temperature increases. Commonly a 10°C increase in temperature doubles or trebles reaction rates, whether the rates be slow or fast. Biological activity is also increased.

Jenny (1941) showed that within belts of uniform moisture conditions and comparable vegetation, the organic matter content of soil decreases exponentially with increasing temperature. His data indicate a doubling of organic content with each 10°C decrease in temperature.

In some circumstances the increased chemical reaction rate due directly to temperature increase may be offset by the distinct decrease in solubility of oxygen and CO_2 in water with increased temperature. Solution of calcite is particularly complicated. Its solubility depends on the solubility of CO_2, and is less at higher temperatures because the solubility of CO_2 decreases. However, its solution is largely controlled by the production of CO_2 in soil atmospheres, and low temperatures cause less vegetation and so lower CO_2 contents in soil atmospheres, and hence less solution.

It is generally thought that silica is more soluble at higher temperatures, and so leaching of silica should be more prevalent in tropic regions. This has been questioned by Davis (1964) on the basis of analysis of the silica content of rivers, but the general principle seems to be valid, and accounts for the abundance of desilicified clay minerals in freely drained tropical soils.

Carrol (1951) suggests that in tropical weathering, silica is removed from rocks while the pH of leaching water remains high; if drainage is impeded, this silica cannot be removed from the soil (or weathering rock) and resilicification of primarily formed gibbsite to kaolinite occurs. This may help to explain the presence of siliceous layers in the same horizons of Australian fossil laterites where silica was apparently mobilised but not removed. Where little clay mineral formation can take place, because of the nature of parent rock, the silica forms layers of sheets around quartz grains already present, and silcrete or 'grey billy' is formed.

FROST

Frost is of profound importance in mechanical weathering, and the details of its relationship to climate have been admirably discussed by Visher (1945), who prepared the frost maps of the United States shown in Figs 50-53. Fig. 50 shows wide regional contrasts in the average number of nights during which freezing temperatures are recorded in Weather Bureau shelters. At ground level in the open, frost occurs more often than shown, and the map is also based on low-altitude stations, not mountains or even hilltops.

Fig. 51 shows the average depth of frost penetration. The maximum depth of frost penetration is about twice as great as the normal penetration, except in the coldest areas where it is about 50% more than the normal (Fig. 52). These maps throw evidence upon how deep the formation of ice in cracks may split buried rocks, and reveal something of the regional contrast in the strength of frost heaving.

The effectiveness of frost is dependent on the frequency of temperature fluctuations about the freezing point in the presence of water. Minima therefore occur where it is too hot for freezing and too cold for thawing, and also where water is too scarce.

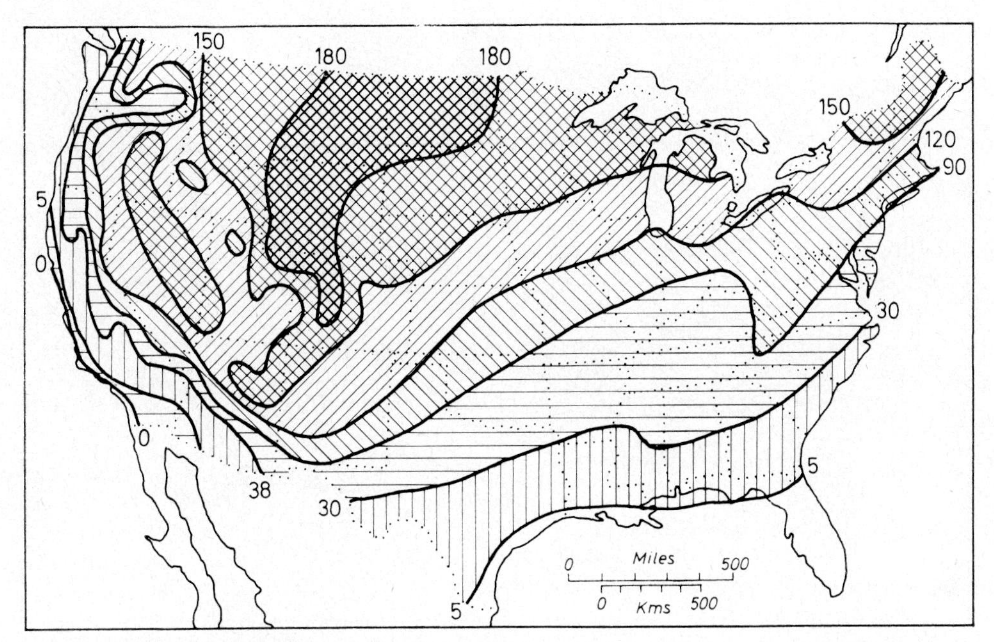

FIG. 50. Frost at night, U.S.A.; average number of times per year (after Visher, 1945).

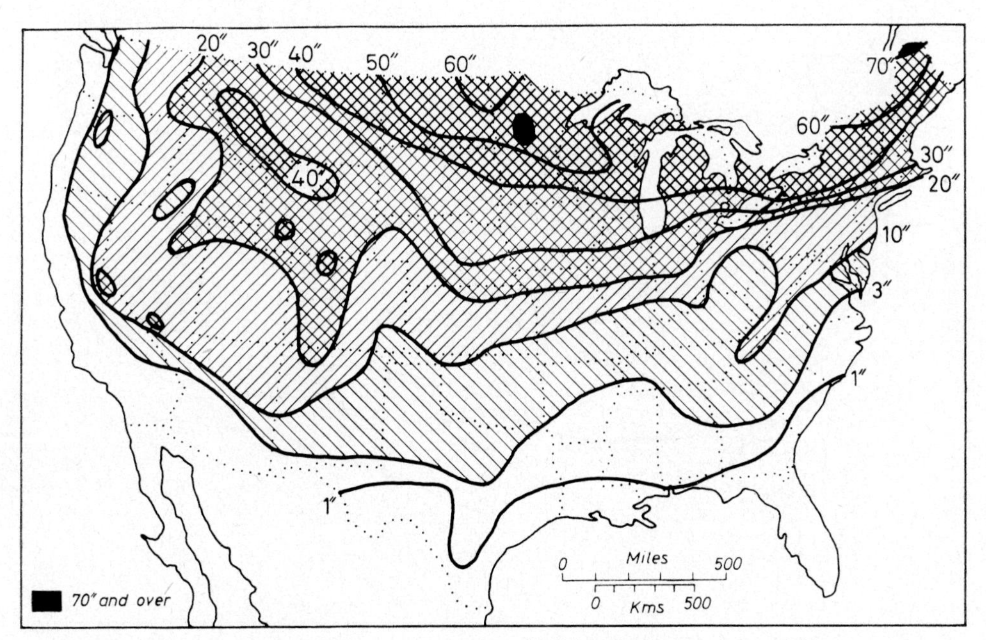

FIG. 51. Average depth of frost penetration, U.S.A.; in inches (after Visher, 1945).

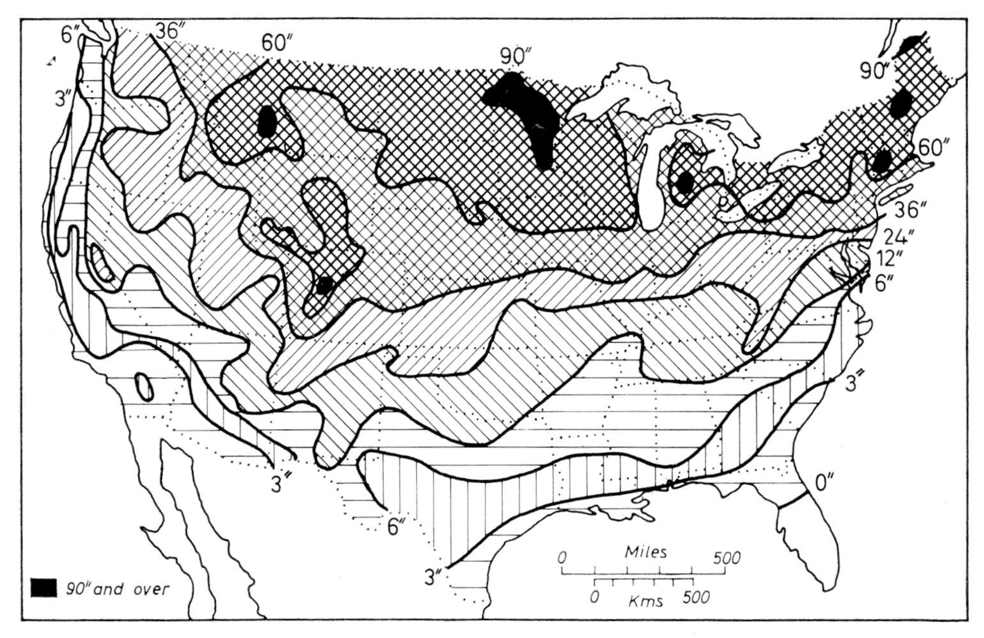

FIG. 52. Extreme depth of frost penetration, U.S.A.; in inches (after Visher, 1945).

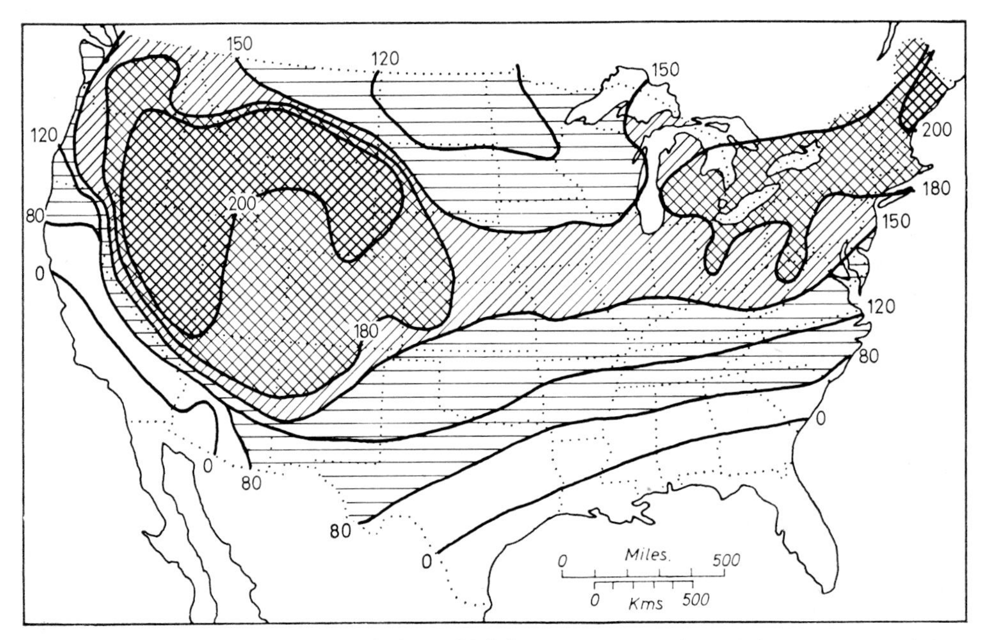

FIG. 53. Frequency of freeze and thaw, U.S.A.; average number of times per year (after Visher, 1945).

conditions. This will perhaps reduce biotic weathering, but by allowing increased removal of weathering products it will altogether increase weathering rates.

Microclimate, or the climate in and close to the ground, would be a better basis for considering climatic effects on weathering than 'coarse' climatic data, but not enough work has been done on this so far. The subject is outlined comprehensively in Geiger's *Climate near the Ground* (1965), but the large amount of data he presents cannot be easily related to weathering.

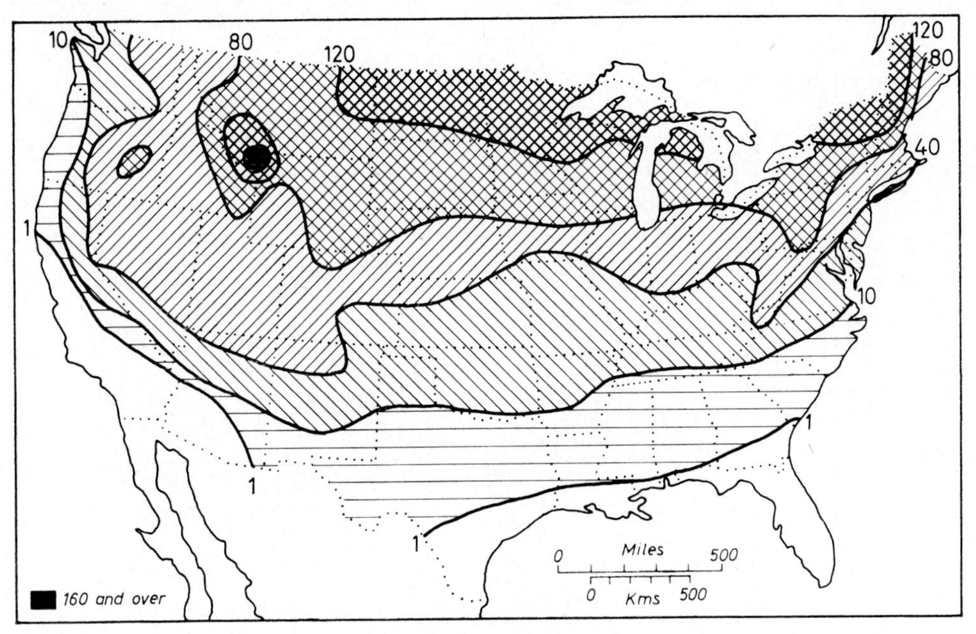

FIG. 55. Duration of snow cover, U.S.A.; average number of days per year (after Visher, 1945).

CLIMATIC REGIONS AND WEATHERING

Climates are classified into types with consistent relationships of the different climatic factors so far considered. It seems reasonable to suppose that different weathering phenomena will be favoured by some climates and inhibited by others, and there may well be a correlation of type and intensity of weathering with different climatic regions.

In arid regions, whether hot or cold, there is likely to be less chemical weathering. Judd (1886), for instance, attributed fresh microcline and orthoclase found in Nile mud to the arid climate of the source area. In tundra regions percolation is ineffective due to frozen ground, and so soil forming processes are weak, though even here there is weak podzolisation, and the translocation of iron and depletion of carbonate and bases. There is much physical weathering, often producing large angular debris, but it has been reported that basic rocks may succumb to granular disintegration (Rudberg,

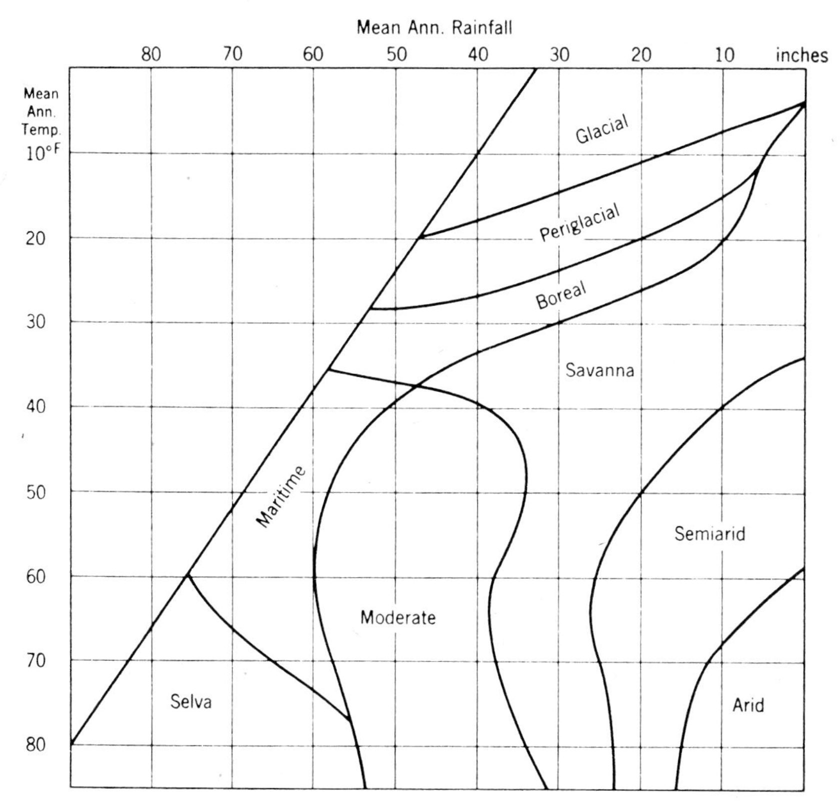

FIG. 56. Diagram of climatic boundaries of morphogenetic regions (after Peltier, 1950).

1963), and in Spitzbergen Czeppe (1964) found that mechanical weathering of certain sandstones was entirely subordinate to chemical exfoliation, although *a priori* frost might have been thought very important. However, most cold region research has demonstrated that the hydrolytic action of water and associated chemical weathering are of only subordinate importance in Arctic climates, on sandstone, quartzite, clay and calcareous shales, phyllites dolerites and many other rocks. It is clear that even the silicate rocks in a cold, arid climate undergo only meagre chemical weathering. Soil moisture, though not insignificant after snow melt, is never sufficiently high for the initiation of effective levels of chemical weathering. In Magdalen Bay, jointed granite blocks remain tightly knit and covered by a horizon of sharp-edged fragments formed entirely by mechanical attack, the smallest of them the size of nuts.

Some Russian writers, such as Stepanov (1965), have emphasised that weathering goes on even within a glacial environment. Snow and ice contain very fine (loess-like) particles. The glaciers of Central Asia alone give, on melting, 160 million tons of fine dusty earth annually and 9 million tons of salt, of which 960 thousand tons are calcium carbonate. The breakdown of these affects chemical properties of ice; for instance, the reaction of Central Asian ice is neutral to weakly alkaline, while Greenland ice is acid. Algae, fungus and bacteria, which may even tint the snow, participate in the weathering of minerals.

A number of geomorphologists have attempted to set up a series of climatic regimes within which the intensity and relative significance of the various geomorphic processes are, so far as we know at present, essentially uniform. These regions are called 'morphogenetic regions' by Peltier (1950), 'Formkreisen' by Büdel (1944, 1948), and to indicate the relation of climate to landforms have also been termed 'climato-morphic regions'.

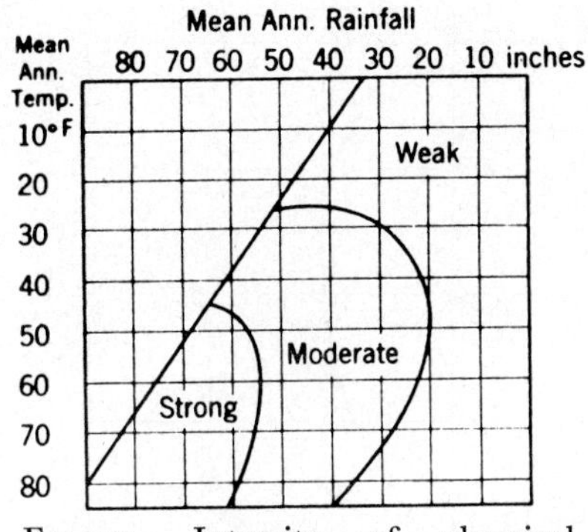

FIG. 57. Intensity of chemical weathering in relation to rainfall and temperature (after Peltier, 1950).

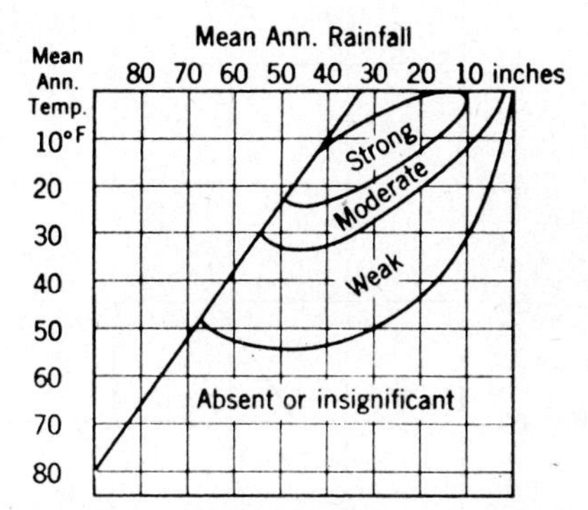

FIG. 58. Intensity of frost weathering in relation to rainfall and temperature (after Peltier, 1950).

Peltier (1950) has set up seven morphogenetic regions, shown on Fig. 56. These regions are hypothetical realms, representing the diagrammatic exposition of a concept, because accurate means of measuring and comparing all the relevant processes are not yet available. In using Peltier's diagrams, this point must always be kept in mind. Nevertheless, the morphogenetic regions provide a useful framework for thought, enabling a qualitative assessment of weathering and the different process active in any particular region under any particular climate.

Peltier provided three diagrams concerning weathering.

Fig. 57 attempts to show the relationship between climate and chemical weathering. Water is essential for most chemical weathering, and in so far as rainfall controls its abundance, an increase in rainfall leads to greater weathering. Temperature mainly affects the rate of chemical reactions, so warmer climates in general are expected to give rise to stronger weathering than cold climates. Thus weathering is greatest in hot, wet climates, and becomes less as rainfall or temperature decreases.

Peltier did not provide a diagram for biological weathering, but suggests that increases in both rainfall and temperature give rise to increased density of vegetation, and so chemical reactions related to organic waste are most rapid in warm, humid regions. Thus biological weathering might follow the same general trend as chemical weathering.

Fig. 58 is Peltier's figure for frost weathering. Two minima occur where conditions are too warm for freezing or too cold for thawing. By superimposing the rainfall

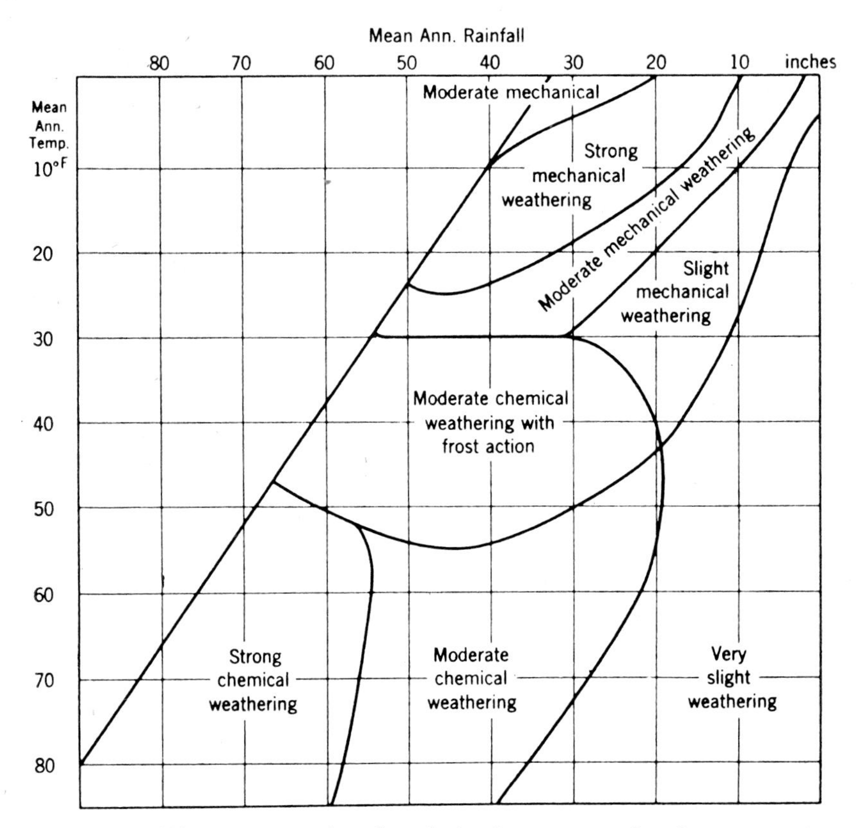

FIG. 59. Diagram suggesting the relative importance of various types of weathering under various temperature and rainfall conditions according to Peltier (1950).

pattern, with minima in polar and sub-tropical zones, upon the thermal pattern the final picture is obtained. Peltier believes that the intensity of frost is of no importance, as large falls in temperature do not cause corresponding increases in pressure. The duration of frost is also regarded as unimportant, for no further weathering is achieved until the next thaw. As stated above, the frequency of freeze and thaw is paramount, and those regions will suffer most frost weathering which have long transitional periods between the season too cold for thawing and the season too warm for freezing.

Peltier equated frost weathering with mechanical weathering, and following Blackwelder (1933) he discounted insolation weathering completely. In view of what was said in Chap. II about insolation weathering, it would now seem that a diagram of physical weathering should also have a high in areas of large diurnal temperature variation, that is in the hot dry zone.

Fig. 59 is Peltier's diagram of weathering regions, derived by adding his diagrams for chemical and frost weathering together. If insolation weathering is important, a patch of moderate or strong mechanical weathering should be added to the bottom right corner.

Another attempt to show the climatic and vegetational control of weathering zones is shown in Fig. 60 from Strakhov (1967). The zones shown are described on p. 126, Chap. XI, and the diagram can be compared with Fig. 79 which shows a comparable zonation of soil profiles. Fig. 60 shows clearly the major features of climatic zonation: maximum leaching occurs in the tropical zone and in the taiga-podzol zone, where precipitation maxima occur; because of high temperatures the alteration in tropical areas is more intense and deeper than in the podzol zone; the tundra and the semi-desert and desert have little chemical alteration because of low temperature and water shortage respectively.

The geographical distribution of the zones of weathering is shown in Fig. 61. This is somewhat confused because of the inclusion of tectonic factors, which Strakhov believes to be of equal importance to climate in weathering.

In this book much emphasis is placed on removal of weathering products, for it is thought that if they are preserved further weathering is inhibited, whereas if they are removed further weathering can proceed uninhibited; weathering proceeds in the manner of a chemical reaction that continues if a product is removed but reaches equilibrium if the products remain in the system. Strakhov on the other hand says, 'For the development of the chemical process the surface layers of rock must not be entirely removed, or, if so, eroded at such a rate that will not inhibit the change of the eluvium from an alkaline to an acid state and the removal of the soluble derivatives', and further, 'with greater variation and range of relief, mechanical denudation becomes more intense and, finally, reaches a value at which the rate of slope wash is greater than the rate of chemical weathering; the latter process inevitably begins to weaken under such circumstances and is ultimately suppressed', and 'In regions of even more rugged mountainous relief and very rapid denudation, chemical weathering is almost entirely absent . . .'.

Certainly in very rugged mountains deeply weathered profiles are unlikely to be preserved, but chemical weathering can still be very active if there is sufficient water and warmth, and few western workers would agree with Strakhov that regions of tectonic activity or regions of uplift (whatever the difference might be) are automatically areas of weak chemical weathering. In the rugged highlands of New Guinea deep weathering profiles are the rule, and it is difficult to find fresh rock.

A really adequate definition of morphogenetic regions must await the collection of more precise data, so that the regions can be derived from a firm factual basis and not by theoretical deduction. At present the morphogenetic regions enable a qualitative assessment of weathering, and of the different processes active in a region, providing a very useful conceptual framework.

PAST CLIMATES

The relationship between climate and weathering is made more difficult to determine because practically all areas have suffered different climates in the past from those

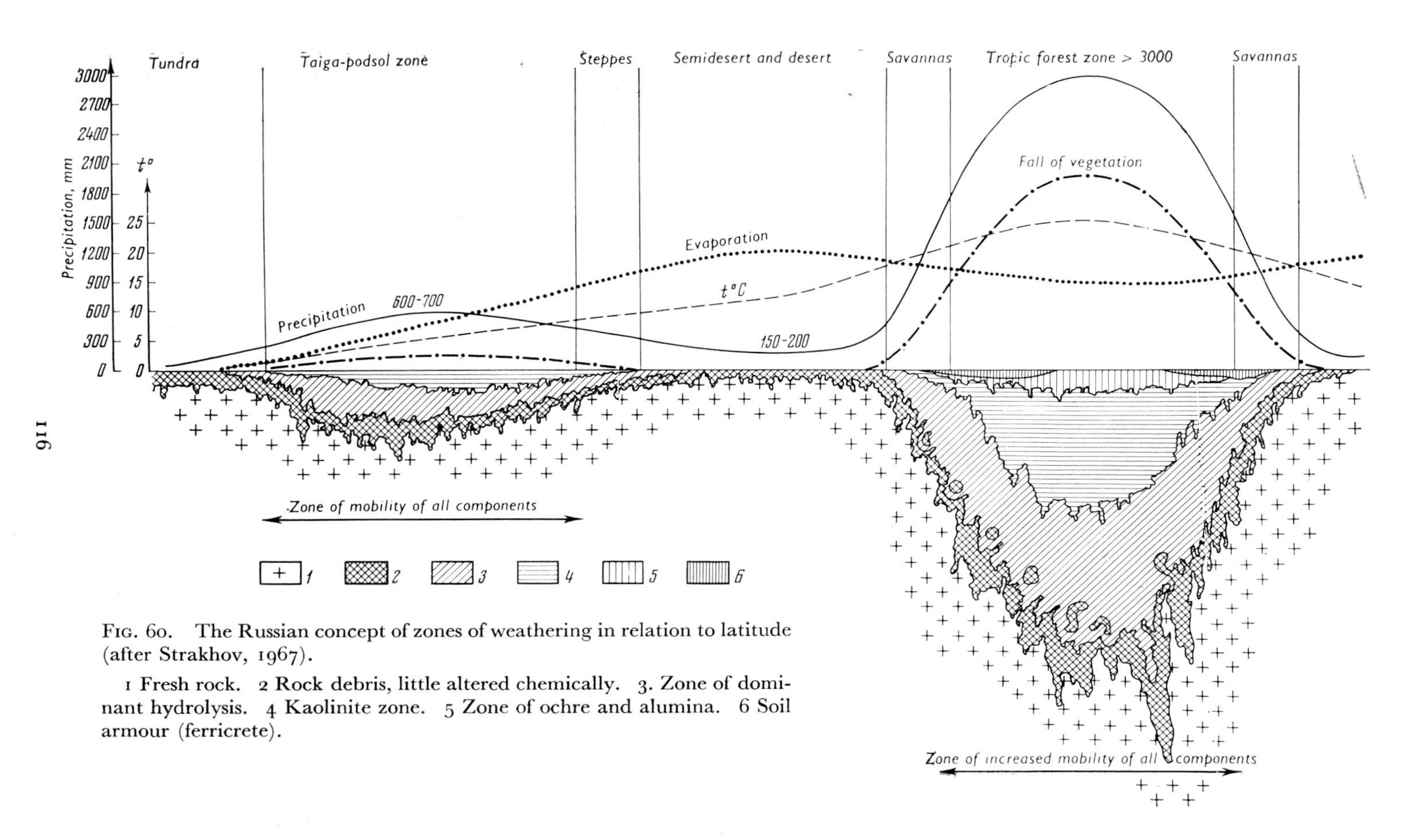

Fig. 60. The Russian concept of zones of weathering in relation to latitude (after Strakhov, 1967).

1 Fresh rock. 2 Rock debris, little altered chemically. 3. Zone of dominant hydrolysis. 4 Kaolinite zone. 5 Zone of ochre and alumina. 6 Soil armour (ferricrete).

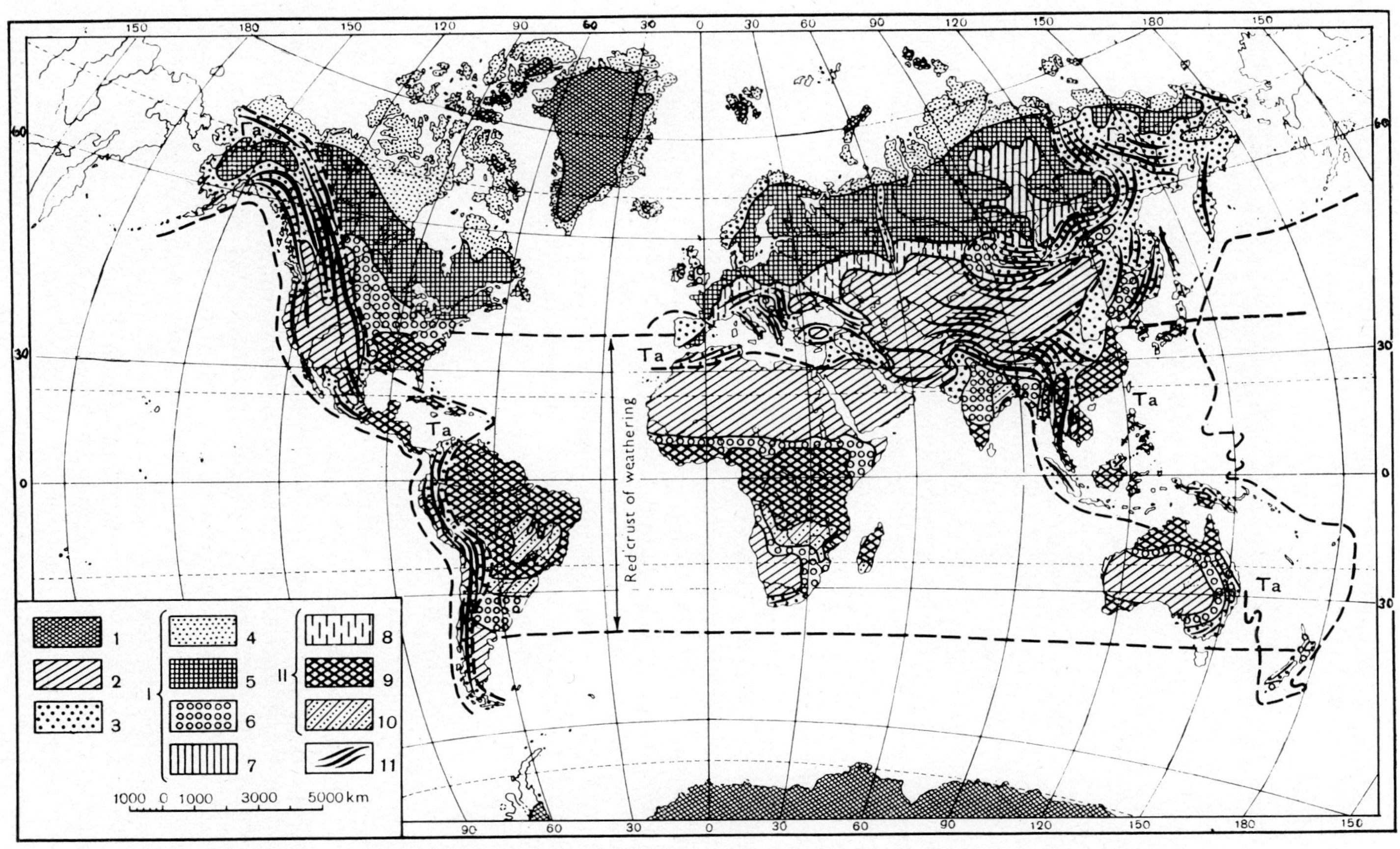

FIG. 61. The geographical distribution of weathering types (after Strakhov, 1967).

1 Glacial sedimentation. 2 Arid sedimentation. 3 Tectonically active—no weathering mantle. 4 Chemical weathering reduced by low temperatures. 5 Normally developed weathering. 6 Chemical weathering reduced by low precipitation. 7 Chemical weathering reduced by high relief. 8 Chemical weathering slight because of low precipitation. 9 Chemical weathering intense. 10 Periphery of zone 9. 11 Mountain ranges. I Region of temperate moist climate. II Region of tropical moist climate. Ta Tectonically active areas.

prevailing at present. It is frequently difficult to determine whether particular processes and landforms belong to the present or former climatic regimes.

It is not necessary to review the evidence for past climatic changes, for it is too well known. Much of the landscape in Europe and North America was created by glacial and periglacial processes, and the evidence has been described many times. In equatorial regions the evidence is not so clear, but there seem to have been shifts in climatic belts, with movement of the desert boundaries, and changes in pluvial conditions.

Nevertheless, although climates have undoubtedly been different in the past, landforms should not be attributed to past climates without thorough investigation, for there are a number of difficulties that can cause error.

Homologous landforms provide perhaps the main source of error. Patterned ground is commonly attributed to periglacial action, but gilgai produce similar patterns in non-glacial and even arid conditions. Frost wedges may be confused with solution tongues. Frost-riven blockfields should be easily distinguished from corestones boulderfields (see p. 214), but there are many examples of the latter in Europe that were once used as evidence of periglacial conditions. Inselbergs were once regarded as characteristic of arid regions, and are now coming to be taken as indicators of humid tropical climates by some authors, when in fact they probably occur in a wide range of climatic conditions.

There are numerous examples of hypotheses of changing climate to account for different weathering phenomena. For example, in Europe, Oligo-Miocene deposits are rich in quartz and poor in feldspar and quartzite pebbles, whereas deposits of Riss and Würm age have plenty of both. Cailleux (1960) suggests that those deposits poor in feldspar were formed under a hot humid forest climate, which seems to be rather a precise and unwarranted speculation.

At Ayers Rock, Central Australia, Bremer (1965) found evidence for tropical humid climate (the sugarloaf shape and the channels), Mediterranean climate (tafoni) and arid climate (scales). In contrast to this explanation, Ollier and Tuddenham (1962) believed they could account for all the features of Ayers Rock entirely by processes taking place under the present, arid climate.

Another problem is the difficulty found in comparing past climates of one area with the present climate of another. There have been repeated attempts to compare the weathering conditions observed in the Arctic with those of the Pleistocene periglacial region of central Europe, but such a comparison cannot be absolutely correct because of the different latitudinal positions of the two regions. In Spitzbergen, for instance, the land lies in darkness for fully four months and there is no sunshine to cause fluctuations of temperature. In the former periglacial regions the sun rises and sets daily throughout the year. The former periglacial areas therefore had greater and more frequent temperature variations than present periglacial areas, where the sun never sets for long periods, and where the angle of incidence of the sun's rays is almost as acute at midday as at midnight. More rapid temperature changes are bound to give more severe weathering and make direct comparison of climates impossible.

Comparison between climates at different altitudes in the same latitude may also be invalid; it is not possible to compare say a formerly cold lowland with a nearby mountain that has a cold climate at present. Mountains are commonly rather small in area, and some have glacial climates and yet are not glaciated because they have too small a snow-collecting area. Precipitation is possibly less effective in weathering on high ground, where it tends to run off rapidly, than on low ground where it is more easily retained. Highlands induce cloudiness, and so have a different pattern of radiation. In mountains the total range of temperature, the frequency of temperature changes and the number of freeze-thaw cycles, would be different from that in lowlands of the same average temperature.

As a matter of scientific methodology, it is perhaps a good general rule to use 'Occam's razor' or the 'Law of parsimony' in such instances, and adopt the hypothesis that rests on fewest assumptions. This means that the smallest possible number of climatic changes are invoked, and any past climate is only accepted when it is proved that present conditions are quite incapable of accounting for the phenomena observed.

X DEEP WEATHERING

Rocks may be weathered to great depths. Rotten rock in place can be termed saprolite. A more general term is regolith, which covers both residual and transported loose or soft material overlying solid bedrock. In many circumstances regolith is a better term to use in profile description because the amount of mass movement, creep or other sorts of deposition at the top of the profile cannot easily be determined. Many studies of deep weathering concern granite, and the term grus, meaning gritty weathered granite in place, is often encountered.

With a few notable exceptions, early geomorphologists appear to have been largely unaware of the great depths attained by the regolith. Where the regolith is most extensive there are few outcrops or exposures, and the failure to observe regolith was perhaps partly due to the usual process of geological mapping. It is customary to work from outcrop to outcrop of solid rock and interpolate the geology of the areas between exposures. Only when mapping specifically for soil or subsurface features does the extent of the regolith become apparent, and a lot of close borings are needed to get a good idea of the three-dimensional extent of regolith. Fortunately there are occasional engineering projects in regolith-covered sites and regolith-exploiting mines, and the investigations for such work provide a great deal of information.

DEPTH OF WEATHERING

It is commonly observed that while regolith is only a few feet deep in temperate regions it extends to a depth of many feet—tens or hundreds—in the tropics, and in Arctic areas is extremely shallow. This effect is so obvious that no one has bothered to compile statistics to 'prove' it. This would be hard to do, because there is great variation in depth of weathering even within one small area.

Porous, permeable and chemically reactive rocks tend to weather deeply; impermeable and inert rocks not deeply, so that, for instance, a dolerite dyke and a thick quartz vein weather to quite different depths in similar situations.

Where climate and topography permit rapid erosion there is less chance for weathered products to be retained and so regolith will not accumulate to great depths. If however erosion is much reduced, then weathering products can accumulate in place to considerable thicknesses. Conditions are ideal on some African plains, where climate is conducive to rapid weathering, relief is very small, and, as watercourses are commonly choked with reeds, erosion is reduced to a minimum. However, there is not complete agreement on the factors involved in deep weathering, and these points will be discussed further later.

It is difficult to find the average depth of weathering in an area, and for comparative purposes maximum depths are possibly of more use at present, though they can be biased by freak results. Some great depths have been reported.

Thomas reported weathering to 300 ft in Nigeria (1965), and from Uganda Ollier (1960) also reported 300 ft. Demek (1964) reported that kaolinitic weathering reaches a depth of 100 m in parts of Czechoslovakia. Ollier (1965) quoted reports of granite weathering to 148 ft in Queensland, 900 ft in New South Wales, 260 ft in Victoria and 120 ft in Western Australia.

During construction work on the Keiwa Hydro-electric Project, Victoria, deep weathering was one of the major problems of the engineers. The following data have been kindly provided by Dr F. C. Beavis, who was Senior Engineering Geologist at the time of construction. Mylonite was found to be weathered to 400 ft on the Keiwa River. Bores on Junction Spur proved weathered granodiorite to 550 ft, and weathering on this spur was everywhere greater than 400 ft thick. Bores on the Big River at the foot of Mount Bogong proved schists weathered to 600 ft. In the Head Race Tunnel an oxidised copper lode was found 1050 ft below the surface, and weathered gneiss, with biotite weathered and feldspars partly weathered and with joints open and iron-stained, was found 1150 ft below the surface.

As far back as 1930 Vageler reported weathering to 1312 ft. Perhaps the greatest depths claimed so far are by Razumova and Kheraskov (1963), who reported that on the Russian platform weathering processes have penetrated through individual fissure zones to the base of the platform to a depth of more than 1000 m and up to 1500 m.

Further evidence for very deep weathering comes from the investigation of metalliferous ore deposits formed by supergene enrichment, as described on p. 131.

WEATHERING FRONT

In some rocks, especially dense rocks with good jointing such as granite and basalt, there is a remarkably abrupt junction between weathered and fresh rock. The transition may occupy a few millimetres, or may appear completely sharp. This junction was called the basal platform by Linton (1955), but because it is very irregular was termed a 'basal surface of weathering' by Ruxton and Berry (1957). As isolated blocks and corestones provide equivalent surfaces to that at the top of the fresh rock, Mabbutt (1961b) proposed the term 'weathering front', which seems to be the present accepted term, but 'basal surface' can still be used for the top of a continuous mass of fresh rock.

Not all rocks display an abrupt weathering front, and in porous rocks or very fissile rocks like shales there is no clear boundary between weathered and fresh rock.

ZONES OF WEATHERING

Between fresh rock at some depth, and the topsoil at the surface, the regolith is not even and may be divided into a number of zones that occur in fairly constant arrangement. These are not soil horizons, though soil layering may add to or overlap other zones.

Wilhelmy (1958) summarised much work in many areas and believed the following zones were generally present on granite:

1. A red or yellow loam (the soil profile).
2. Weathered granite *in situ*. Structure was lost in the upper part, but frequently present below.
3. Decomposed granite with rounded corestones.
4. Less weathered blocks, angular and locked together, separated by narrow bands of more weathered material.

There is a decrease in amount of solid rock up the profile, and an increase in the angularity of blocks downwards.

Ruxton and Berry (1957) give a rather similar sequence from Hong Kong:

Soil (not counted by Ruxton and Berry as a weathering zone).
Zone I—residual debris of structureless sandy clay or clayey sand.
Zone II—residual debris with rounded, free corestones.
Zone III—angular, locked corestones with residual debris.
Zone IV—partially weathered rock with minor residual debris along structural planes.

A sharp junction separates Zone IV from unweathered rock.

A more detailed, less generalised sequence is given by Mabbutt (1961a) from Western Australia, and includes the lower parts of a lateritic soil profile.

1. Mottled zone. Little rock structure retained.
2. Pallid zone. Original rock structure retained.
3. Lower pallid zone, with rounded corestones.
4. Joint blocks.
5. Exfoliation plates.
6. Sharp weathering front, below which is fresh granite.

Brunsden (1964) gives a similar section from Dartmoor, England, complicated only by a periglacial, disturbed layer at the top:

Migratory layer—disturbed, structureless sand and granite fragments.
Zone 1. Undisturbed, structureless.
Zone 2a. Rotted granite that crumbles in the hand, but retains structure.
Zone 2b. Coherent rotted granite, with some rounded corestones.
Zone 3. Decomposition along joints. Angular, locked corestones. Spheroidal scaling.

This merges into solid rock with brown staining. Elsewhere on Dartmoor, Brunsden finds a sharp weathering front, as in the other sequences given above.

Writing of granite weathering in Australia, Ollier was unable to report the regular occurrence of zones in many places, and found rapid transitions and considerable irregularity in the disposition of the features, especially corestones, on which zones are based (Fig. 62). Even in Hong Kong the zones are by no means perfect, as can be seen in Fig. 109, p. 199. Ruxton and Berry believed that pockets of corestones high in the weathered profile represented partially abandoned portions of the weathering profile.

Nevertheless, it seems that as a generalisation the concept of zoning is true, and the scheme of Wilhemy above is perhaps the most straightforward expression of the general succession. Other sequences are given as local examples, and there is no reason why

Fig. 62. Juxtaposition of various degrees of weathered granite, Tarago, Victoria. Corestones of fresh and crumbly rock in a matrix of granite completely weathered to clay.

these should not have variations and peculiarities not found in other places. It is certainly helpful to have the 'Zone' idea in mind when describing weathering sections, although in some instances it may not be applicable, and though there are no sequences that can be expected to apply in all cases.

The most generalised sequence of zones is:

1. Soil (horizons possibly overlapping lower zones).
2. Structureless regolith.
3. Saprolite retaining rock structure.
4. Structured regolith with rounded corestones.
5. Structured regolith with angular, locked corestones.
6. Unweathered rock.

The zones described above are based on preservation of rock, in pieces of various sizes, or preservation of rock structure in the regolith, but zones can be based on many other criteria.

One possible division is into a lower zone, in which regolith is preserved undisturbed, and an upper zone where the regolith has been disturbed, largely by organic activity; thus separating the soil layer from the underlying regolith.

It will be shown (p. 174) that considerable chemical alteration of rock takes place under conditions of constant volume alteration. This provides another basis for division between a lower zone where volume remains constant, and an upper zone where expansion takes place.

From a chemical point of view the weathering layer could be divided into zones on the dominant kind of reaction. In deeper layers hydrolysis is probably paramount; near the soil carbonation and reaction with organic acids may be important. In limestone country the effects of the latter are apparently important, and are probably concentrated near the water table. In upper layers oxidation is very important, and an oxidised zone, marked by the red and brown colours of iron oxides and hydroxides, is one of the most obvious zones to separate.

Chemical zones are important in commercial weathering profiles, such as nickeliferous laterites or oxidised ores.

The presence of water has a profound effect on weathering, and so hydrological zones can also be incorporated into weathering zones, and a separation of a weathering profile into a phreatic zone, and a vadose zone, with the various subdivisions mentioned in Chap. VIII is also feasible.

Zones within the regolith can also be recognised on a mineralogical basis. The weathering series listed in Chap. V provides the basis for the method. Close to fresh rock the regolith usually contains numbers of weatherable minerals, and with increasing distance from fresh rock the more weatherable minerals disappear. It may be possible to distinguish several zones by the successive disappearance of different minerals, but more commonly there is only a distinct basal zone, with a large number of weatherable minerals present, and an upper zone in which the weatherable minerals have largely disappeared. Such a situation was described by Radwanski and Ollier (1959) in Uganda (see p. 60).

Yet another zone indicator is colour. Cores at Tarago were red near the surface, but lower portions were pink and finally grey.

Ruxton and Berry characterised the zone of weathering further by reference to chemico-mineralogical change and the state of physical disintegration, which were represented by the symbols in Table 13.

By using capitals to indicate normal presence in a zone, small letters to show minor amounts only and parentheses to emphasise rarity of occurrence, the various zones can be described symbolically as in Table 14. Zones of weathering are of importance in engineering works, and schemes for describing the zones have been devised by such bodies as the Snowy Mountains Authority and the State Rivers and Water Supply Commission of Victoria, whose work frequently involves deep weathered zones. The terminology used would not conform with geological practice, but 'weathered' means

Table 13. Features of weathered material and their symbolic expression (after Ruxton and Berry, 1957)

Chemico-mineralogical change	Effect	Symbol
Reddening and argilisation	Formation of reddish-brown silt and clay	D
Complete decomposition of feldspars and biotite	Formation of light-coloured kaolinitic debris	C
Partial decomposition of feldspars and biotite	Formation of grus	B
Partial decomposition of biotite	Formation of brown margin to joint blocks and core stones	A

State of physical disintegration	Cause	Symbol
Differentiated debris	Further disaggregation, illuviation or eluviation	Z
Residual debris	Disintegration and disaggregation	Y
Grus	Spheroidal scaling	X
Core Stones	Penetration of weathering agents inward, normal to the open structural surfaces	W

altered but not disintegrated, and the material can range from crumbly to hard, while 'decomposed' means there has been alteration to clay minerals, or accumulation of clays, which are often wet and unctious. The zones used are: decomposed; weathered; fresh, with iron-stained joints; fresh with closed joints.

Table 14. Description of Zones of Weathering (after Ruxton and Berry, 1957)

Zone	Weathering stages present	Characteristics
I	D, (c, b, a); Z, (y, x, w)	D-Z, corestones generally absent
II	C, B, A, (d); Y, X, W, z	C-Y, corestone generally common
III	B, A, c; X, W, y, (z)	B-X, dominant core stones
IV	A, b, or a, b; x	Partially weathered massive jointed rock

* Capitals indicate normal presence, small letters minor amounts only, and parentheses are used to emphasise rarity of occurrence.

The frequency of joints or other fissures is used to indicate the brokenness of the rock, and a further distinction is drawn, where present, when joints are clay filled or iron stained.

Fig. 63 shows an example of bore records for an engineering scheme in weathered rock.

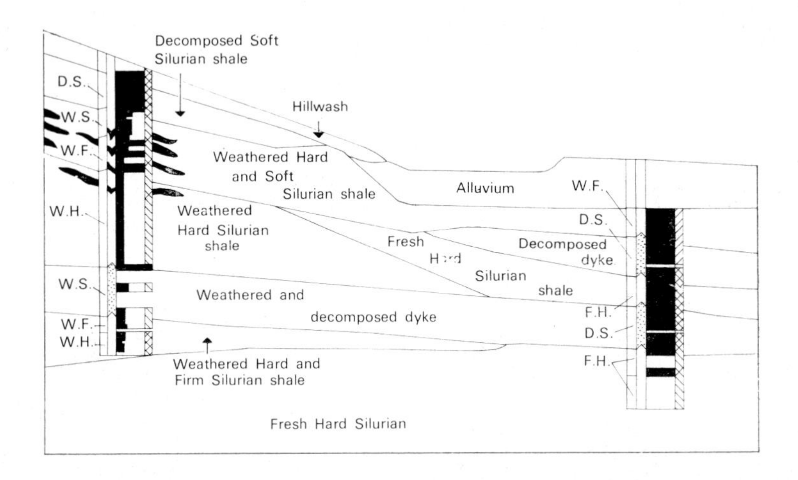

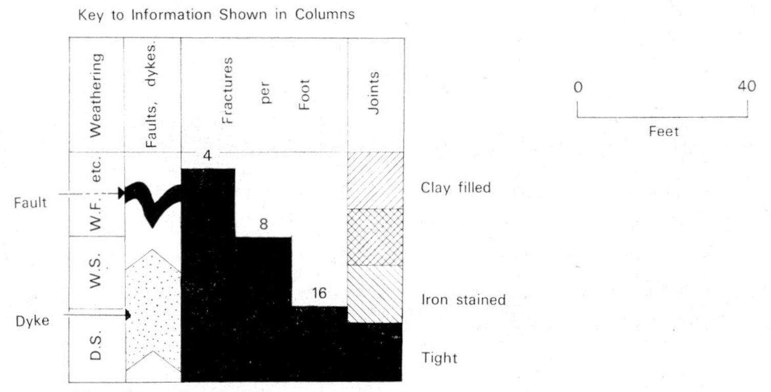

FIG. 63. Semi-diagrammatic representation of weathering data from bore hole records. Diagram based on an actual example from the State Rivers and Water Supply Commission of Victoria.
F=fresh; W=weathered; D=decomposed; H=hard; F=firm; S=soft.

Russian workers such as Ginzburg (1957), Nikitin (1965) and Strakhov (1967) adopt a zonation somewhat different from that used by Western workers. The Russians refer to the entire weathering profile as the crust of weathering, and point out that the upper zones are only present in moist tropical areas. At the base of a deep weathering profile there is said to be rock rubble, mechanically broken but scarcely touched by decomposition. This zone is quite absent from the description of western writers, and in my experience is absent from weathering profiles. Upwards follows the 'horizon of decomposition', which is probably equivalent to a zone of dominant hydrolysis, and equivalent also to the pallid zone. It is said to be characterised by hydromica, hydrochlorite, montmorillonite and beidellite, with occasional opal or carbonate. Structural features of the rock are preserved in this zone. Next comes the mottled zone, and finally the ocherous zone; these are clearly equivalent to the mottled zone and ferrallitic soil zones of western writers.

DEGREE OF WEATHERING

The zones of weathering refer to positions within a recognisable and often consistent series, but do not necessarily indicate the degree of weathering. For the latter, new terms are needed. In everyday usage such terms as 'weathered', 'rather weathered' and 'nearly fresh' are common, but they are not really adequate for description, and do not allow any genuine comparative judgements to be made.

Ollier (1965) suggested the following five-point scale based on friability:

1. Fresh. A hammer tends to bounce off the rock.
2. Easily broken with a hammer.
3. The rock can be broken by a kick (with boots on), but not by hand.
4. The rock can be broken apart in the hands, but does not disintegrate in water.
5. Soft rock that disintegrates when immersed in water.

A similar sequence was given by Melton (1965):

Class 1: A completely fresh fragment showed no oxidation stain on surface and no visible alteration or weakening of the rock.

Class 2: The surface of the fragment was stained or pitted, but the interior was not visibly altered.

Class 3: The surface of the fragment was deeply pitted or a thick weathering rind was present; the interior showed some staining. The fragment broke after repeated hammering, but was definitely weaker than fresh rock.

Class 4: The fragment was partially decomposed throughout but still cohesive and could be broken by hand or by a single light blow of hammer.

Class 5: The fragment was thoroughly decomposed and not strong enough to resist rough handling or being dropped a foot or two.

TOPOGRAPHY AND DEEP WEATHERING

Many examples of deep weathering are found in areas of considerable relief. These may be relics of once much thicker layers of regolith, and it is possible that weathering occurred when the areas had much less relief than at present.

If the association of deep weathering with peneplains could always be assumed, it provides a method of estimating the original thickness of regolith. At Tarago, for instance (p. 100), the top of a bore with 260 ft of regolith was itself about 250 ft below the projected erosion surface, which would suggest an original thickness exceeding 500 ft.

In this area the hilltops are occupied by lava flows of Tertiary age. These originally flowed down valleys in a landscape that was already deeply weathered. The present scenery is produced by inversion of relief. Thus there is every reason to believe that the weathering did occur beneath a higher and flatter surface than that of the present day, and the original depth of weathering was much greater.

Great depths of weathering are also found in the Snowy Mountains of Australia, where relief is of several thousand feet. Completely weathered rock has been found at depths of 900 ft below an undulating ground surface.

If a former plain is cast over this area, the highest in Australia, then the depths of weathering postulated become staggering, possibly up to 5000 ft. The alternative

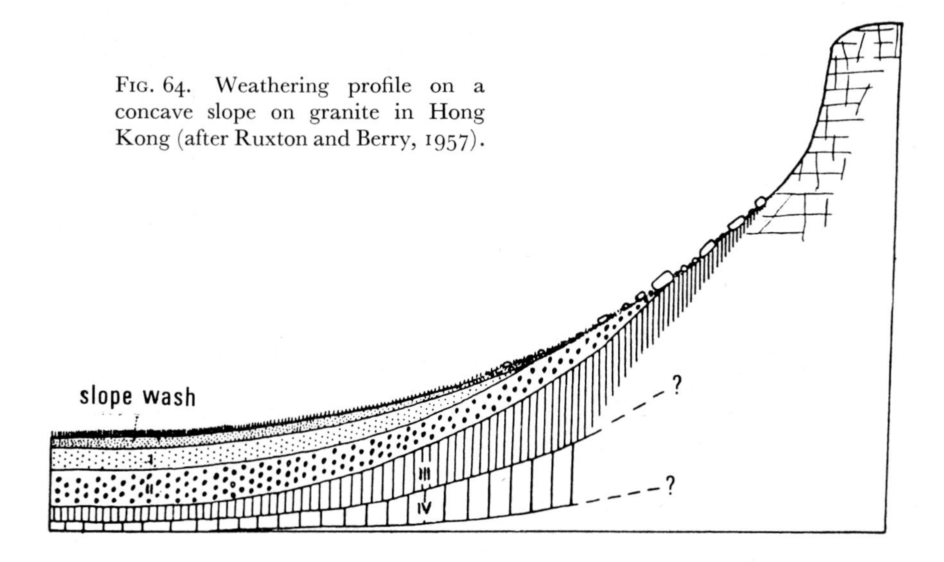

FIG. 64. Weathering profile on a concave slope on granite in Hong Kong (after Ruxton and Berry, 1957).

explanation is that weathering up to 1000 ft thick took place and was preserved on some of the most rugged mountain slopes in the country, which also seems very unreasonable. A similar situation exists at Kiewa, discussed on p. 121. Whatever the final answer is, conventional ideas of deep weathering need to accept a change of scale; either weathering can go on to depths undreamt of, or deep weathering can occur and be preserved on very steep slopes.

In the examples given above, deep weathering has been assigned to earlier geological times; deep weathering took place below a former plain, and the regolith is now being eroded. Of course, some weathering is taking place at the present time, and some writers, such as Ruxton and Berry, believe that deep weathering is a contemporary process operating at a rate comparable with rates of erosion. In a comparative study of weathering profiles in Hong Kong and the Sudan (1961b) they write 'Since, in general, the basal surface retreats at the same rate as the slope . . .' and also 'Retreat of the basal surface precedes retreat of slope. Mass-movement and erosion act above (often far above) the basal surface; only deeply entrenched stream courses operate below it.'

The normal profiles described by Ruxton and Berry from Hong Kong are said to occur on slopes less than about 15°. On concave slopes complete profiles are found near

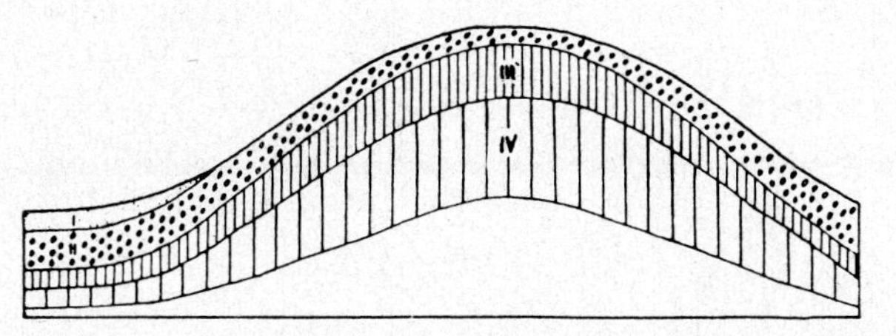

FIG. 65. Weathering profile on a convex hill in Hong Kong (after Ruxton and Berry, 1957).

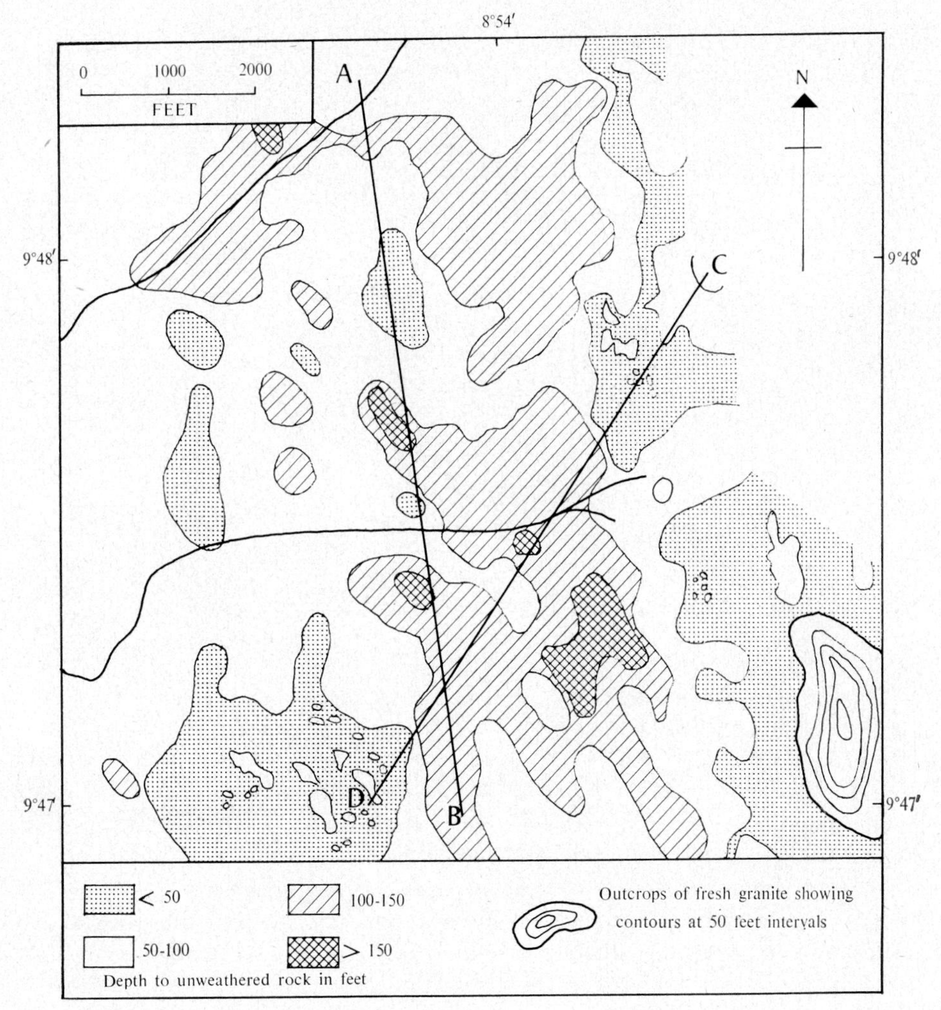

FIG. 66. Deep weathering patterns on granite near Jos, Northern Nigeria (after Thomas, 1966).

the base (Fig. 64). Some convex hills have the thickest profiles beneath the hilltops (Fig. 65).

Figs 62 and 109 show a more irregular arrangement of weathering zones in relation to slope, and suggest that the weathering zones were formed relative to some long-disappeared slope, and that the present slope cuts obliquely across the zones.

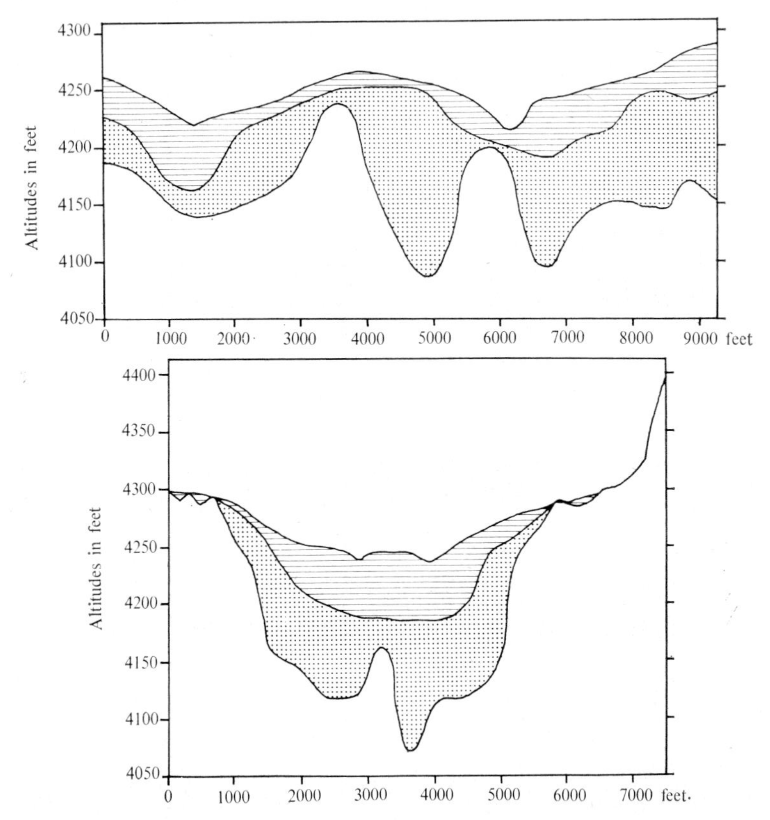

FIG. 67. Cross sections along lines *AB* and *CD* of Fig. 66. Horizontal shading indicates superficial deposits, stippled shading indicates weathered rock *in situ*.

Thomas (1966) has provided some of the best evidence about the three-dimensional form of the regolith. From an area in Nigeria he collated boring records and seismic results to produce a map showing depth of weathering (Fig. 66), from which numerous sections may be drawn, two of which are reproduced in Fig. 67.

There is no relationship between the present topography and the pockets of deep weathering, and the deep weathering is related to variations in fissility, petrology and other features of the bedrock, not to the disposition of streams. Thomas's observations bear out the statement of Enslin (1961), who observed that in South Africa the weathered rock 'tends to vary in depth and lateral extent, and to form a large number of isolated basin-shaped or trough-like groundwater compartments. These areas of deep decomposition show practically no surface indications but can be differentiated by geophysical methods.'

CLIMATE AND DEEP WEATHERING

It will be noticed that not all the examples of deep weathering are found in tropical areas. Dartmoor, Germany, Czechoslovakia and the Sierra Nevada, in California present widely different climates. It is possible, however, that the deep weathering in

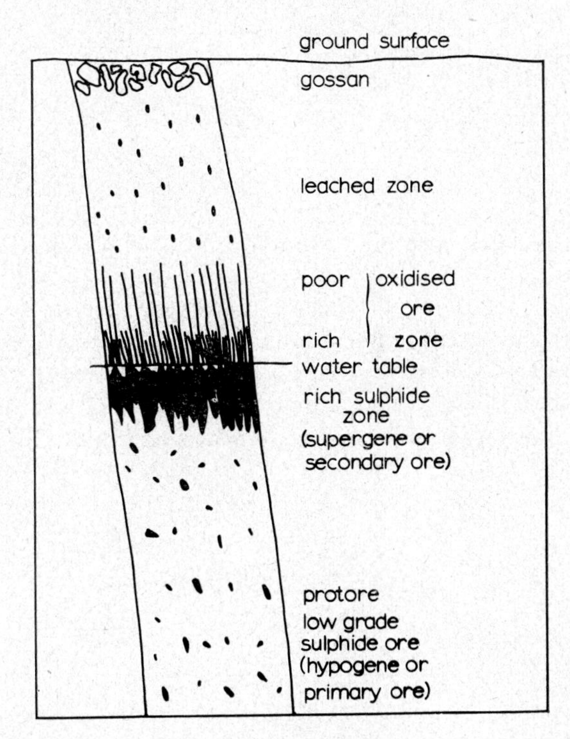

FIG. 68. Zones in an ore body formed by supergene enrichment.

these areas did not take place under the climatic conditions of the present day, and indeed it is commonly assumed that some sort of more tropical climate, possibly in Tertiary times, prevailed when the weathering occurred. In such cases it is important to work out the geomorphic history in sufficient detail to determine the age of the weathering, after which the influence of past and present climates can be assessed with greater certainty.

EVIDENCE FROM ORE DEPOSITS

A vast amount of information about depth and zones of weathering is available from the records of mines working commercial deposits formed by weathering. Much of this is recorded in textbooks of economic geology such as Emmons (1940), Bateman (1950) and Lindgren (1933).

The term primary and secondary are often misleading if applied to ores, so hypogene and supergene are preferred. Deposits formed by ascending water are termed hypogene. These may be affected by weathering, and rearranged by descending water to form supergene ores. In many areas mineralised but worthless deposits have been thus concentrated into valuable ore. Low-grade material that may be concentrated into

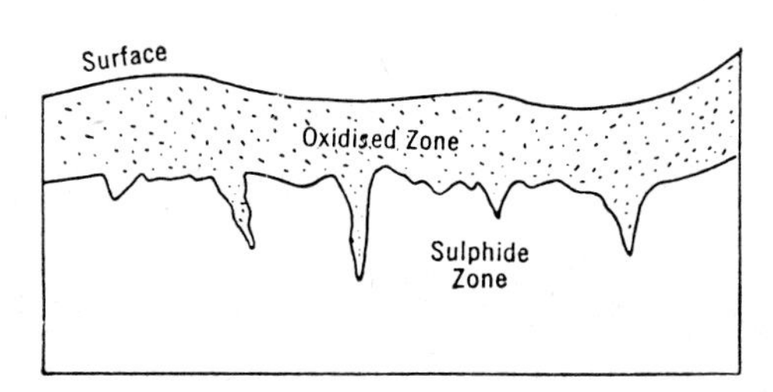

Fig. 69. Contact between oxidised and sulphide zone in the ore deposits at Cananea, Mexico (after Bateman, 1950).

ore is called protore. Supergene enrichment is especially important in the sulphide ores.

Fig. 68 shows the zones in a supergene ore body. Residual ores may be concentrated at the surface by removal of valueless material; some iron ores and bauxites are formed in this way. Alternatively a gossan may be formed, which is a mixture of limonite and quartz, often an indicator to prospectors of rich ores below.

The upper part of the weathering profile, above the water table, is the oxidised

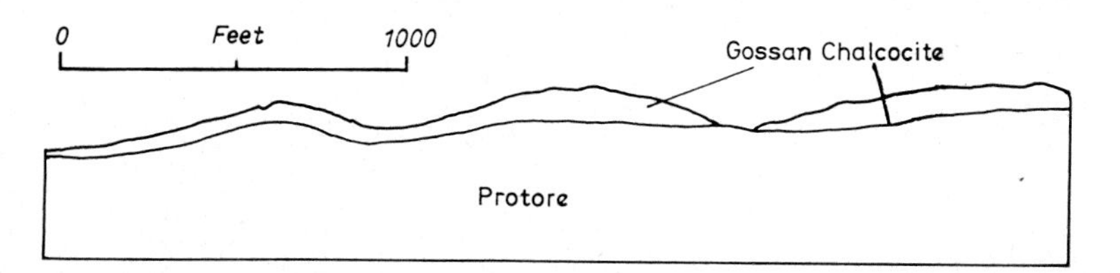

Fig. 70. Thin band of sulphide enrichment at Ducktown, Tennessee (after Lindgren, 1933).

zone. This in turn may be divided into a leached zone at the top, and a zone of oxide ores below, normally getting richer to the base.

The lower part of the weathering profile is the enriched sulphide zone, where the richest ore is usually found. This has been derived from the weathering of material above, and precipitated in the reducing conditions below the water table. The rich sulphide zone will often merge into protore below, but occasionally has a sharp boundary.

It is evident that the boundary between the oxidised and sulphide zones corresponds to the position of the water table at the time of ore genesis, but of course water levels may now be different. Thus at Cripple Creek, Colorado, oxidised ores are found 200 ft below water level, and in Zambia the oxidised zone is submerged to depths of 2000 ft.

In contrast, lowering of the water table has left sulphide zones stranded above the water table at many places, including Bingham (Utah), Bisbee (Arizona) and Rio Tinto (Spain).

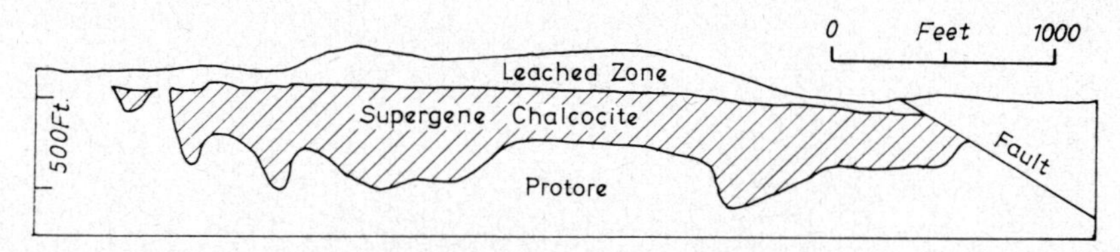

FIG. 71. Thick zone of sulphide enrichment at Ely, Nevada (after Lindgren, 1933).

There are special circumstances under which oxidation may take place up to scores of feet below the water table, as at Cananea, Mexico (Fig. 69), where the top of the sulphide zone is in general flat, corresponding to the water table, but is penetrated by long pendants of oxidised ore, corresponding to areas of high permeability such as fracture zones and faults.

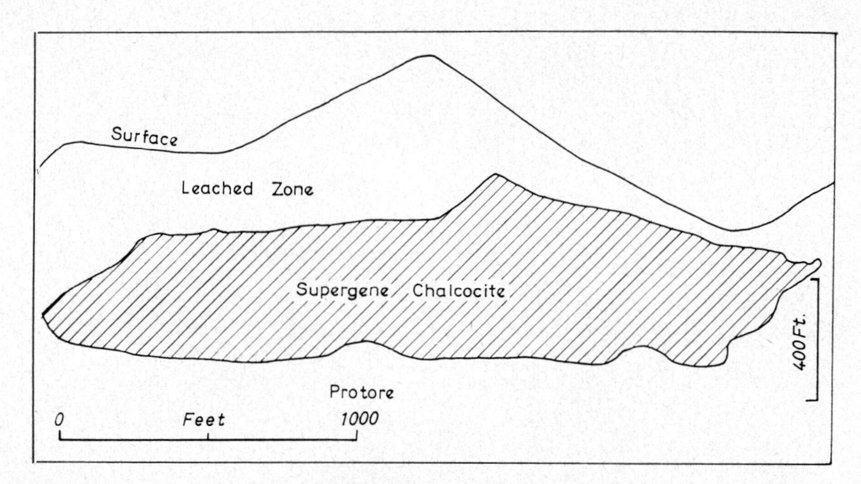

FIG. 72. Thick zone of sulphide enrichment at Morenci, Arizona (after Lindgren, 1933).

All the material above protore can be regarded as part of a weathering profile; thicknesses of oxidised and sulphide zones are measures of the thicknesses of zones of weathering, and the total depth to protore can be regarded as the thickness of the weathering profile.

The oxidised zone normally ranges from a few tens to a few hundred feet deep, but is known to be greater than 2000 ft in a number of places, including Kennecott, Alaska

(2000); Tintic, Utah (2000); Zambia (2000 and deeper); and in the Lonely mine, Southern Rhodesia, it reaches a depth of 3000 ft.

The zone of sulphide enrichment may be confined to a layer with a thickness of only a few feet, as at Ducktown, Tennessee (Fig. 70), or may be very thick as at Ely, Nevada (500 ft, see Fig. 71); Butte, Montana (1000 ft); Morenci, Arizona (average thickness 450 ft with a maximum thickness of 1000 ft, see Fig. 72); and Bingham, Utah, which is over 1400 ft thick.

XI SOILS

There is no generally agreed distinction between weathering and soil formation. Some seem to regard them as the same thing—a far-fetched idea perhaps, for unloading at least can hardly be regarded as a soil-forming process, and weathered rock at a depth of 200 ft can hardly be called soil. Reiche has said that no useful purpose can be served by making a distinction between soil formation and weathering, but I shall make a distinction here which is useful at least in organising the book, and, it is hoped, reflects views commonly held about the distinction in question.

Weathering is taken to mean any alteration of rocks and minerals; soil formation is the production of layers or horizons in weathered material near the earth's surface. (More formal definitions would have to take into account diagenetic and metamorphic alteration of rocks and minerals, and the possible existence of soils with no horizons.)

However, although a distinction may be drawn between soil formation and weathering, it is necessary in a book on weathering to include a study of soils because the two processes are related in many ways; it is not possible to understand soil formation without some knowledge of weathering, and weathering cannot be studied adequately without taking into account the soil zone at the top of the weathering profile. Soil horizons may be formed by various processes, some of which involve weathering. Another important relationship is that soils are often formed on weathered material rather than on fresh rock. The rate of soil formation will often be dependent on the weathering rate of the parent material, and conversely the rate of weathering of a rock will be affected by the type of soil-forming processes taking place near the surface. In many instances the weathering profile consists of nothing more than the soil profile.

SOIL MATERIALS AND SOIL PROFILES

For many purposes we are interested in soil as material, and can describe it by various properties such as mineral composition, shear strength, pore space, nutrient content or whatever interests us. In this regard it is like any other earth material, and standard descriptive techniques are available. From the point of view of weathering studies the main properties of interest are the grain size distribution and mineral content, and a brief account of the techniques of soil description methods is given in Chap. XV.

Soils are seldom uniform through a considerable thickness, but consist of a number of fairly distinct layers or horizons which differ from one another as soil materials.

Within any one horizon the soil material remains reasonably uniform. It is found furthermore that the vertical sequence of different horizons is often consistent over wide areas. The stack of soil horizons in any one place is called a soil profile, and extends from the ground surface to the parent material. The topsoil and subsoil, without weathered rock, are sometimes known together as the solum.

When pedologists (soil scientists) talk of soils they usually talk of soil profiles. There are many kinds of soil profile, but certain types seem to predominate, and the commonest profiles have been given names and classified in various ways.

Before soil profiles can be described, however, it is necessary to know how soil horizons are named, and before either profiles or horizons can be discussed profitably one needs to know something of the soil-forming processes.

SOIL FORMING PROCESSES

There are very many processes that take place in soils and may to some extent affect the formation of soil profiles, but the major soil-forming processes are:

1. Organic accumulation.
2. Eluviation.
3. Leaching.
4. Illuviation.
5. Precipitation.
6. Cheluviation.
7. Organic sorting.

Organic accumulation takes place mainly at the ground surface due to accumulation of decaying vegetable material. In podzols there may be secondary accumulation of organic matter in the B horizon (see p. 143).

Eluviation is the mechanical translocation of clay or other fine particles down the profile, a washing down the profile of mineral particles.

Leaching is the removal of material from a horizon in solution, and its movement down the profile.

Illuviation is the accumulation in the lower part of the profile of material washed down (eluviated) from above.

Precipitation is the formation of solid matter in the subsoil from solutions washed from above.

Cheluviation is the downward movement of material, akin to leaching, but under the influence of organic 'chelating agents'.

Organic sorting refers to the separation of material, usually of different grain size, by organic activity. For instance, termites make mounds of fine-calibre material, and coarse fragments tend to accumulate in the subsoil at the base of termite activity; earthworms make casts at the surface, eventually making a surface layer of fine material. Clearly many processes, working at different rates on different parent materials, can give rise to many combinations, and it may seem remarkable that soils present such a rational picture as they do.

Soil formation can also be regarded as a result of gains and losses, transformations and translocations (Millar, Turk and Foth, 1966). If we consider the topsoil, it gains water from precipitation, from condensation, or by flow from elsewhere; it gains oxygen and carbon dioxide from the atmosphere, and minor amounts of other gases; organic matter is gained from biotic activity; energy from the sun; and mineral matter may be washed on as a sediment. At the same time the topsoil is losing water by evaporation, transpiration and seepage; losing nitrogen by denitrification; and carbon dioxide from organic respiration; energy may be lost by radiation and mineral matter by erosion. If all these gains and losses are balanced the topsoil will remain in a steady state equilibrium and be unaltered through time. If there is not a balance then the topsoil will change its properties due to gains or losses, and will change towards another equilibrium position.

However, we must realise that the picture is in reality very complicated. Not all changes will be towards equilibrium but may be progressive, for some losses can never be replaced, and some changes are irreversible.

This theoretical approach may seem unduly complicated, but it is even harder in practice to examine a soil profile and work out whether or not it is in equilibrium, or exactly what soil-forming processes gave rise to the properties of the different horizons. It is difficulties such as these that have led many soil scientists to avoid wherever possible genetic implications when they describe soil profiles or horizons. Nevertheless, an interest in soil genesis is still there, and the soil scientist wants to work it out. His caution is to prevent him begging the question by assuming a knowledge of genesis in his first descriptive approach to a profile.

Soil horizon nomenclature

The assemblage of soil layers or horizons is known as a soil profile. Ever since the days of Dokuchaiev, the father of pedology, whenever soil scientists have studied and classified soils they have used the concept of the soil profile. Russian pedology not only provided the valuable profile concept, but also provided a system for labelling the component horizons within the profile.

Dokuchaiev originally used the letters A, B, and C for the layers in a chernozem soil formed on loess in Russia (Fig. 73). The upper layer, rich in organic matter, he called the A horizon. The mineral matter of the loess—perhaps the 'mother rock' of the soil—was called the C horizon, and the layer of mixing of the two was called the B horizon.

The simple idea of the B horizon being a mixture of A and C was not maintained, and in describing a podzol the upper sandy layer of the profile was called the A horizon, including any slight admixture of organic matter and the heavier clay horizon beneath, was called the B horizon. The parent material beneath was called the C horizon. Now the formation of this soil appeared fairly straightforward: the parent material was weathered and then clay was washed out (eluviated) from the A horizon and washed in (illuviated) and deposited in the B horizon. 'A' and 'B' horizons then came to be generally regarded as the designations for the eluvial and illuvial horizons (Fig. 74).

Later difficulties concerned the definition of the C horizon, which, it was found, was not always the 'parent rock'. For example a rock, say a granite, was deeply weathered, and a podzol formed on the upper part of the regolith. Then A and B were the eluvial and illuvial horizons, the weathered granite, i.e. the parent *material* was the C horizon, and the solid granite, i.e. the parent *rock*, was the D horizon.

The original ideas of soil horizons were applied to freely drained soils, and waterlogged soils were a complication that developed special and distinctive features due to reducing conditions in the anaerobic, waterlogged horizons. These grey, sticky often mottled horizons, are called gley, or glei, and they do not have a constant relationship to

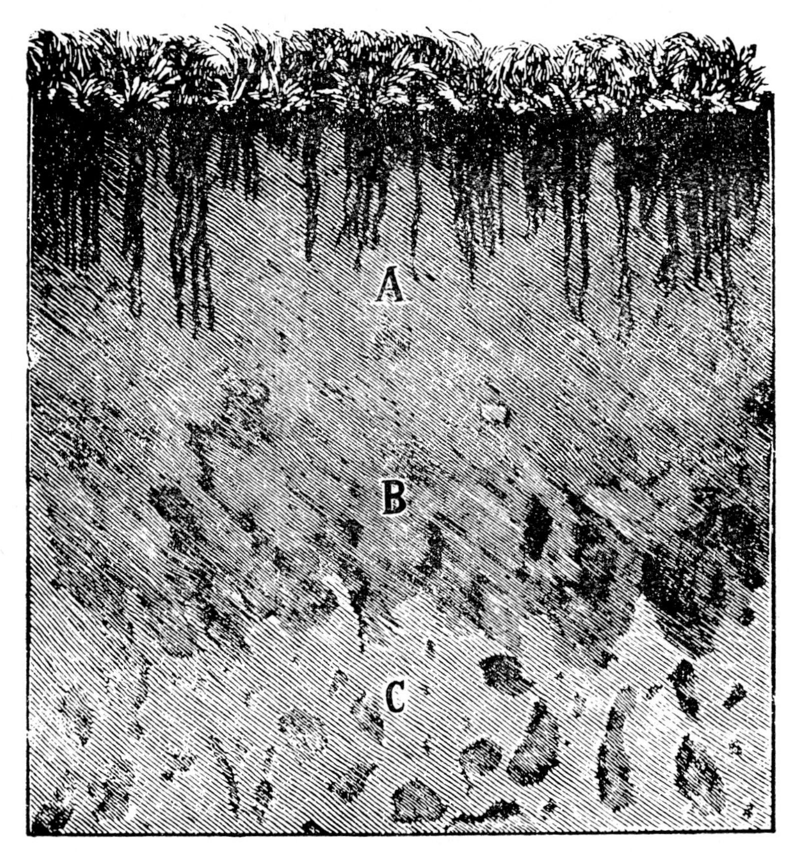

FIG. 73. Dokuchaiev's illustrations of the chernozem. Note the irregular pocketed *B* and the crotovina in the *C* horizon.

other horizons. Some gleys are groundwater gleys, due to waterlogging at depth, and others are surface water gleys due to perched water over deeper soil. The gley horizon may be symbolised by G.

In a similar manner other horizons and horizon characteristics may be symbolised either by a horizon symbol or by a suffix or prefix to another horizon symbol.

Difficulties often arise from assumptions regarding genesis of horizons. How do we know that layers are due to eluviation and illuviation? In the first place and with no prior knowledge could it be assumed that the layers are due to a depositional origin—that apparently leached layer may be a layer deposited by wind, by hillside creep, by alluvial deposition, or volcanic ash accumulation, depending of course on the environment. And could not lower horizons also be due to deposition, before the deposition of the A horizon deposit? Or perhaps the 'A'+'B' horizons are a deposit quite unrelated to 'C', in which profile differentiation has taken place to provide A and B horizons. Even more tricky possibilities can be imagined. For instance, a layer may be added to a bedrock, eluviated, and illuviation take place in both the base of the added layer and the top of the underlying material. The situation could be further complicated if the underlying material in question was already layered—say a stratified alluvium for instance, or an old soil with well-marked horizons.

The recognition and distinction between soil horizons and depositional strata, that is between pedological and geological layering, is a major problem in many studies of soil genesis, and not as easy as is often supposed. In early studies, geological additions

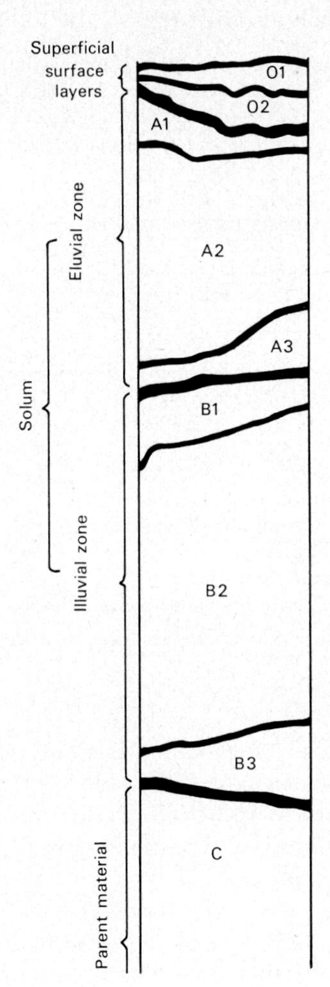

FIG. 74. The *A*, *B*, *C* nomenclature of soil horizons on the basis of eluvial and illuvial zones (from Millar, Turk and Foth, 1966).

were often overlooked and attributed to pedological processes. On the other hand, in some modern profile descriptions from Australia hardly any pedological layering is recognised, and almost every horizon is attributed to a geological event (e.g. Churchward, 1963; van Dijk, 1959; and examples in Butler, 1967).

It is fairly common at present for soil profile descriptions to go down to the depth of a soil auger or soil examination pit, yet lower horizons in weathered material may be very significant. It is to be hoped that for weathering studies a method of profile description will be devised that enables description of entire weathering profiles, however large, and not merely the 'soil profile' at the top.

Happily there appears to be a trend towards uniformity in horizon nomenclature at present, but there are still considerable differences in usage between nations, organisations and individual researchers.

The A, B, C nomenclature is fairly standard for main horizons.

The French also distinguish the following:

(A) simple disaggregation of parent material,
A organic accumulation and/or leaching,
(B) structurally differentiated subsoil horizon,
B texturally differentiated subsoil horizon,
G gley with red mottles in grey soil (permanently saturated),
g pseudogley, with grey mottles in brownish soil,
Ca carbonate.

The Americans use several other horizon symbols:

O superficial organic layers.
L litter layer of forest soils (O1),
F structure evident in organic matter } names
H amorphous organic material } for O2,
R bedrock.

The Americans also use suffix symbols as follows:

p plough layer,
t illuvial clay,
ir illuvial iron,
h illuvial humus,
hir both iron and humus illuviated,
b buried soil horizon,
ca lime accumulation,
cs calcium sulphate (gypsum) accumulation,
ca accumulation of concretions,
f frozen soil,
g strong gleying,
m cementation, induration,
sa salt accumulation,
si siliceous cementation,
x fragipan (firmness, high density).

These symbols are used in conjunction with the horizon symbol, thus, B2t, Bhir, etc.

SOIL CLASSIFICATION

Soil profiles can be classified in many ways, and many classifications are in existence. Some of the earlier ones, founded by the Russian school of pedologists, are dominated by a geographical approach. The zonal soils are thought to reflect zones of climate and vegetation, and those types that do not fit are placed in the outcast groups of Intrazonal soils and Azonal soils.

At the time such soil classifications were evolved, classifiers were much impressed by the success achieved by biologists in the classification of living things, and a hierarchical system was set up with Orders, Suborders, Great Soil Groups and other units corresponding roughly to the biological Phylum, Order, Class and so forth. There was no agreement on which factors should have priority in this hierarchy, and no consistency even within one classification, and so a multiplicity of classifications arose, all based on the same loose general principles, but differing in details.

Arguments have also arisen as to whether soils should be classified according to genesis (which is not an observable property of the soil) or on profile morphology, which is observable but to many not so fundamental. The modern American, French and Australian classifications attempt to be based more on morphology than genesis.

Another possibility is now becoming apparent. It is possible with a purely morphological classification to use the methods of numerical taxonomy that are already being applied in several branches of classification, especially in biological taxonomy (Sneath and Sokal, 1963). In this type of classification all characters that one desires to use are given equal weight, rather than arranged in a hierarchy at the whim of the classifier. Mathematical treatment is used to find how the soils are best grouped together, and then the properties of the 'clusters' can be used to define soil groups. This method promises to bring together the mathematically rigid classifiers and those who use what has generally been called a 'natural' classification. Rayner (1966) applied numerical taxonomy to a number of English soils, and found that the clusters he obtained agreed well with the field classification of the surveyor who described the profiles, but, of course, the numerical methods are more objective.

THE WORLD PATTERN OF SOILS

The world picture of soil development can now be described. Besides giving an account of soil geography, this exercise will provide a vocabulary of common soils, indicate the varying importance of several soil-forming factors, and present a possible method of soil classification.

In this section a traditional classification of soils is used. The major divisions are into Zonal, Intrazonal and Azonal soils.

Zonal soils reflect climate and vegetation to a large degree, hence their name. They are formed on well-drained sites on non-extreme parent material.

Intrazonal soils are well-developed soils formed where some local soil-forming factor is dominant.

Azonal soils have poorly developed profiles because some factor such as parent material or lack of time has inhibited soil formation.

ZONAL SOILS

Polar desert soils. General characteristics of the Antarctic polar desert soils are dryness, low temperature, low organic matter content, mildly acid to alkaline reaction, and commonly the presence of salt efflorescences, desert pavement and patterned ground. In the Arctic there are extensive wet areas, but some well-drained sites carry zonal soils called Arctic Brown soils.

Podzols. The group of soils known as podzols has probably received more detailed attention than any other group of soils. In Europe they occur in humid regions, where

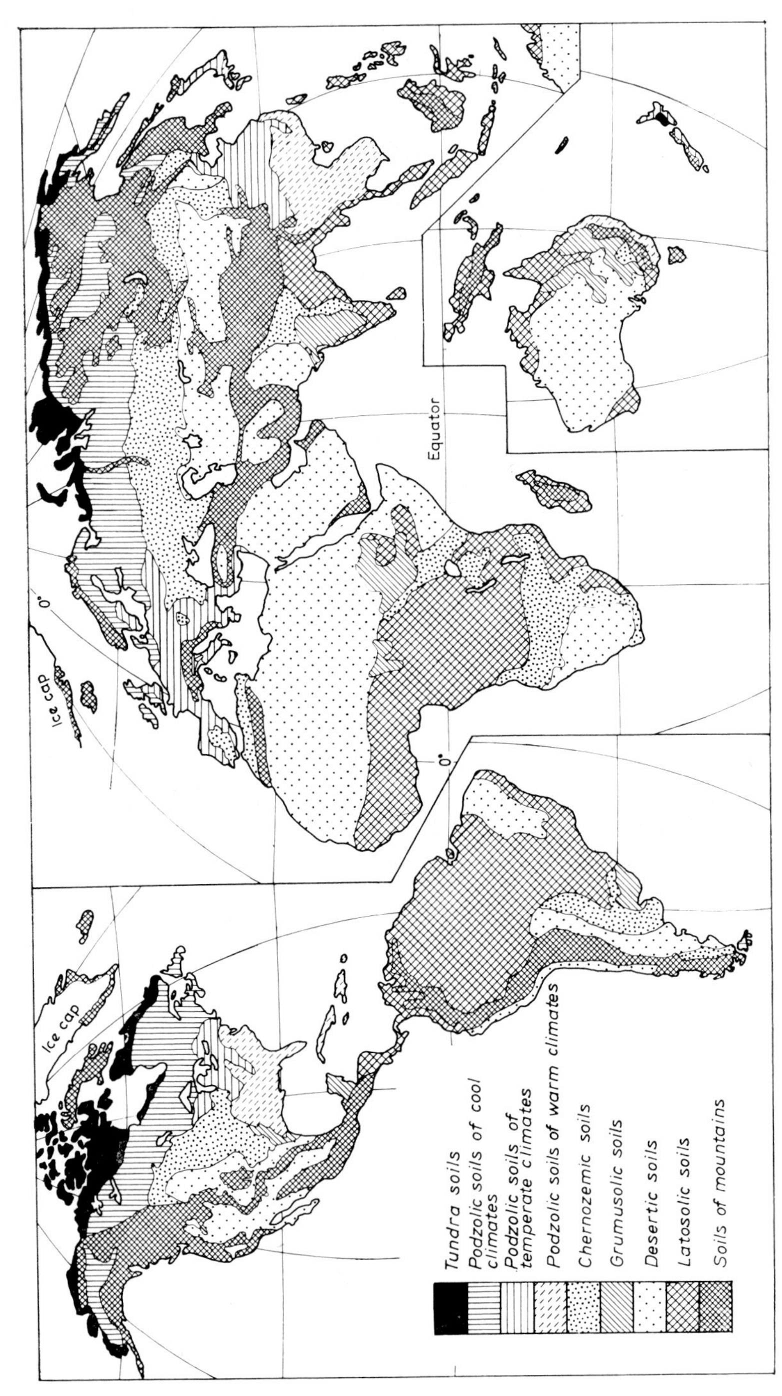

FIG. 75. Broad schematic soil map. (Map prepared by Soil Conservation Service, United States Department of Agriculture.)

excess precipitation over evaporation leads to eluviation and leaching. They are found south of the tundra region; typically under coniferous forest though also present under heath; both these vegetation types have little demand for nutrients. There is considerable evidence that the vegetation *causes* podzolisation, largely by the action of chelating agents produced by decaying leaves (see p. 51).

The typical horizons of a podzol are a bleached A horizon, poor in humus, clay and sesquioxides, overlain by a raw humus layer and underlain by a B horizon rich in clay, humus and sesquioxides. The profile may be subdivided as follows:

O1 litter,
O2 raw humus (mor),
A1 mineral soil with humus infiltration,
A2 bleached layer,
B1 humus enriched layer
B2 } sesquioxide enriched layer (subdivided on
B3 } structure, etc.),
C parent material.

In some podzols, humus is washed down the profile and accumulates as a humus enriched B horizon, as in the profile above. Such soils are known as humus podzols. In others there is a marked accumulation of iron oxide, forming the iron podzols. The iron oxide can make a distinct band of hard material known as orstein. This can be impermeable, and its development may lead to impeded drainage and a gley podzol may result.

Sandy, base-poor parent materials give rise to the best developed podzols. The soils are acid, too acid for earthworms, and so horizons remain sharp.

There is considerable literature on tropical podzols, that is soils with podzol types of profile, but formed in tropical areas. These are frequently found under particular types of vegetation, such as the bracken-covered podzols of Madagascar, which suggests that vegetation rather than the direct action of climate is responsible for podzolisation.

Some soil profiles appear to be visually homogeneous, but chemical analysis reveals that they are 'podzolic' in character. Such soils are called cryptopodzols.

The name podzol is a Russian term meaning 'ash soil', and refers to the ash-grey colour of the bleached A horizon.

Podzolic soils. There is a whole range of soils between the true podzol and the typical brown earth described later. Soils in this range are known as podzolic soils, and are extensively developed in North America.

Grey-brown podzolic soils have a well developed grey to yellowish brown A2 horizon, and a conspicuous grey to reddish brown B horizon, generally darker than the parent material. The solum is acid throughout.

Grey wooded podzolic soils are found on base-rich parent materials in association with podzols. They have a platy, light grey A2, and the B horizon is thick (50 cm), grey brown and blocky.

Red-yellow podzolic soils are found in regions of warmer and wetter climate than other podzolic soils, and extend into tropical areas. Soils are characterised by low accumulation of surface organic matter, a deep horizon of eluviation and a thick illuviated zone in which oxidation and hydrolysis of iron produce red and yellow colours.

Brown earths (Brown forest soils, braunerde). These soils typically occur under deciduous forest in the areas of milder climate to the south of the main podzol coniferous forest zone. The soils still exhibit leaching and eluviation and are typically acid, but all these features are less intense than in the podzols.

Carbonates (which may be present in the parent material) are all leached, but there is no downward movement of sesquioxides, and the silica sesquioxide ratio remains constant down the profile in the typical brown earth. Humus is not raw, and is well distributed through the profile. Free sesquioxides are present, giving rise to the typical brown colour of the soil. Soil structure is granular, and the soils have a better base status than podzols.

Brown Earth, Braunerde and Brown Forest Soil are terms used in Europe where this sort of soil seems to be common though podzolic soils predominate in North America.

Prairie soil (Brunizem, Degraded Chernozem). These soils merge into the Brown Earths, but occur in the drier areas beyond the range of natural forest. They have a typical three-part profile:

A dark, organic rich horizon;
B transition horizon (often brown);
C parent material.

The B horizon is a zone of clay accumulation, but horizon boundaries are diffuse. These soils are approaching the Chernozem type (see below), have a favourable base status, but do not have the accumulation of carbonates typical of chernozems.

Chernozem (Black Earth). These soils develop under steppe, prairie or semi-desert in semi-arid climates. The moisture deficiency of these areas leads to incomplete leaching of lime, and a layer of carbonate concretions is formed. Nevertheless, the areas with chernozems are too wet for accumulation of salt. Over the lime concretion layer is a deep dark layer of soil, typically 70 to 100 cm thick, but up to 150 cm or more. The change to parent material is a rapid transition, over only a few centimetres. There are usually many infilled animal burrows (crotovinas), and there is no eluviation of clay, sesquioxides or humus. The humus content is usually less than 10%, but grassland humus is generally dark and this is sufficient to account for the colour. The soils have a well-developed crumb structure. In the United States and in Russia chernozem is commonly formed on loess, which seems to be the ideal parent material for this soil.

Since dryness of climate seems to be a major control on chernozem distribution, it is not surprising that the geographic extent is great, with a wide latitudinal range. Continental interiors, as in central North America, Argentine Pampas and the Ukraine, form favourable sites, and so apparently do plateaus in South and East Africa, and western India.

Chestnut soils. On the arid side of the chernozem belt, under a natural vegetation of low grass-steppe, is a belt of reddish brown soils known as the chestnut soils. Compared to the chernozems these chestnut soils have carbonate closer to the surface, lower organic matter content, and a prismatic or platy structure instead of the crumb structure. Gypsum may be present as well as carbonate.

Brown and Grey soils of semi-desert (Sierozem). These soils can be regarded as extreme forms of chestnut soils, where lime and gypsum come even nearer to the surface, and organic matter is still lower.

FIG. 76. A chernozem profile, Narracoorte, South Australia.

Grumusols (Regur, Black Cotton Soils). These are clayey soils of grass or savanna-covered areas with humid to semi-arid, tropical to temperate climate with marked wet and dry seasons. They are dark, made of expanding clay, and gilgai are commonly present. There are no eluvial or illuvial horizons.

Red earths (tropical red earth, red loam, possibly similar are the red podzolic soils of the United States). These soils occur in tropical regions, with adequate but not excessive rainfall, and probably are most typical of savanna areas. They are red in colour, freely drained, and friable though frequently nearly structureless. Acidity increases down the profile, and due to loss of silica there is a relative increase in sesquioxides giving the red colour. The soils are often mottled at depth in red, yellow and grey, and there is usually a gradual transition to parent material. Organic matter is reasonably high, although it does not show as a dark colour.

Lateritic Soils (Latosols, Ferrallitic Soils). The original laterite was defined by Buchanan as a subsoil which goes hard on exposure to the atmosphere, the name coming from the Latin *later*, a brick. The term has been greatly abused, but generally means

FIG. 77. A lateritic soil profile, Casterton, Victoria, Australia.

a tropical soil with a concretionary iron-oxide horizon which is normally red in colour. Other parts of the profile may not be present, but include a surficial horizon over the ironstone, a mottled or vesicular zone and a pallid zone.

There appears to have been upward movement of iron, giving accumulation in the upper horizons and depletion in the pallid zone. This migration was once thought to be due to upward capillary movement of groundwater caused by evaporation during the dry season, and laterites are indeed common in savanna countries where marked alternation of wet and dry seasons is found. However, it seems more probable that the iron moves in solution through groundwater towards a precipitation site at the surface of the groundwater, rather than that the groundwater itself moves, for modern knowledge of soil physics indicates that extensive upward movement of groundwater by evaporation is most improbable. Some soils that have the 'lateritic' type of profile, with pallid zone, etc., have a bleached surficial horizon from which iron appears to have been leached downwards to the ironstone horizon. These profiles thus exhibit some features of podzols, and are known in Australia as lateritic podzolic soils.

In tropical regions movement of silica is enhanced compared with that in temperate soils. Silica is leached, away and there is a relative increase of sesquioxides.

Australian silcrete soils are of similar general profile to laterites, but have a hard layer of silica concretions, usually chalcedonic or opaline. In these profiles it seems

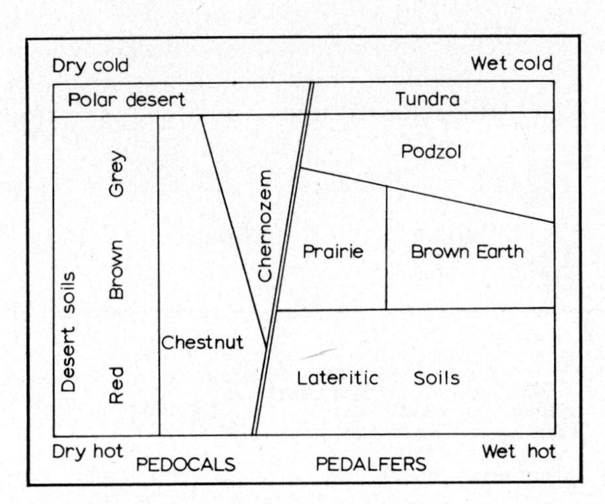

FIG. 78. Climatic relationships of the zonal soils.

that weathering released silica which migrated within the profile but was not removed by through drainage. Such soils are found in the more arid parts of Australia, and, if their accumulation is indeed due to lack of through drainage, the areas have had a fairly arid climate for a long time.

There is a vast literature on laterite: a comprehensive summary is given by Sivarajasingham *et al.* (1962).

Fig. 78 shows the climatic relationships of the zonal soils, and Fig. 79 shows the profile relationship of a number of them. There are no sharp boundaries between the different groups and all sorts of intergrades are possible, some of which have been given names although they are not included here. From north to south, ice disappears as podzols come in; then podzolisation decreases, organic accumulation increases and horizons become less sharp; next carbonate appears and gradually comes closer and closer to the surface, but in arid areas carbonate is replaced by more soluble salts; towards the equator red earths and laterites appear, but are separated from the arid soils by carbonate soils similar to those to the north of the deserts. Some of these relationships are touched on again on p. 115.

One interesting point that comes from such a diagram is the division that can be drawn on it separating two major groups of soils, the pedocals, that is soils with calcium carbonate, and the pedalfers, soils with aluminium and iron but no calcium carbonate.

AZONAL SOILS

In the zonal soils described above it was assumed that climatic conditions had acted long enough to produce the typical soil for the particular climatic zone. If for some reason this does not happen, azonal soils may occur. These include three types:

Lithosols. These have an A horizon directly on parent rock. They are commonly formed on steep slopes where erosion removes soil almost as fast as it is formed. With increasing development such soils form a thicker A horizon and turn into ranker or other soil types. On freshly exposed rock it may simply be lack of time that gives rise to lithosol, and with sufficient time a zonal soil may form.

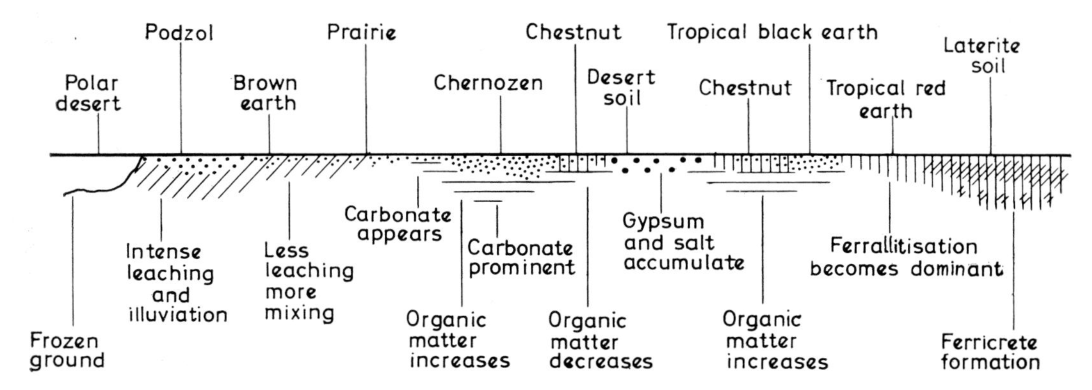

FIG. 79. Diagrammatic representation of profile relationships of zonal soils in a traverse from pole to equator. Increases and decreases are from left to right.

Alluvial soils. These soils may have a weak development of an A horizon but commonly they have only inherited stratification. In actively aggrading valleys the periodic addition of new material prevents full profile development. Old alluvium, such as on a river terrace, may develop a zonal soil, and on badly drained sites alluvium may give rise to hydromorphic soils (see p. 149) rather than azonal soils.

Dry sands. There is very little soil formation on either desert, beach or blow out sand. The parent material is poor in nutrients, and movement of sand prevents any but the most limited soil formation.

INTRAZONAL SOILS

In some circumstances the normal 'climatic' controls are not able to produce typical 'climatic' soils, because of the dominance of some particular feature or condition. When some local conditions are especially active and interrupt or prevent zonal soil formation the soils produced are called Intrazonal soils. There are four main groups:

Saline soils, in which the presence of salt dominates soil profile characteristics.
Hydromorphic soils, where excessive moisture dominates.
Calcareous soils, for a calcareous parent material seldom produces the same sort of soil as does a 'non-extreme' parent material under the same climatic conditions.
Organic soils, where organic matter is so high that it dominates the profiles.

Saline Soils

These are mostly found in deserts, and are distinguished by either an excess of sodium salts, or exchangeable sodium. There are three common soils in this group.

Solonchak (white alkali soils). These have excess salt, usually chloride or sulphate. They usually occur in depressions and exhibit a white efflorescence in drought periods. The subsoil is coloured grey or brown and is structureless or nearly so.

Solonetz (black alkali soils). These soils are characterised by the presence of sodium carbonate. They have good structure when dry, but lose structure when wet.

Solodic soils (soloth, degraded alkaline soil). In these soils leaching in the presence of excess sodium causes deflocculation of clays, and clays and sesquioxides are leached down the profile, forming a bleached, eluviated horizon (provided no calcium carbonate is present). The process is rather similar to podzolisation, but even stronger. The resultant soil profile looks much like a podzol, except for relics of columnar structure in a solodised solonetz.

If salt were to act directly on a parent material it might give rise to profiles that look exactly like podzols and some authorities believe Australian podzolic soils are formed by the action of cyclic salts.

Hydromorphic soils

These soils have profiles dominated by the effects of impeded drainage.

Tundra soils. These have frozen ground below the surface, which forms an impervious subsoil. There is little chemical weathering or organic decomposition, and there is therefore a tendency for peat formation.

Gley (glei, meadow soil, Wiesenboden). In climates without permafrost impeded drainage gives rise to gley soils which have three typical zones (belts) of groundwater, reflected in soil horizons.

1. A surface, aerobic zone, coloured grey or brown and with a high humus content.
2. A zone of fluctuating hydrological conditions, grey with iron oxide mottles in yellow, red or brown, and even concretions. There may even be the development of 'bog iron ore'. Secondary gypsum, manganese dioxide and carbonates may also occur.

3. A permanently anaerobic zone, coloured blue grey and containing iron sulphide.

The three horizons listed above are best developed on a porous parent material. In soils on impermeable parent materials the same zones occur but they are less distinct.

Planosol. In areas where the zonal soils are somewhere in the range from podzol to chernozem, flat sites or slight depressions with impeded drainage may develop an intense clay pan—a B horizon high in clay and very compact, with an abrupt contact with the A horizon. Such soils are called planosols.

Groundwater podzols. These soils develop on sandy parent material in sites with a high water table. The profile consists of a strongly bleached A horizon and a B horizon with marked illuviation and precipitation, the top of the B corresponding to the groundwater level.

Tropical gley soils (vlei, mbuga). These soils can be regarded as a hydromorphic variant of tropical black earths. They are black, grey or brown, with iron oxides and sometimes carbonate at depth.

Calcimorphic soils

Soils developed on calcareous parent material are known as calcimorphic soils. There are two main types, one a dominantly organic soil, and the other a dominantly mineral soil.

Rendzina (humus carbonate soils). These soils have the following typical horizons:

A1 grey to black, crumb structured, organic rich horizon, up to 30 cm thick.
A2 a whitish grey horizon, containing much limestone.
C the parent limestone rock.

There is no B horizon. Some insoluble residues from limestone solution may be present, but they are masked by the large amount of organic matter present.

Rendzinas may be subdivided into mor-rendzinas, which have raw humus, and mull-rendzinas in which earthworms are present and the humus is not raw.

Terra rossa. These soils are red, usually of heavy texture, and overlie parent limestone rock with a sharp junction. Solution pipes, lapies and other solutional features may cause the junction to be very irregular. The soils often present problems when studied in detail as they are not just insoluble residues, and frequently the iron content appears to be too high to be derived from the underlying rock.

Terra fusca is a brown version of the same sort of soil.

Organic soils

Organic soils include peats, bog soils, fen soils and similar soils that consist almost entirely of organic matter. Since they form simply by accumulation of organic matter and have no importance in weathering they will not be discussed further.

SOME OTHER SOIL CLASSIFICATIONS

THE AMERICAN SYSTEM

The U.S. Department of Agriculture has developed a new system of soil classification which has appeared over the years in a number of 'approximations' as the system evolved. It is claimed that the system is entirely morphological, so that genetic considerations (and controversies) will not affect the classification of any given soil profile. Nevertheless, a conventional horizon nomenclature is used, and the mere identification of a layer as a B horizon presumes some knowledge of profile genesis.

The system has a completely new terminology, which while not exactly euphonious at first hearing, at least avoids the danger of attaching yet more definitions to old terms. Some of the new terms are listed in Table 15, which also shows the main features of the classification.

Table 15. American Soil Orders and approximate equivalents in Great Soil Groups.

Order	Meaning	Approximate equivalents
1. Entisol	Recent soil	Azonal soils
2. Vertisol	Inverted soil	Grumusol
3. Inceptisol	Young soil	Some brown forest, gley
4. Aridisol	Arid soil	Desert, sierozem, solonchak, etc.
5. Mollisol	Soft soil	Chernozem, chestnut, prairie, rendzina.
6. Spodosol	Ashy soil	Podzols
7. Alfisol	Pedalfer (Al-Fe) soil	Gray-brown soil, degraded chernozem
8. Ultisol	Ultimate (leaching)	Red-yellow podzolic and reddish brown lateritic of U.S.
9. Oxisol	Oxide soil	Lateritic soils
10. Histosol	Tissue (organic) soil	Bog soils

Since 1965 this system has been the official classification system of the United States Department of Agriculture. The system is described in the U.S.D.A. *Soil Classification, Seventh Approximation* (1960).

THE FRENCH SYSTEM

This system concentrates on profile characteristics and the hydromorphic, calcimorphic, azonal and other variants are incorporated in the main system rather than separated at a high level in the classification. This is a great advantage in realistic soil mapping. The following symbols and terms are among those used:

(A) simple disaggregation of parent material.
A leaching and/or accumulation of organic matter.
(B) subsoil horizon distinguished on structure only.
B subsoil horizon distinguished on texture.

Sol lessivé is equivalent to a Gray-Brown Podzolic Soil.

The following table reveals the main features of the classification. The classification goes much further than shown here, and is periodically amended or revised. For further details see Duchaufour (1960) and Aubert (1963).

The French system of soil classification.

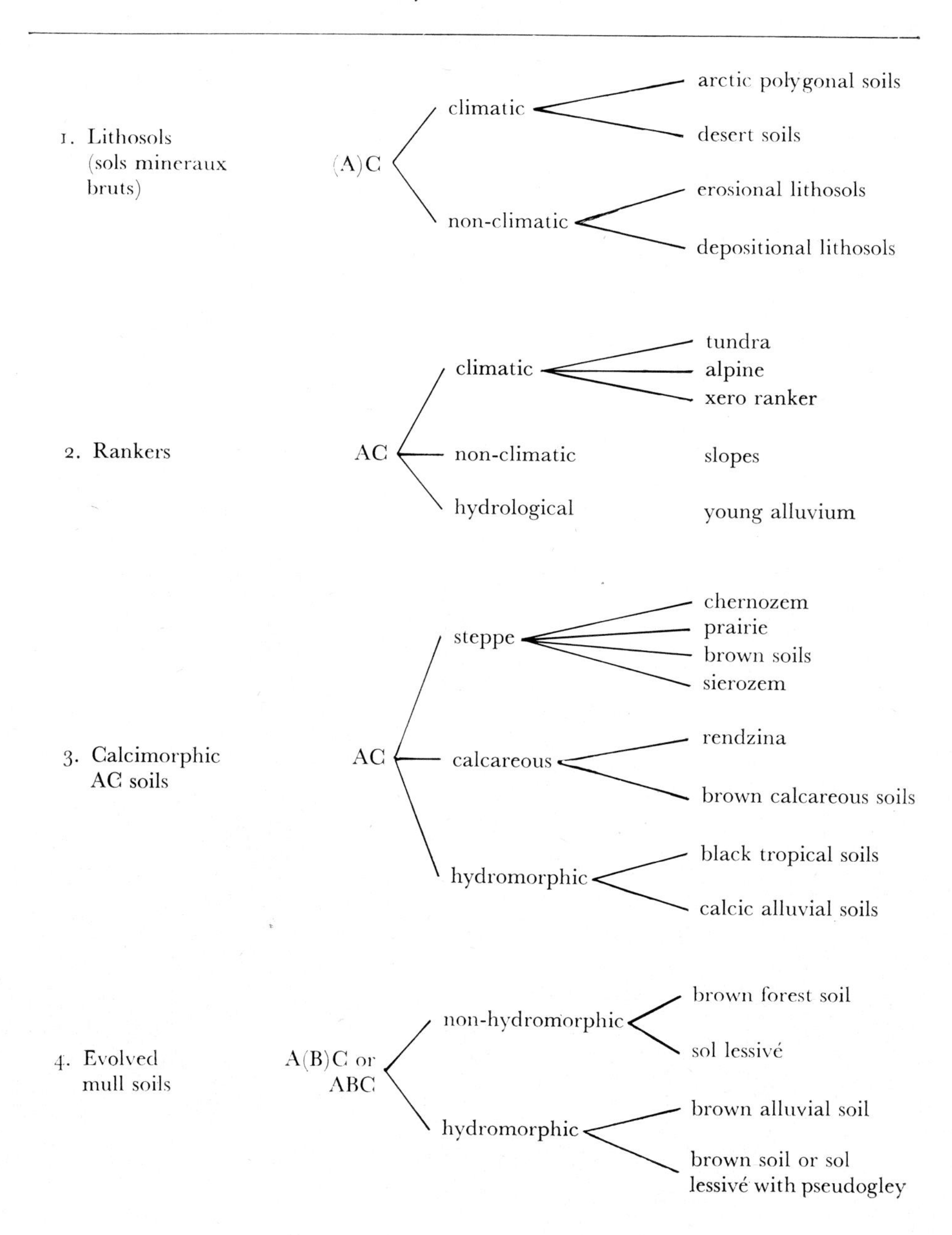

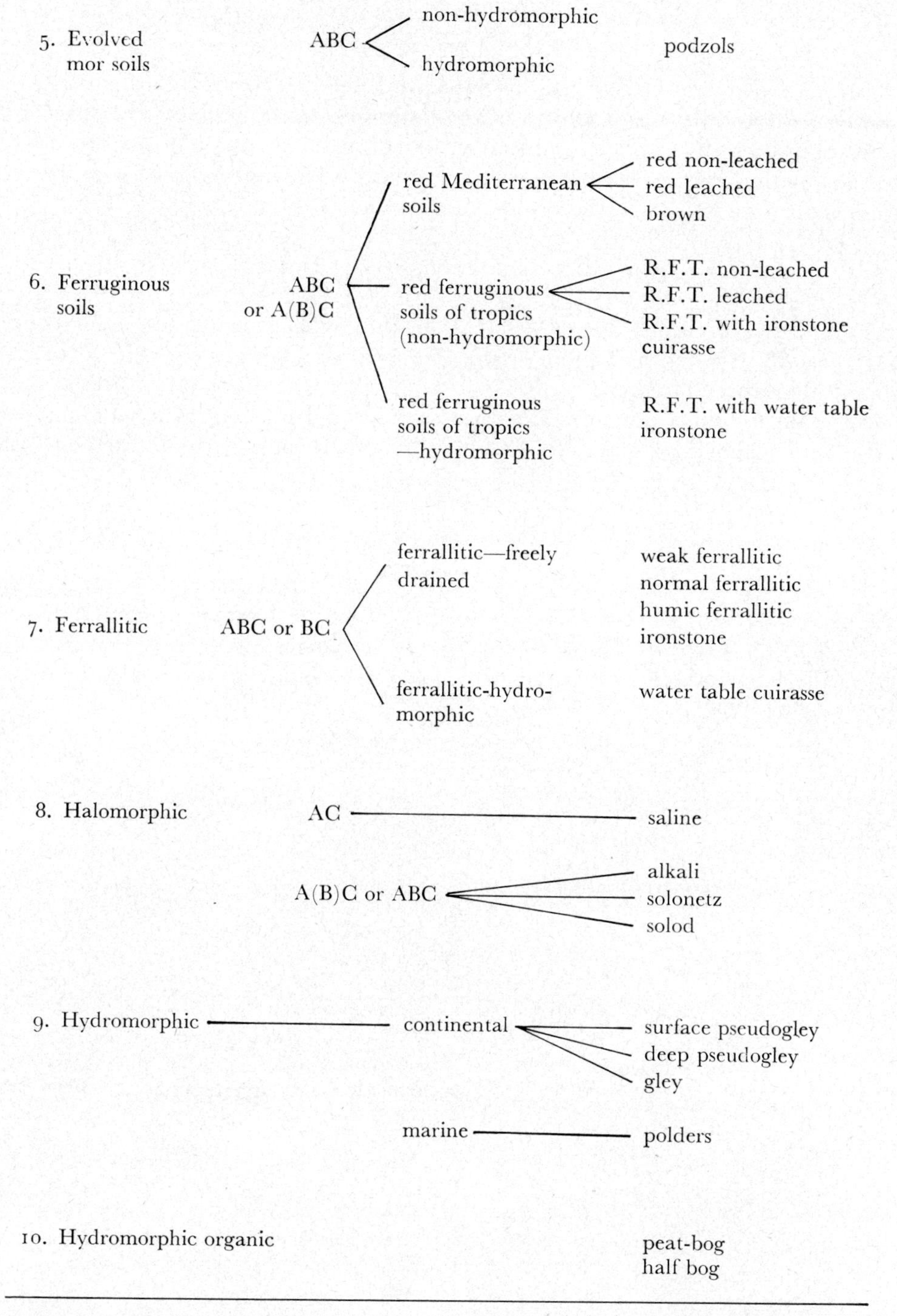
5. Evolved mor soils
ABC
non-hydromorphic
hydromorphic
podzols
6. Ferruginous soils
ABC or A(B)C
red Mediterranean soils
red non-leached
red leached
brown
red ferruginous soils of tropics (non-hydromorphic)
R.F.T. non-leached
R.F.T. leached
R.F.T. with ironstone cuirasse
red ferruginous soils of tropics —hydromorphic
R.F.T. with water table ironstone
7. Ferrallitic
ABC or BC
ferrallitic—freely drained
weak ferrallitic
normal ferrallitic
humic ferrallitic
ironstone
ferrallitic-hydromorphic
water table cuirasse
8. Halomorphic
AC
saline
A(B)C or ABC
alkali
solonetz
solod
9. Hydromorphic
continental
surface pseudogley
deep pseudogley
gley
marine
polders
10. Hydromorphic organic
peat-bog
half bog

NORTHCOTE'S CLASSIFICATION OF AUSTRALIAN SOILS

In Australia Northcote (1962) has developed a different classification for the *Atlas of Australian Soils*, and details are given in the letterpress accompanying the *Atlas*. This classification attempts to be very objective, based on the profile morphology, and not genetic, climatic, or otherwise based on assumptions. The highest priority in the classification is given to texture of the profile, and the major divisions are:

Uniform (U); Gradational (G); Duplex (D).

The first two divisions are self-evident; in duplex soils there is a sudden texture change, usually at a depth of a foot or so. The change takes place within a distance of not more than three inches.

The main divisions are subdivided on such properties as presence or absence of carbonate, cracking clay, sand, etc. Peat and some alluvial soils are treated separately.

KUBIENA'S CLASSIFICATION OF EUROPEAN SOILS

Kubiena (1953) formulated a classification for the soils of Europe, presented in outline below. Like the French classification, this uses the diagnostic horizon sequence of the total profile, such as ABC soils, (A)C soils, etc. Kubiena tries to avoid the 'artificial'

Table 16. Kubiena's classification

A	Sub-aqueous	A	soils not forming peat,
		B	peat,
B	Semi-terrestrial soils	A	raw soils,
		B	marsh,
		C	high moor peat,
		D	salt soils,
		E	gleys,
		F	alluvial soils.
C	Terrestrial soils	A	climax raw soils,
		B	rankers,
		C	rendzinas,
		D	steppe soils,
		E	altered soils on calcareous rocks,
		F	red and brown loams,
		G	latosols,
		H	brown earths,
		I	pseudogleys,
		J	podzols.

classification systems in which selected properties are arranged in order of significance to establish a classification hierarchy. He claims his classification is a natural one, ordered by *all* soil properties. If all properties were given equal weight it would be a kind of freehand attempt at the methods of numerical taxonomy (see p. 141), but it is clear that drainage is given first priority, and parent material receives considerable attention. To avoid the introduction of many new terms, Table 16 is slightly simplified from Kubiena's original.

AVERY'S CLASSIFICATION OF BRITISH SOILS

It is possible to make a classification in which emphasis is placed on the overall moisture status of the profile, with the form of humus giving a basis for early differentiation. Drainage status and humus form are characteristics of importance in agriculture, and a classification that stresses them is possibly most useful for a soil survey designed for agricultural use rather than fundamental pedology. B. W. Avery presented the following scheme for classifying British soil in 1956. Of course, many of the important soils of the world are not represented, but this classification may be taken as an example of a classification built up for a specific purpose, not evolved from a grand theoretical idea.

Table 17. Avery's classification of British Soils

A Terrestrial soils	I	Raw Soils
	II	Montane Humus Soils
	III	Calcareous Soils
	IV	Leached Mull Soils (brown earths, etc.)
	V	Podzolised (Mor) soils
B Hydromorphic Soils	VI	Alluvial Soils
	VII	Gley Soils
	VIII	Gley Podzolic Soils
	IX	Peaty Soils
	X	Peat Soils

SOIL-FORMING FACTORS

A valuable concept in pedology is that of the soil-forming factors which are:

parent material; time; climate; topography; living organisms.

These are of course factors of weathering too. Any given soil can be regarded as the product of the inter-action of all these factors.

PARENT MATERIAL

Granite can give rise to soils as different as laterites and podzols, and basalt can give rise to bauxites or brown forest soils. This might suggest that parent rock is unimportant compared with other factors, but its influence must not be overlooked.

In temperate climates, there is usually a change in soil type at every geological boundary. An area may have one 'great soil group' dominant, but there will be differences within the group at geological boundaries. In tropical areas the influence of rock seems to be rather less and ferrallitic soils from quite varied parent materials come to be very similar. Even here, however, the more extreme parent materials, such as quartzite, give rise to distinctive soils.

The features of the parent material that affect soil formation include texture, porosity, drainage, mineral composition, stratification—in fact all the factors that control rock weathering (Chap. VII).

The parent material provides chemical elements and mineral grains, and these may exert varied influences in soil formation. Thus a pure clay or a basic igneous rock provides no sand grains, and therefore could never give rise to a true podzol. A very pure quartzose sandstone on the other hand would produce no clay, and therefore could not form a podzol either. Chemical aspects of the parent material have most effect through the formation of clay minerals and their subsequent ion exchange. A rock rich in K and poor in Mg and Ca is likely to give rise to illites. A rock deficient in K and rich in Mg is likely to give rise to montmorillonites.

Development of soil profiles is generally faster in an acid environment, and so if the parent material is rich in lime, soil formation may be retarded. This is often noticeable in soils developed on relatively young glacial drifts.

The texture of the parent material affects the rate of soil formation; freely-drained parent materials can develop soils much faster than dense impermeable parent materials, and the soils on coarser parent materials can respond much faster to changes in conditions.

The nature of the parent material will have a very marked effect on young soils, and its influence will become less as the soil becomes older.

It is useful to distinguish between non-extreme and extreme parent materials. Some parent materials such as pure sand, ore bodies, or limestones give rise to special soils very much dominated by the parent material. Others, such as granites, shales, and all rocks with a wide variety of chemical content, are non-extreme, and on such rocks the influence of other factors is likely to be more important.

The parent material factor is frequently complicated by the superficial addition of layers of different parent materials that can simulate soil horizons. It is frequently a considerable problem to distinguish geological layering from pedogenic horizons.

As an example of a composite profile we can take the Batcombe soil series of south-east of England (Fig. 80). The bedrock is Chalk. This is overlain by red-brown clays with numerous flint pebbles and nodules, and the topsoil is lighter and siltier. At first sight this looks like a deep insoluble residue from chalk solution (the chalk contains flint nodules) with some pedological differentiation to give the lighter topsoil. However, detailed investigations show that most of the clay with flints is weathered Eocene material, partly reworked by Pleistocene periglacial processes. The silty topsoil is largely due to incorporation of wind-blown loess, and the base of the clay with flints is a distinct layer that has originated by solution of chalk *beneath* the Eocene deposits. These different materials have been somewhat mixed by pedological processes but have retained sufficient individuality to be identified (Avery, Stephens, Brown and Yaalon, 1959).

TIME

The longer a ground surface persists, the more chance there is for ultimate soil development. A lava flow from a new volcanic eruption has no soil; freshly deposited alluvium on a floodplain has no soil.

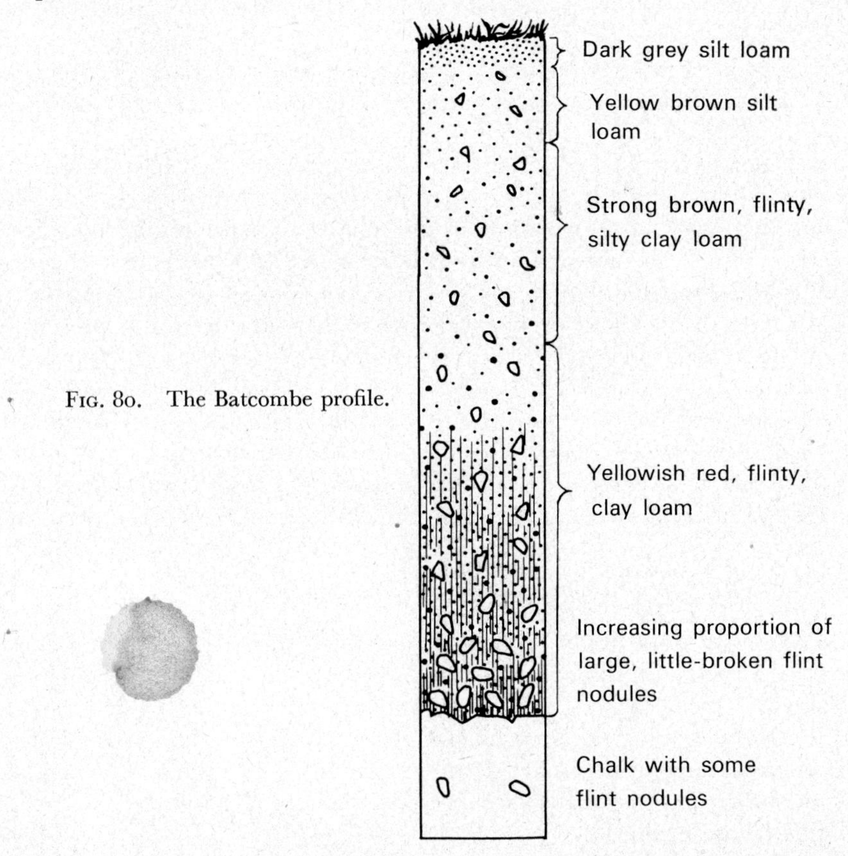

FIG. 80. The Batcombe profile.

Our first theoretical question might be 'When does a parent material become a soil?' From the detailed studies considered in Chap. XIII we shall see that soils form over a period of time at varying rates. The curves of Salisbury (p. 245) and Hissink (p. 245), which focus attention on a particular soil property (leaching), may be taken as examples. Both these curves come smoothly right to the present day, with no distinct break suggesting the start of pedogenesis: the only possible place to consider as the start of pedogenesis is the present. Thus we can regard recently exposed or deposited parent material as soil at its very beginning. At time zero, parent material and soils merge together.

With the passage of time, the soil will evolve further, and new concepts are then needed to consider what happens. Does the soil reach a certain state of development beyond which it cannot go, or could it be that a stage is reached when removal of material and addition of material (from weathering) are just balanced and there is no change in the soil profile? On the other hand it may be that soils continue to change, but at an ever-decreasing rate, so that after a while they are virtually stable. We can conceive of soils evolving through various stages, which may be called incipient, immature and mature soils. Is there then a further evolution to something further, a post-mature soil?

Such discussion is rather idealistic. In the real world changes in other factors—vegetation, climate, erosion rate and so forth—are likely to upset the balance before a true equilibrium is reached.

Numerous examples of rates of soil formation are given in Chap. XIII, but a few main points may be given here. On porous parent materials where leaching can operate from the start, soil formation is rapid at first and gradually slows down. On such materials periods in tens or hundreds of years are sufficient for a distinct soil to develop. On impermeable parent materials pedogenesis is slow at first, but porosity increases as soil forms, so the process accelerates until a mature soil is present, when it slows down. On glacial tills 300 years may be enough to form a soil. On dense basalt or a fresh granite surface very much longer would be required. Furthermore some soils, such as rendzinas and rankers, are mature very quickly; others such as laterites may take so long that the whole concept of maturity is difficult to apply.

Thus there is a difference of perhaps a thousandfold in the rates of soil formation, but most soils on non-extreme parent materials can form in less than a thousand years.

CLIMATE

Climate is of major importance in soil formation. Rainfall, temperature and their seasonal and diurnal variations affect soil directly, and also act through vegetation and hydrology.

It has long been realised that certain climates give rise to certain soils. For instance, in equatorial areas ferrallitic soils are dominant on all well-drained sites; in cold humid areas of coniferous forest podzols are present; dry areas have soils with carbonate accumulation in the profile.

A number of parameters of climate have particular significance.

Temperature is important, high temperatures speeding up most chemical reactions and therefore chemical weathering. Of particular significance perhaps is the greater solution of silica in tropical regions. In cold areas there is little chemical weathering, organic activity is reduced, and consequently soil formation is limited. Frost and ice affect soil moisture conditions and physical aspects of soil very considerably.

Rainfall is clearly important for leaching. In soils with lime accumulation there is frequently a direct correlation between the depth to the calcium carbonate horizon and the amount of rain.

The precipitation/evaporation ratio is also of great importance; if precipitation exceeds evaporation there will be loss of ions in drainage waters, if evaporation is in excess there will be accumulation of salts either at the surface or within the profile.

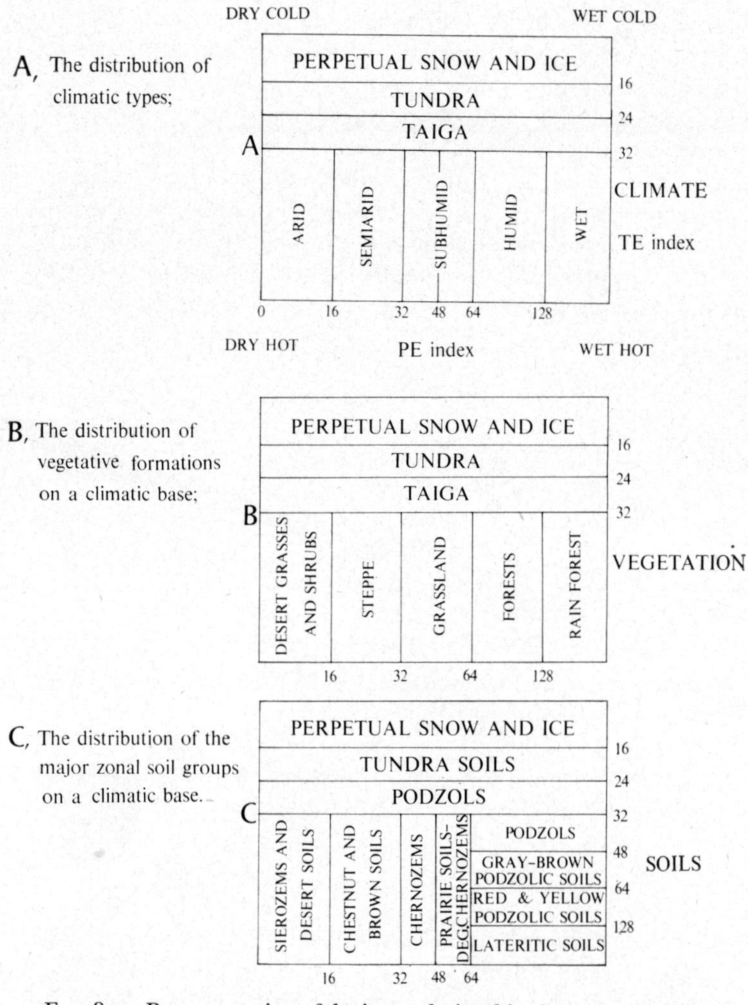

FIG. 81. Representation of the interrelationships between climate vegetation and soil (after Blumenstock and Thornthwaite, 1941).

Seasonal variations are important for, in drying, the soil may suffer irreversible precipitation of material, especially iron oxides.

The main aspects of climate can be plotted on axes of a diagram such as Fig. 78, and on this diagram the zonal soils can be placed in approximate position. A series of more elaborate diagrams showing the interrelationships of climate, vegetation and soil are shown in Fig. 81. The zonal soils, which occupy particular climatic zones, appear to be largely controlled by climate, either directly or indirectly, and the particular soil shown on the diagram should occur under the climate indicated on any freely drained parent material of non-extreme composition provided there are no other 'extreme' complications of slope, vegetation or other factors.

The climate changes with altitude as well as latitude, and a series of climatically controlled soils may form an altitudinal series, as shown in the well-known diagram reproduced in Fig. 82.

To what extent zonal soils really do occur in their appropriate zonal climates is not clear, and to some extent the zonal classification *assumes* that climate is the most important soil-forming factor. In areas complicated by high relief and rapid changes of parent material many complications occur, and many areas reflect in their soils not just the present climate but a history of climatic and physiographic change.

In spite of these limitations the climatic or zonal classifications and concepts will probably remain popular for they provide a simple, idealised basis to which further

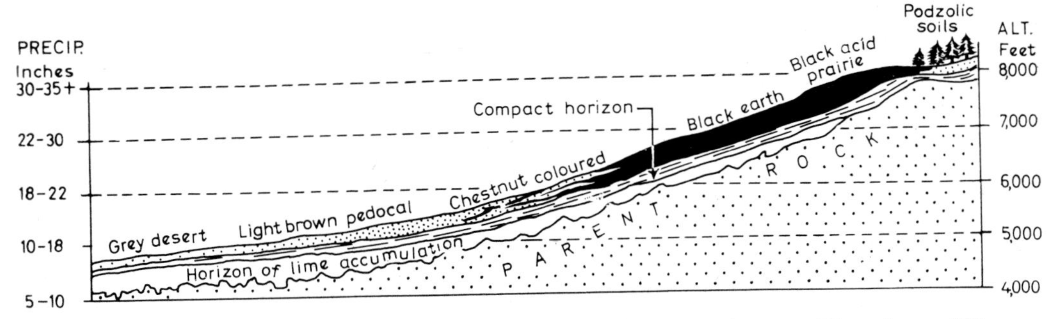

FIG. 82. Gradation of soil profiles from desert to humid mountain top. West slope of Big Horns, Wyoming (after Thorp, 1931).

concepts can be added. Zonal concepts possibly provide the best simplification of the complex multi-variate problem of soil relationships.

TOPOGRAPHY

Topography exerts an influence on soils and weathering in several ways.

One is the effect of pure altitude, and it is found that zones of different soils occur at different altitudes, just as they occur at different latitudes, because of different climate.

Another aspect of topography is that of the actual landform on which soil is formed, as soils on river terraces, eskers, beach ridges, alluvial fans and many other landforms, will have their soils controlled to some extent by the shape of the landform and its internal structure, composition and origin. Because of this, the geometrical aspects of topography cannot really be isolated because their effects are compounded with those of parent material, time and other factors. Nevertheless, with care, some general conclusions can be reached about the effects of some aspects of topography on soils.

A very simple relationship between a soil property and slope steepness was demonstrated by Norton and Smith (1930). They plotted angle of slope against thickness of A horizon (Fig. 83) and, as might be expected, showed that the horizon is thickest on level topography and least on steep slopes.

Erosion is likely to be more severe on steeper slopes, and they may carry truncated soils. The removal of material promotes more rapid weathering and renewed soil formation.

Slope angle does not always correlate exactly with soil properties, for there may be differences in what is above or below the slope, and different parts of a single slope of constant gradient may have different properties. This is particularly true for drainage relationships—a soil on the lower parts of a slope will receive, in addition to rainfall, water draining from higher up the slope. In Uganda, Radwanski and Ollier (1959) found that on hillslopes of about 10° there was a red soil on upper parts of the slope and a brown soil with considerable iron concretions on the lower part.

Slope erosion may also affect the distribution of soil types, as shown by Ellis's (1938) example from Manitoba (Fig. 84). Note that even without erosion there are different

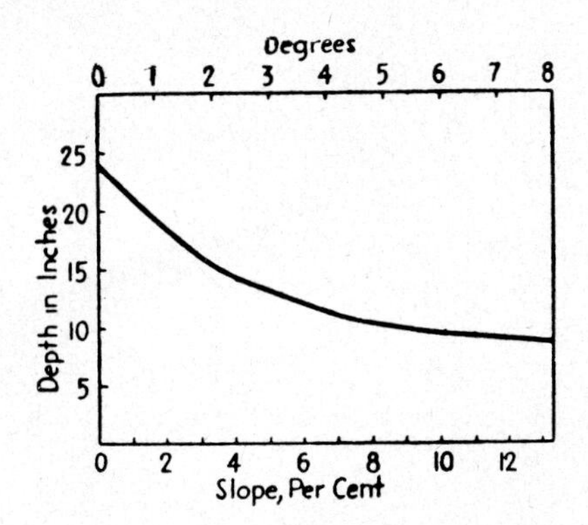

FIG. 83. Relationship between slope and thickness of A horizon of timbered soils derived from loess (after Norton and Smith, 1938).

soils in different parts of the slope. The hilltop, although of gentler slope than the mid-slope, has shallower soils. A nearly flat site on a topographic high can be an exporting site, and an equally flat site on a topographic low is an accumulating site. This applies not only to material washed over the surface but to material leached and eluviated downslope through the subsoil, such as the carbonate in Fig. 84.

In a similar way, Maignien (1959) emphasised the mobility of sesquioxides downslope. Fossil laterite on high sites is dissolved, the dissolved iron drains downhill, and new sheets of laterite are formed at low levels. The soil/topography relationship is even more complicated in this instance, for Maignien believes that the formation of the low-level laterite reduces infiltration of water, increases run-off, and causes incision of the streams or rivers; soil formation is affecting topography rather than the reverse.

Ideas similar to those of Maignien are incorporated in Chenery's hypothesis (1960) that as hillslopes retreat the laterite sheet at their base extends backwards.

Whatever the genetic relationships may be between soil and topography, it is found in many areas of the world that there is a frequent repetition of topographic and soil sequences. Milne (1935) used the word catena (Latin = a chain) to describe 'a regular repetition of a certain sequence of soil profiles in association with a certain topography'. Unfortunately this word has become very confused because of further debate as to whether it referred to soils on a single parent material or several, one climate or more, and so on; the arguments are summarised by Watson (1965). Catena is still a much used term, but to avoid the arguments mentioned above the word toposequence may be substituted. Fig. 85 shows typical toposequences, and also demonstrates that the units of a toposequence may not be symmetrically disposed on opposite sides of valleys. Some toposequences on one parent material are virtually sequences of soils differing in hydrological conditions. Ellis (1938) considers that hilltops, losing water

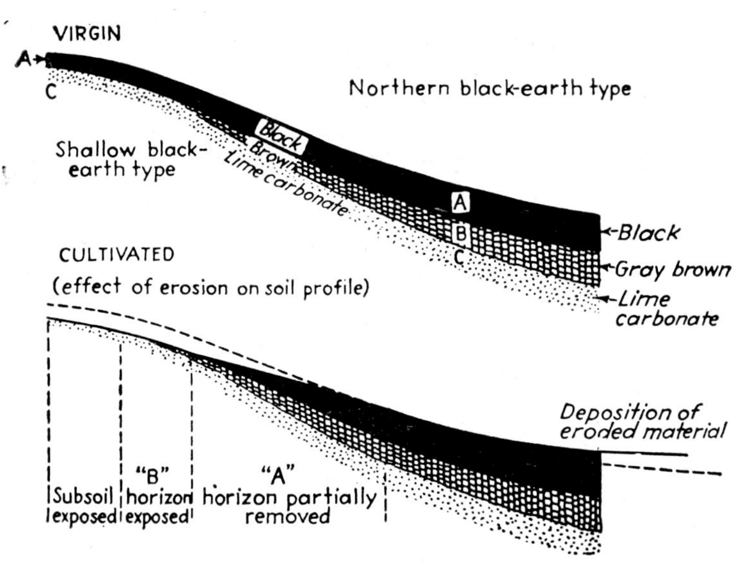

Fig. 84. Effect of erosion on soil type in Manitoba (after Ellis, 1938).

by run-off, are locally arid, whereas depressions or footslopes, receiving water from upslope as well as their own precipitation, are locally humid. In the Scottish Soil Survey, series are grouped into larger units called soil associations, which consist of a number of series derived from similar parent material but which differ in drainage.

In generally flat-lying areas with high water tables, quite small topographic differences can have a profound effect on soils. Mattson and Lonnemark (reported in Jenny, 1941) describe a sequence of podzols from Lake Unden in Sweden. On small hills a few metres high and not affected by the water table an iron podzol is formed. On lower slopes where the water table is within the profile, humus podzols are formed, and on the lowest sites a non-podzolised bluish-gray bog soil is formed.

A toposequence of soils may include palaeosols, especially on remnants of erosion surfaces preserved either as plateaus or as benches. Some plateau laterites of Africa, and silcretes of Central Australia, are of this kind. The various horizons of the palaeosol outcrop on the hillsides provide a variety of parent materials, and complicate the soil topographic relationships very much. Studies of such situations have been presented by Mulcahy (1967), Hays (1967) and Wright (1963).

Some further examples of the interaction of soil, weathering and topography are given in Chap. XII.

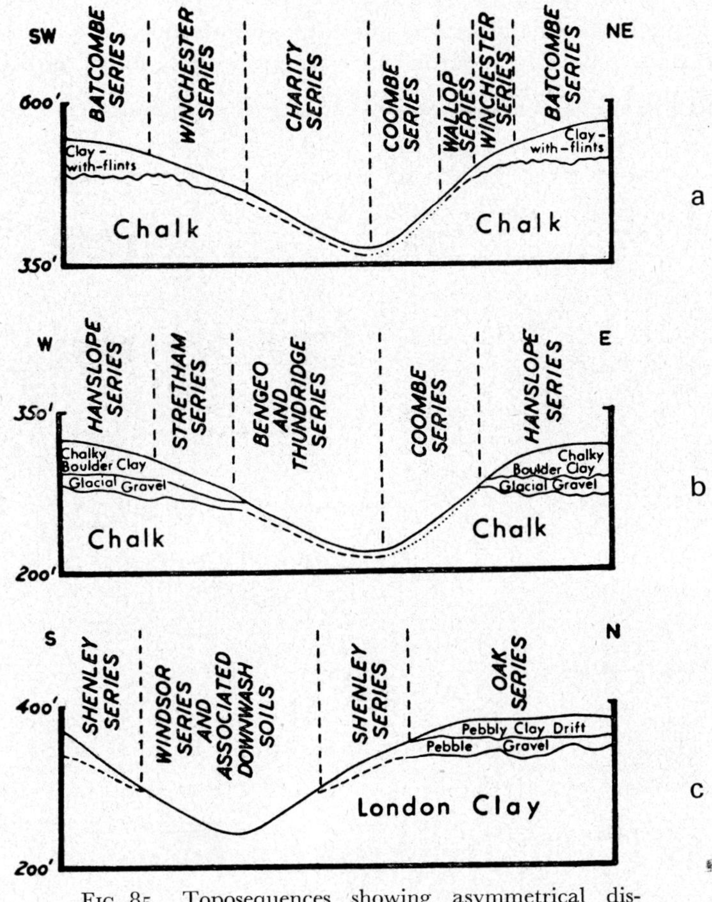

FIG. 85. Toposequences showing asymmetrical distribution of soils.

a The Clay-with-Flints area of West Hertfordshire. *b* The Chalky Boulder Clay area of East Hertfordshire. *c* The London Clay area of South Hertfordshire. All sections are one mile long and the vertical scale is greatly exaggerated (after Thomasson and Avery, 1963).

THE ORGANIC FACTOR

The organic factor is important in weathering, but even more so in soils. Organic matter itself is the basic feature of some horizons. Soil fauna may make or destroy horizons, by sorting or mixing of soil particles.

Vegetation is not always an independent variable, and distribution of vegetation may be controlled by soil, distribution of soil controlled by vegetation, or both soil and vegetation may be controlled by other factors.

Many trees produce a leaf litter from which may be extracted a powerful chelating agent that can quickly form a podzol in laboratory conditions. However, many do not in fact form podzols, so the effects of the entire organic population of the soil are probably involved. Whatever the controls are, it is clear that soil and vegetation distributions are closely related.

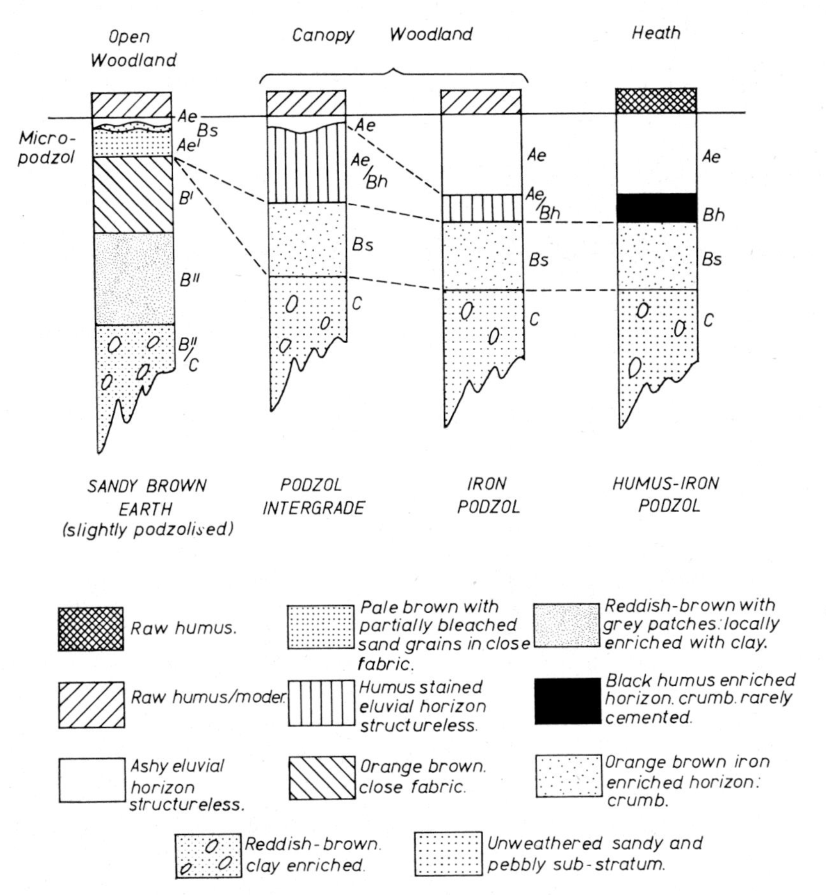

FIG. 86. A genetic sequence of soils on sandy sub-strata under different vegetation (after Mackney, 1961).

Certain major soil types have specific vegetation associated with them, and a change in vegetation may cause a change in soil.

Compared to forest soils, grassland soils are darker, and contain about twice as much organic matter which is more evenly distributed through the profile. Forest soils show greater eluviation and leaching, which gives more strongly developed B horizons.

Mackney (1961) described a very close relationship from Warwickshire, England, where on coarse parent material a variety of podzolised soils were formed. Under open oak woodland the soil was a slightly podzolised sandy brown earth; under canopy woodland there were iron podzols; and under heath the soil was a humus-iron podzol (Fig. 86).

In Europe podzols are associated with humid cold climates and coniferous forest; brown forest soils with humid temperate climates and deciduous forest. The two soil types merge in southern Sweden, where Tamm (reported in Jenny, 1941) found that vegetation controlled the direction of the reversible reaction between podzol and brown forest soil.

In areas of beech and oak forest, brown forest soils are the regional soil. Where the deciduous forest is replaced by conifers or Calluna heath, the soil becomes strongly leached, high acidity develops, raw humus accumulates and true podzols are formed.

In generally podzolic areas it was found that where birch (and associated grass cover) invaded coniferous plantations, the podzol was converted to a brown forest soil. Acidity dropped from pH 4 to pH 5·5, earthworms came in, and the raw humus and the leached A2 horizon were converted to mull.

Organic matter in the soil is itself weathered or altered, very largely by biological activity. Part of it turns into humus, by a process called humification, and part becomes gaseous (NH_3, NO_2, CO_2) or goes into solution, a process which for some reason is called mineralisation.

A summary of the classification of soil humus is given below:

Mull	complete organic-mineral complex	aggregated clay-humus
Moder	partial organic-mineral complex	clay and humus not aggregated
Mor	raw humus	fibrous humus not aggregated with clay

SOME CONCEPTS RELEVANT TO SOIL AND WEATHERING STUDIES

MATURITY

A number of different concepts have come to be regarded as concepts of soil maturity.

Shaw (1928) proposed the following degrees of maturity:

Raw soil;
Young soil, only slightly weathered;
Immature soil, only moderately weathered;
Semi-mature, already considerably weathered;
Mature soil, fully weathered.

Although they do not actually use the term maturity, the stages proposed by Mohr and van Baren (1954) are very similar:

Initial stage, unweathered parent material;
Juvenile stage, weathering has started, but much of the original material is still unweathered;
Virile stage, easily weatherable minerals have largely decomposed, clay content has increased;
Senile stage, decomposition arrives at a final stage, only the most resistant minerals have survived;
Final stage, soil development complete, soil is weathered out under the prevailing conditions.

Both Shaw and Mohr and van Baren are listing intensity of mineral weathering; horizon formation and profile development are not mentioned. These are not really tables of degree of maturity.

Miller, Turk and Foth (1966) write of immature soils, mature soils and old-age soils.

Immature soils are characterised by organic matter accumulation at the surface and little weathering, leaching or eluviation. These are AC soils.

Mature soils are characterised by the development of a B horizon.

Old-age soils have very marked differences between the properties of the A and B horizons.

They give as examples of each: Regosol (an azonal soil on soft unconsolidated mineral deposits with few if any soil characteristics), as an immature soil; Prairie soil (Brunizem) as a mature soil; and Planosol as an old-age soil.

It may indeed happen that under some circumstances a parent material may give rise to these soil types in succession. However, in some situations a Regosol or a Prairie soil may be the ultimate stage of soil development. In such circumstances is not the Regosol or Prairie soil the mature soil? It is not clear whether all Regosols are meant to be regarded as immature, or only those where there is evidence that further development is possible with the passage of time.

Marbut (1935) defines a mature soil as one 'whose profile features are well developed'. As a definition this is hard to use. Presumably a Regosol, Prairie soil or Planosol could all be regarded as mature. A skeletal soil has a profile in which skeletal soil features are well developed. It may or may not evolve further, depending on other conditions, and if it does evolve it may go through a succession of stages, each with a well developed profile.

Yet another concept is expressed by Jenny (1941) as follows: 'mature soils are in equilibrium with the environment'. This is very similar to Nikiforoff's statement (1942) 'A mature soil represents the steady state of the dynamic system comprising the soil and its environment, the latter including the climate and the organic world'. Such definitions have no reference to profile features at all, but emphasise soil-forming processes, the environment, and the balance between factors. On this basis it is impossible to be sure whether most soils are mature or not. The concept of equilibrium or steady state soils is further discussed on p. 167.

Jenny also discussed the idea of maturity as a soil property-time function and this is perhaps the most valid and useful of all the maturity concepts. From the graph of Salisbury (p. 245), we can see that although change is rapid at first, it eventually falls off until the curve is almost level. Further increments of time produce negligible change in the soil property. When this stage is reached the soil may be called mature, with respect to that property. But not all soil properties approach maturity at the same rate. When all major properties have reached the point of negligible further change with time, the soil as a whole may be called mature.

A mature soil of this kind would be in equilibrium with its environment, but this equilibrium is not the basis for calling it mature.

It should be noted that a mature soil is not necessarily inert, with pedogenic processes at a standstill. Soil properties may vary around a mean due to daily, seasonal or other fluctuations in the environment.

STEADY STATE SOILS

Steady state soils are those that are in equilibrium with their conditions, especially topographic conditions, and although not inert they retain the same morphology.

The simplest picture of a steady state soil is to imagine that it is losing material by erosion, but gaining material by weathering. Such losses affect the top of the profile, and the gains are in the subsoil, so for the profile to remain constant the different horizons move down through the profile. Many features of the profile may be in a steady state equilibrium. Organic matter may be added to the profile at the same rate as organic matter already in the profile decays; clay may be produced by weathering at the same rate as it is eluviated. If the profile is considered as part of a slope, gains from upslope are equal to losses downslope, both in material passing through the profile such as ions in solution, and colluvium mechanically traversing the surface.

A more elaborate statement of the steady state soils system has been presented by Nikiforoff (1942). He believes that in a steady state soil, production of any material is equal to the losses, and expresses the relationship by the formula

$$S_n = A\left[\frac{(1-r) \cdot 1-(1-r)^n}{r}\right]+A,$$

where S_n is the amount of any substance present at the end of n years, A is the amount of this substance synthesised during one year, r is the rate of decomposition expressed as decimal portion of the amount present, n is the number of years.

It must be noted that this formula is based on a number of assumptions that are questionable. It is assumed that the amount A is the same each year throughout the period characterised by the steady state, and that r is the rate of loss and is a constant proportion of A. A is an *amount*, r is a *rate*. It would seem an equally valid assumption that the amount lost is constant, and the amount added is a proportion of it. As for the constancy of the proportion, that appears to be brought in to make the formula work.

It would seem much simpler to avoid these assumptions, and to treat both additions and losses as rate phenomena, and when the rate of addition equals the rate of removal there is a steady state or equilibrium. In fact, further in his paper Nikiforoff uses this approach, as when he writes '[the carbonates] synthesis and retention in the soil take place only if the rate of their solution and leaching is lower than the rate of production. The ratio between these rates is not the same under different climatic conditions.'

Besides the steady state within the soil profile due to the normal soil-forming factors, we must also consider the relationship between a soil and the topographic position, especially soils on slopes.

Let us assume first that debris accumulates at the foot of a cliff, a certain amount being added each year. Some will be lost by weathering, creep or wash. When the amount added equals the amount lost, the slope will be in equilibrium. The soils on such a slope will be formed on debris, and will receive annual increments of fresh debris to balance their losses due to wash, leaching, etc. They will then be steady state soils on a steady state slope.

Quite complex slopes with very different soils on different slope elements may be in a steady state.

Thus in Fig. 87, the plateau soil is a sedentary soil in equilibrium. On the convex slope there is soil creep, the plateau soil thins out, and at the edge becomes incorporated with chalk. The rates of creep, soil formation, and loss to the lower slopes may be in balance. On the steep, straight slope (free face) little mineral debris is retained and

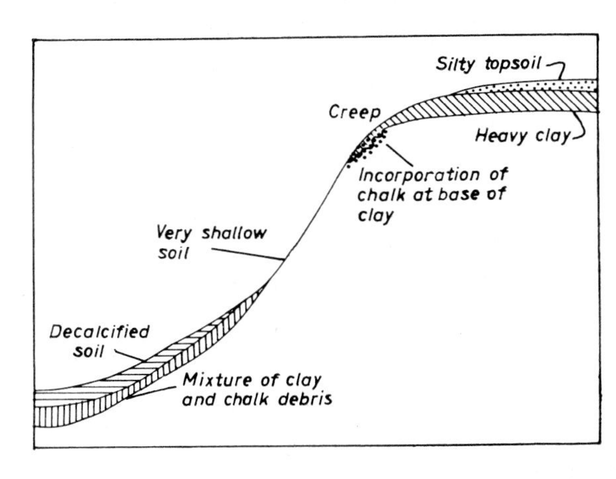

FIG. 87. Diagrammatic representation of soils in relation to slope in the Chiltern Hills (simplified from Ollier and Thomasson, 1957).

there is only a shallow rendzina. On the debris slope at the bottom there is accumulation of a mixture of clay and chalk, partly decalcified, which again can be in equilibrium with supply of material and pedologic processes.

It would be possible for slope retreat to take place, with the slope profile moving to the right of the position shown, and still retain identical slope components with the same soil profiles.

Another example is provided by the slope shown in Fig. 88, which is an East African catena. The soils on the simple, straight slopes have mottles and iron concretions at depth which are never found at the surface, and Radwanski and Ollier (1959) suggest that the process of iron movement and precipitation is taking place all the time, and the mottled subsoil is gradually destroyed from above, but develops downwards by encroaching on the underlying deeply weathered rock. The profile is in equilibrium with the present environment and the development of all its horizons is keeping pace with all the processes responsible for their removal. The profile is lowered as a whole and its morphology is not affected.

PERIODIC SOILS

In contrast to steady state soils are periodic soils. These have been described in particular from Australia, and a summary of present ideas concerning them is given by Butler (1967).

A well-developed soil takes time to form, and so represents a period of stability at the ground surface. Alternating with such stable periods there may be periods of activity when slopes are eroded, and eroded material is deposited on lower slopes or in valleys. The two phases together—stable and unstable, soil forming and erosion/deposition—make a cycle which has been called a K-cycle (Butler 1959). A number of Australian

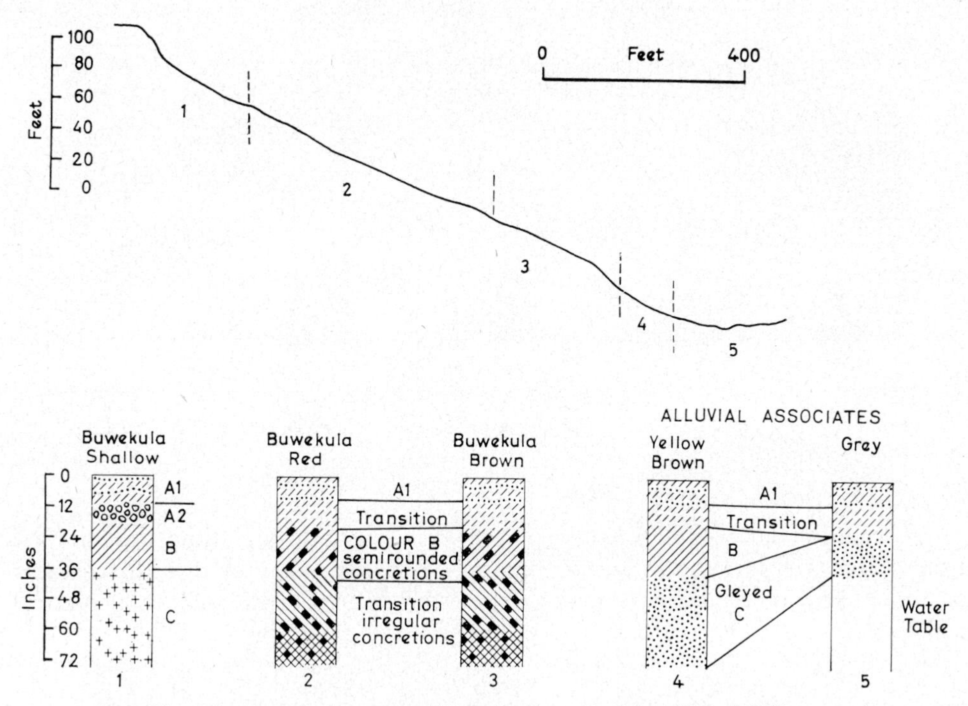

FIG. 88. The Buwekula catena, Uganda (from Radwanski and Ollier, 1959).

studies, summarised by Butler (1967), indicate that there have been several successive K-cycles in parts of Australia, and furthermore each stable period was characterised by a different dominant soil, as shown below:

K1 (most recent)	minimal prairie soil,
K2	red earth or red-brown earth,
K3	red podzolic soil.

Some K-cycle soils are preserved at the surface as palaeosols, other old ground surfaces are deeply buried and their soils would be fossil soils on the definition of p. 173. Yet others are at shallow depths or partially buried, and present the complications described on p. 173. Figs 89 and 90 show in a diagrammatic way examples of ground-surface relationships.

It should be stressed that the theoretical basis of K-cycle thinking is not yet proven. An alternative set of assumptions is equally feasible, if not more probable, namely

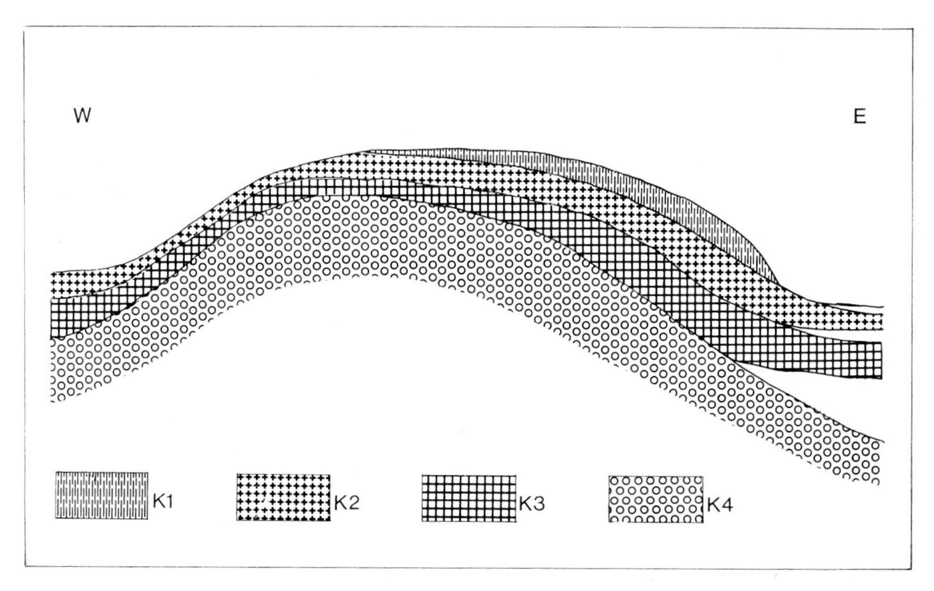

FIG. 89. Ground-surface relationships in the Riverine Plain, New South Wales (after Butler, 1967).

that soil formation and weathering take place continuously through time, though erosion and deposition are periodic at any one place. Periodic soils may owe their existence to exceptional types of mass movement on slopes, or to peculiarities of local geomorphological history.

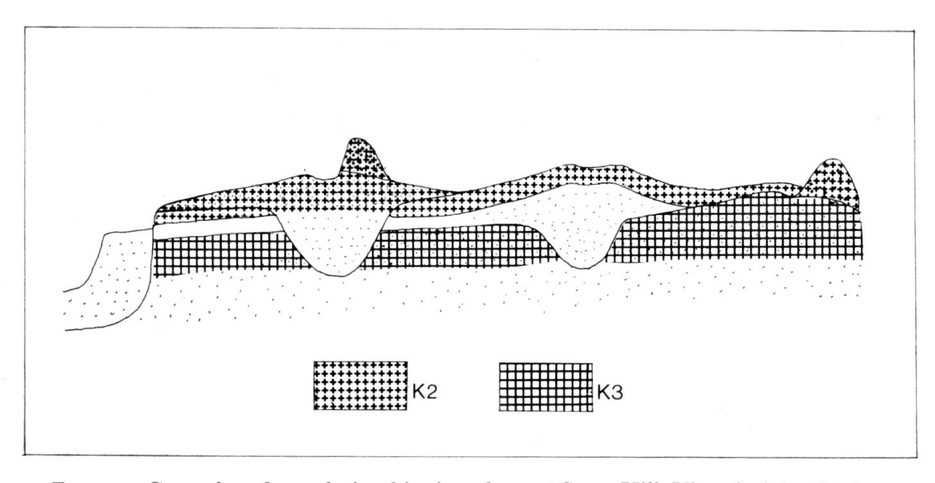

FIG. 90. Ground-surface relationships in a dune at Swan Hill, Victoria (after Butler, 1967).

NORMAL SOILS AND NORMAL RELIEF

It is all too easy to talk of 'normal' soils, but it is necessary to be precise as to what normal means, especially since some soil scientists follow Marbut (1928, also see Jenny, 1941) in using 'normal' in a specialised and (to geomorphologists) a very strange way.

Marbut considers normal relief may be described as smooth, undulating, or rolling, with a relation to drainage such that the permanent water table lies entirely below the bottom of the *solum*. This idea may have arisen from a misunderstanding of the 'normal cycle of erosion' of W. M. Davis. Modern geomorphologists no longer restrict 'normal' to humid temperate landscapes, let alone restricting it to subdued undulating relief. Arid landscapes, high mountain landscapes, and steep or flat landscapes can all be considered equally normal.

Marbut defines a 'normal profile': 'The profile features are of the same *kind* as those in other soils of the same region and all the (profile) features are present which are present in any of the associated soils.' He further believes that *normal profiles are formed exclusively on smooth, undulating topography*.

When the concepts of maturity are compounded with concepts of normalcy, the situation becomes further confused, and the conflicts that can arise may be illustrated by an example provided by Jenny. The Putnam silt loam and the Marshall silt loam occur together in the prairie region, and have the following properties.

Table 18

Putnam	very heavy B horizon	flat sites
Marshall	feeble horizon development	undulating sites

Local soil workers regard the Putnam as the mature soil, for with further illuviation of clay into the B horizon the Marshall could mature into a Putnam-like profile.

Marbut believes that the Marshall is the mature prairie soil, because it is on undulating topography. Because the Putnam is found in flat sites Marbut believes it is a transitory soil, and in the course of time will become a Marshall. Presumably he also believes that the Putnam sites will become undulating in the course of time, which is presupposing quite a lot about geomorphic evolution of the prairie landscape.

I suggest a more useful concept of geomorphic soil relationships might be to regard all topographic variants of soil as normal if they are maintained under natural conditions. In this way all slope soils are normal, not only soils on smooth undulating slopes as in Marbut's concept. In high mountain regions scree slopes are normal landforms, and the skeletal soils associated with them are normal soils. In many topographic catenas the soils at the foot of the slope or in valley bottoms are waterlogged at times, and the hydromorphic soils in such sites are normal soils. On flat sites soils are as normal as on slopes, and factors other than slope determine their maturity. Even actively developing soils are normal, if they are maintained under natural conditions.

Thus on the outside of a river bend where the meander undercuts the river bank, skeletal soils will be normal.

On this basis the only abnormal soils are those that are not maintained under prevailing natural conditions. Palaeosols would come into this class, but most abnormal soils are artificial. Soils on mine dumps, artificial soils put over open-cast mines and quarries, and many garden soils would be in this class. So would the Great Orme soil series of North Wales, which was solemnly reported to have a parent material of broken bottles and old tin cans.

REVERSIBLE AND IRREVERSIBLE CHANGES

Some soil profiles can be converted into others, and back again to the original, if conditions are right. Thus a podzol can be converted to a brown forest soil, and a brown forest soil can be converted to a podzol, as described on p. 165. Such changes consist mainly of separation or mixing of horizons, and changes in organic matter that can be accomplished by vegetation and organisms.

Other changes are more drastic. Thus a weathering limestone may have a skeletal or rendzina soil. If a thick layer of insoluble residue accumulated through extensive limestone solution, a terra rossa might develop. This could only be converted to a rendzina by complete erosion of the soil profile and a fresh start on the limestone parent material.

Many other changes in soil property are irreversible.

Leaching removes bases, and this can go on as long as weatherable minerals can release further bases. Eventually all the weatherable minerals will be altered, and no pedological process can replenish them.

The precipitation of iron oxides in soils is often virtually an irreversible process, which accounts for the persistence of many iron-rich palaeosols.

How do these irreversible reactions affect one's concept of maturity? It seems that the maturity concept is valid for most soils, but over vast lengths of time a soil can become so depleted of extractable material and suffer so many irreversible changes that it can become an almost inert weathering residue. Such soils are rare, for on such a time scale geomorphic processes of erosion and deposition will affect most of the earth's surface, providing new ground surfaces and new parent materials.

PALAEOSOLS AND FOSSIL SOILS

Some soils were formed long ago under conditions that are no longer present at the location in which they occur; these are palaeosols and fossil soils.

Palaeosols are still at the earth's surface, but it is usually evident that they could not be formed under present-day conditions. Usually they were formed under a different climate and different vegetation, and for some reason have resisted conversion to what would be a 'normal' soil in their present situation. Frequently they are very depleted soils, with accumulations of the most resistant minerals such as quartz, kaolin and iron oxides. They have suffered considerable irreversible changes, and therefore cannot be converted into other soil types very easily. If they are really out of their environment they can be regarded rather as parent materials for the present-day soil-forming processes to act on, but being more or less inert they suffer little change.

Palaeosols are often associated geomorphically with old erosion surfaces.

Fossil soils can be used to describe soils that are buried, to distinguish them from palaeosols that are at the surface. They may be buried under alluvium, under lava flows, under glacial tills or even preserved under sedimentary rocks like the fossil soils

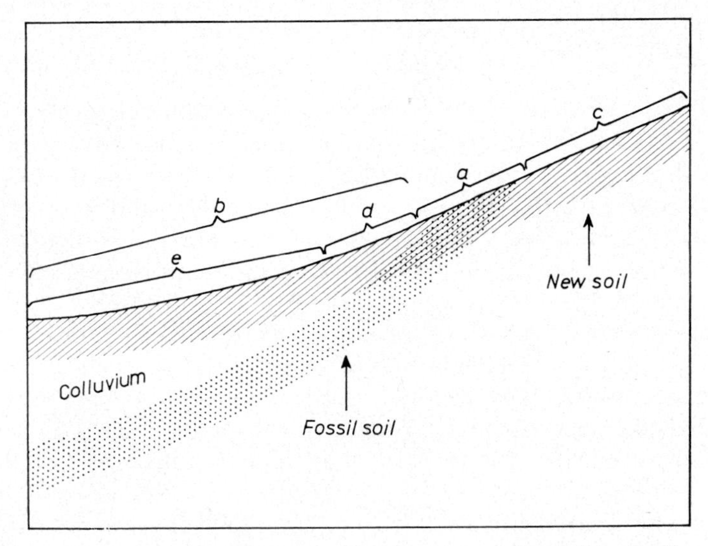

FIG. 91. Relationships of an eroded and partly buried fossil soil.

present below some coal seams. Buried soils may or may not be similar to soils forming at the surface at the present time, and similarities and differences between fossil soils and present-day soils in any locality largely depend on the amount of time that has elapsed since they were buried.

A special problem is presented by fossil soils buried beneath colluvium (Fig. 91). There will be one area where the fossil soil will outcrop at the surface (*a*) and another where it is buried (*b*). A new soil forming on the later ground surface will have various parts of the fossil soil as parent material at (*a*), bedrock parent material at (*c*), colluvium as parent material at (*e*), and the fossil soil will be incorporated into the subsoil of the profile at (*d*). Similar complications can arise by other means, but the erosion and deposition on hillsides present the commonest situation.

XII WEATHERING AND THE EVOLUTION OF LANDFORMS

CONSTANT VOLUME WEATHERING

As mineral alteration results in the formation of new minerals of lower density than the originals, it is usually assumed that the mineral expands on weathering. Rocks, being composed of minerals, are assumed to expand likewise when weathered. Commonly when a rock weathers at the earth's surface it does expand, the change in volume causing distinctive weathering features. However, a great deal of weathering occurs *without volume change*. Constant volume weathering takes place at some depth below the earth's surface.

In many areas where there is extensive deep weathering it is found that structures of the original rock are perfectly preserved despite extensive chemical and mineral alteration. Small quartz veins are often the best indicators, but joints, small faults, original bedding or gneissic banding, xenoliths, and many other sedimentary, igneous or structural features may be preserved as 'ghost' structures in the weathered material.

The outline of original joint blocks can be seen, and the joints remain straight. It seems very unlikely that joint blocks could expand without deforming or destroying the joint pattern, yet joints between thoroughly weathered blocks remain straight, and furthermore in attitudes that match the jointing in associated unweathered rock.

Regolith and saprolite

The use of these two terms was discussed on p. 120. It was noted that a deep weathering profile consists of a number of zones: the lower part, of weathered rock in place, may be termed saprolite; the entire profile, with redistributed, mixed and possibly transported material at the top, is termed regolith. In most weathering profiles the upper part has a volume change, even if the regolith is entirely sedentary, and perhaps saprolite could usefully be restricted to mean weathered rock with no volume change. A term is needed then for the upper part, and perhaps expanded saprolite would suffice.

In this way regolith means the entire weathering profile, and the regolith consists of: transported material (which is not always present), expanded saprolite, saprolite. A soil profile may be superimposed on the regolith, and may involve all the profile or only the upper part.

The saprolite on granite, rich in quartz grains, may be known as granite sand, or grus, growan or by other local names.

Soil horizons of constant volume

Some soil horizons may extend into the saprolite zone of constant volume alteration. Perhaps the commonest soil horizon which shows constant volume alteration is the pallid zone of thick laterites, which frequently displays rock structures very clearly. Many other soils have constant volume alteration material in their profiles, and it is not uncommon for constant volume saprolite to come within three feet of the ground surface. Frequently there is not sufficient structure in the parent rock to recognise saprolite unequivocally, and it is necessary to generalise from those localities where rock structures are preserved and are distinct.

Desert varnish

A feature of weathering in arid regions is the production of desert varnish, a shiny coating on quartz pebbles, rock fragments and occasionally boulders. It usually consists of iron oxides, manganese dioxide, and sometimes silica. The coating is usually too thin to be seen by naked eye in cross-section, but in some siliceous varnish there is a zone of about 1 mm thick behind the varnish, from which the silica may have been derived, with no volume change. Scheffer *et al.* (1963) have shown that, in some instances at least, colonies of blue-green algae on rock surfaces mobilise Fe ions and produce the concentration of oxides that constitutes the varnish. It often seems impossible for the iron coatings on quartz pebbles to have been derived from the interior of the pebbles, and the varnish may be a surface accretion in this case. There may be several kinds of desert varnish, produced by different mechanisms. Blackwelder (1948) in Egypt, and Hunt (1961) in the U.S.A., have shown that desert varnish is over 2000 years old. Chemical data on desert varnish have been summarised by Engle and Sharp (1958).

Patina

Boulders or pebbles may be depleted by weathering at the surface to form a skin distinguished by colour, porosity or other feature, but with no change in volume. Flint nodules and pebbles often have a distinct patina, due to leaching of silica from the surface. The thickness of the patina may be an indication of age, and archaeologists can use the patination of stone implements as a rough means of dating.

Spheroidal weathering

Although it is commonly supposed to be due to expansion, Ollier (1967) made a case for using spheroidal weathering as a term for a constant volume alteration type of exfoliation. The layers in a spheroidally weathered block are caused by chemical migration of elements within the rock, with periodic deposition of the Liesegang ring type. Augustithis and Ottemann (1966) have shown that the bands are due to chemical enrichment and depletion involving both major and minor elements. Hydrolysis is the main weathering agent, and volume is kept constant by removal of weathering products in groundwater. Spheroidal weathering is found in rocks with all the features of constant volume alteration described above.

Alteration is of course most rapid at edges and corners of blocks, so with advancing exfoliation an originally angular joint block is reduced to a rounder, often spheroidal, shape, surrounded by spheroidal shells. This term 'spheroidal weathering' has been

FIG. 92. Spheroidal weathering in granite, Heathcote, Victoria. The corners of the joint block appear to have suffered little more than colour banding, but in the centre there is parting along the planes of spheroidal weathering.

used loosely, and it is here suggested that the term should be restricted to those instances where weathering is truly spheroidal, and attacking the rock from all sides and not only from above. This then restricts it to weathering undergound. There are however two possible sources of confusion that complicate spheroidal weathering landforms.

1. Small boulders weathering at the surface may decompose by 'flaking' (see p. 186), but being small they may be attacked from all sides and simulate spheroidal weathering. In fact such boulders, being unconfined, would probably expand.

2. Spheroidally weathered boulders formed at some depth may be exhumed by subsequent erosion so that they appear at the ground surface. These do not constitute proof of the formation of spheroidally weathered boulders at the surface. Once exposed at the surface, their course of weathering is likely to change to a volume-increase alteration.

Colour banding

Rather similar to spheroidal weathering, but much less regular in appearance, is colour banding (Fig. 93). This is due to alternating enrichment and depletion, mostly of iron oxides, and like spheroidal weathering is probably due to periodic precipitation. The

FIG. 93. Colour banding in steeply dipping Silurian mudstone, Keilor, Victoria.

colour bands are usually irregular; they may go through rock structures, and do not necessarily follow the edges of joint blocks or other regular features. The bands are often very variable in thickness, whereas those in spheroidal weathering are often remarkably constant. Colour banding is common in porous rocks such as sandstones, and the banding may result from the patchy and periodic drying up of groundwater, with precipitation at the surface of diffuse water bodies.

Weathering rinds

Protective rinds of iron oxides are found on the surface of some boulders and may have a thickness of a few millimetres to about 10 cm. One kind is developed as coatings from downward percolating water (possibly in a humid climate) and the other by capillary rise of solutions to the surface of a boulder (possibly in an arid climate). The first kind is formed while a boulder is at some depth below the earth's surface; the second could be formed with the boulder at the surface, but not necessarily. It is not easy to distinguish a weathering rind formed at depth and subsequently exhumed from a rind formed at the surface. Probably most rinds are formed at depth.

FIG. 94. Joints impregnated with iron oxides stand in relief on a shore platform, in weathering granite, Cape Patterson, Victoria. (Photograph by A. A. Baker.)

Although iron oxides are the commonest addition in a protective rind, silica accumulation and perhaps other materials such as manganese may give rise to rinds.

Shell, crust, skin, coating and pellicle are terms sometimes used in the same way as rind.

Case hardening

A weathered rind formed from simple addition of material to a boulder surface does not affect the core of the boulder, though the rind may become more resistant to abrasion due to its impregnation. If there is movement of ions to the surface of the boulder from the inside, a hard surface forms at the expense of weakening of the inside, which often becomes poorly cemented, and prone to granular disintegration. The formation of a hard rind outside a softer core is known as case hardening.

Joint hardening and softening

Joints and other fissures in rocks provide a meshwork of ready channels for the movement of water and dissolved ions. There is frequent depletion or impregnation along the joints, which come to have a different resistance to erosion from the inter-joint blocks. If the impregnated joints are harder than the blocks, later erosion may give rise to a boxwork type of feature (Fig. 94). Conversely, if the joints are softer, erosion attacks them faster and on exposed faces, the boulders rather than the joints stand out in relief.

FIG. 95. Weathered out joints in sandstone. Baby rocks, Arizona.

GENERAL SURFACE LOWERING

In considering the evolution of landscape it is often loosely said that weathering will cause general surface lowering. In most landscapes this is in fact unlikely.

Mineral expansion

From a theoretical point of view, we have seen that weathering of minerals tends to produce minerals of lower density than the originals. Because of this we might expect them to have a greater volume than the originals, and indeed the swelling of minerals on weathering is well known. Some of its consequences have already been considered;

some kinds of granular disintegration or exfoliation appear to be due to expansion of minerals upon weathering.

If we now consider a plane surface of fresh rock and deduce what happens to it upon weathering, does it not seem likely that the expanded minerals on the surface will cause a *rise* in general surface level? Added to the expansion of individual minerals is the pore space that develops as grains fall apart, so it seems most likely that weathering would cause general surface rising if there were no concomitant erosion. Even on isolated hills, such as tors and rounded inselbergs, the process of weathering causes lowering only so long as weathering products are removed by erosion.

Limestone

An exception to this 'general surface rising' might be found on limestone or other soluble rock, where material is lost from the surface in drainage water. The loss might be counteracted to some extent by the accumulation and swelling of insoluble residues, but in general surface lowering seems feasible on limestone. The limestone pedestals described on p. 248 could hardly have formed otherwise than by the general surface lowering of the limestone surfaces that were not protected by overlying boulders.

In South Wales, Thomas (1952) was able to show by a survey of outliers of the Millstone Grit which overlies the Carboniferous Limestone that solution subsidence had lowered the outliers by a vertical displacement of up to 600 feet (Fig. 96).

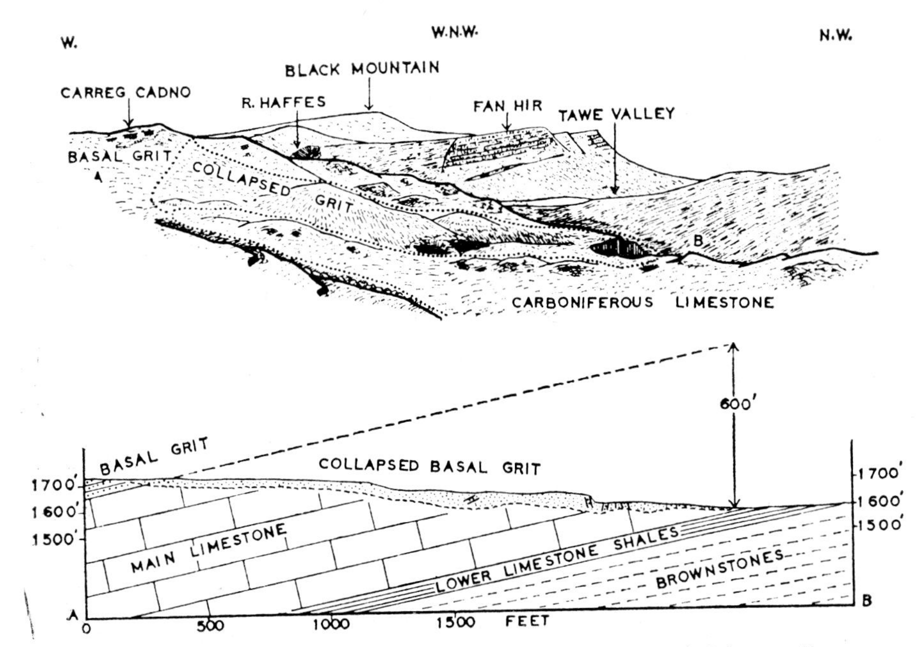

Fig. 96. Panoramic view and section across collapsed Basal Grit over limestone in South Wales (from Thomas, 1963).

Significance of constant volume alteration in general surface lowering

Where weathering profiles are deep, it is found that constant volume alteration features are present to within a few feet of the top of weathered rock. If the volume remains the same there can be no change in general surface level. Lowering can only be by erosion, and rising can be due only to the expansion of the top few feet of soil and expanded weathered material, which would be negligible on the scale used when considering landscapes as a whole. The widespread preservation of constant volume weathering profiles indicates that, for most places at least, there is no general surface lowering due to weathering.

Trendall's hypothesis

One type of surface lowering has been in vogue in recent years and requires more detailed examination. Trendall (1962) put forward the hypothesis that laterite formation caused general surface lowering. As the laterite sheet and land surface sink together the general surface remains flat, producing what Trendall called 'apparent peneplains'.

This idea was based on calculations of the amount of iron in the ironstone layer of laterite profiles, and the amount present in the parent rock. Assuming simple vertical profiles, it is possible to calculate how much parent rock must weather to provide the amount of iron found in the ironstone layer. Calculations on this basis showed that a column of rock greater in thickness than the column now present between the ironstone and bedrock were required. From this Trendall concluded that the column was originally higher, and that as weathering proceeded the general surface became lower, and the laterite sheet settled to a lower elevation. The loss of volume of the profile as a whole was accounted for by lateral removal of weathering products in groundwater.

The defect in this hypothesis is that beneath the laterite the regolith retains all the features of constant volume alteration! This means that the laterite is not sagging or settling to lower levels by removal of material from beneath. De Swart (1964), noting the evidence of constant volume weathering, put forward an alternative suggestion. He still believed that 'a considerable proportion of the iron in laterites has obviously been derived from rocks removed during the lowering of the surface on which they rest', and suggested this is achieved by 'solution in depth and physical erosion at the surface'. As Thomas has pointed out (1965), it is difficult to see how this could operate, because erosion at the surface would be at the expense of the laterite crust, and would not contribute to the thickening of the laterite from below.

Yet again we can consider the laterite as moving down the profile as a sheet, dissolving at the top and reprecipitating below, a kind of steady-state profile like the Buwekula profile (p. 168). However, laterite profiles commonly have an iron impoverished horizon (the pallid zone) separating the ironstone layer from the bedrock,

FIG. 97. Unloading in medium-grained quartz monzonite, Sierra Nevada, California. (Photograph by N. King Huber of the U.S. Geological Survey.)

and such downward movement of ironstone would not actually gain any significant quantity of iron from bedrock as in Trendall's hypothesis.

WEATHERING WITH EXPANSION

LANDFORMS DUE TO UNLOADING

Unloading is the process of massive exfoliation (dilatation) that gives rise to slabs of rock several feet in thickness, and associated landforms. Such landforms are most commonly seen in granites, but are also found in other igneous rocks, in metamorphic rocks, and in some sedimentary rocks (p. 5).

Unloading sheets

The dominant feature of unloading is the production of large and thick curved slabs of rock, and exposure of the dome-shaped rock beneath (Fig. 97). The detached slabs are several feet thick, too thick to be due to chemical or other physical weathering. The

FIG. 98. Unloading sheets and their disintegration in massive conglomerate at Mount Olga, Central Australia.

partings between slabs tend to be parallel to the ground surface, and have been called 'topographic jointing' although of course they follow a 'generalised topography' and not all the details. V-shaped valleys cut in granite bedrock may exhibit unloading parallel to the ground surface on both valley sides. Bradley described unloading fractures parallel to valley walls in the sandstones of the Colorado Plateau (see p. 7). The formation of steep slopes by erosion can give rise to new unloading planes parallel to the new ground surface, and the new and old unloading planes may intersect as at Vaiont (see p. 6). At depth, the fractures get flatter and the sheets thicker, as can be seen in large quarries and other exposures (see p. 6). Other terms for unloading sheet are exfoliation plate, slab and shell.

Unloading domes

These are convex hills, usually devoid of vegetation, and partly covered by the remains of several unloaded sheets. The sheets break up and migrate downslope, and are dispersed by various agents of weathering and erosion (see Fig. 98). Unloading domes may be a few tens of feet to hundreds of feet high. They are found in all climates, hot and cold, wet and dry. They may occur in isolation or in groups. They may be on a broad plain, or in high mountains.

F. E. Matthes first used the term exfoliation dome in 1930.

In the literature there is some confusion between unloading domes and various other sorts of isolated hills such as inselbergs, bornhardts and monadnocks. Since some

FIG. 99. Kernsprung at Devil's Marbles, Northern Territory, Australia.

of these hill types are often associated with particular climates, type of slope development, stage of cycle of erosion or other relationships, it is best not to use them for unloading domes. If required, unloading dome can be used as an adjective to describe such a hill, e.g. unloading-dome inselberg.

Ruwares

Ruwares are gently domed or almost flat outcrops of bare rock that barely break the ground surface. They are believed by Mabbutt (1952) to represent the late stages of dome exfoliation. However some ruwares may have a different origin.

Kernsprung

Kernsprung or Kernsprünge are boulders that have been fractured into a small number of large fragments. They were formerly attributed to insolation (see p. 15), but this is now thought very unlikely as the mechanism. Ollier (1965) has tentatively suggested that the spontaneous opening of granite joints that cleaves some boulders (Fig. 99) might be due to expansion of a kind similar to unloading.

Arcuate landforms due to unloading

Hack (1966) reported circular patterns and exfoliation traces in the granite country of the Grandfather Mountain area, North Carolina. Air photographs of the area showed

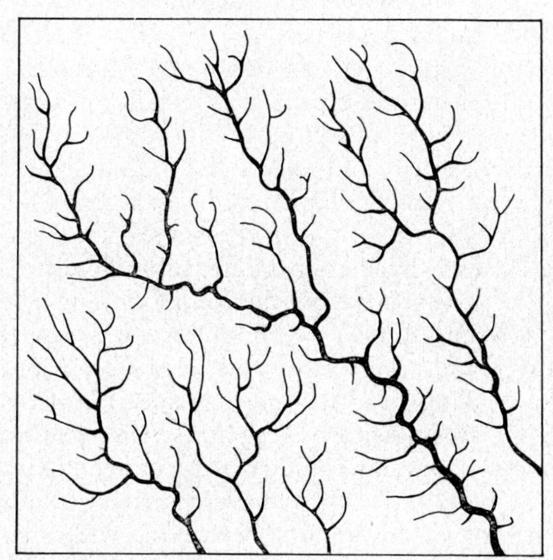

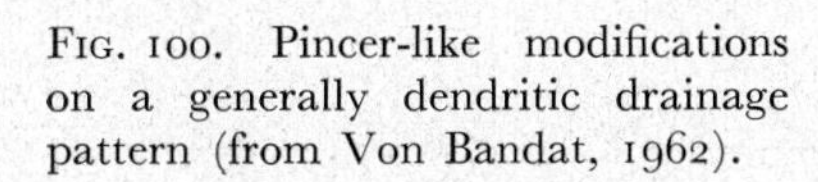

FIG. 100. Pincer-like modifications on a generally dendritic drainage pattern (from Von Bandat, 1962).

numerous large-scale arcuate, circular and elliptical patterns with diameters from 1000 to over 4000 ft. These are commonly defined by curving streams, curving ridges and curving belts of contrasting vegetation. The primary rock structures of the area do not correspond to any of these patterns, and they are probably due to unloading.

The drainage pattern called 'pincer drainage' by Von Bandat (1962) is typical of plutonic igneous rock, and is probably due to the streams following the arcuate traces of unloading sheets (Fig. 100).

EXFOLIATION

Exfoliation is a general-purpose word used to describe several processes and several landforms having in common the possession of more or less concentric shells of rock over an inner core. Exfoliation can usefully be subdivided into a number of divisions, as follows:

Spheroidal weathering—constant volume alteration;
Unloading—expansion due to innate forces in the rock;
Flaking—expansion due to external forces.

Spheroidal weathering and unloading have already been discussed (pp. 175 and 182). It remains to describe flaking (also known as spalling, scaling, desquamation, onion skin weathering, and by other names).

Flaking

1A. Caused by fire. Boulders heated by fire will lose flakes of rock with curved inner surfaces and sharp edges, the flakes being up to several inches thick in their widest part, and fresh. Generally only one fracture plane is formed, and there is no succession of concentric surfaces as in other exfoliation forms. The boulders from which flakes are derived become more rounded.

2A. Caused by salts. The growth of salt crystals in layers or pockets below the surface of porous rock can undoubtedly give rise to flakes. This happens in building, and also on natural surfaces, especially on porous rock. The flakes are found to be backed by a layer of salt up to 2 mm thick. The flakes themselves tend to be of fairly uniform thickness, and only a few millimetres thick. Further details from Fox (1935) are given on p. 13. Flaking due to salt growth has been called exudation (Thornbury, 1954).

3. Caused by chemical decay of minerals. This is the commonest type of flaking, and hydration is probably the commonest type of chemical alteration involved. Flakes produced tend to be a few millimetres or a few centimetres thick and of fairly constant thickness, and are oxidised and otherwise decomposed.

4. Flaking by other mechanisms. Flaking has occasionally been attributed to lightning, frost, rootlets or periodic temperature changes. It is very doubtful if any of these mechanisms could really produce flaking.

Watanabe *et al.* (1954) have described exfoliation due to atomic bombs. One can only hope that this kind of weathering will remain extremely rare.

Flaking gives rise to some other landforms besides the production of flakes. These are particularly common in arid regions where flaking seems to affect many rocks from granites to nearly pure quartzites.

Rounding of boulders

Individual blocks that weather by flaking become rounded because the process attacks corners and edges more than faces. When a boulder is quite rounded it shrinks, and the radius of curvature becomes smaller.

Ground level platforms

Since flaking does not extend below the ground surface, shrinkage of a boulder by flaking stops at ground level. The retreating boulder face generates a more or less horizontal surface across the rock at ground level (Fig. 101), and this has been called a ground level platform (Ollier, 1965). Multiple ground-level platforms are also found.

Flaking hollows

Flaking can occur on concave surfaces, and when it does there appears to be a tendency for the curvature to be exaggerated. Concavities that start off as very shallow depressions grow into deep pockets or, if large enough, caves. These range in size from about one foot across to perhaps fifty feet or more. Some occur in isolation, and at times whole surfaces are almost entirely covered with caves. The formation of hollows by flaking was termed 'negative spheroidal exfoliation' by Wagner (1912), who described examples from South Africa. At Ayers Rock, Australia, Ollier and Tuddenham (1962) found flaking completely unrelated to aspect, shade or position, and the rock (an arkose) seemed to suffer flaking on every exposed surface except the very newest surfaces

FIG. 101. Ground level platform caused by flaking, Cobaw granite, Victoria.

exposed by large rock falls. There appeared to be no difference in type of flake between those in caves and those on the summit of the rock.

Flaking ridges and furrows

At Ayers Rock, Australia, flaking seems quite remarkably efficient at picking out structure, and the bedding, which has no bedding plane parting, is weathered into remarkable ridges and furrows. In other instances joints provide lines of weakness giving rise to grooves or furrows when flaking attacks a rock surface.

Other landforms due to flaking

The convexities and hollows formed by flaking may sometimes be so arranged as to give rise to distinctive landforms.

Pedestal rocks are frequently attributed to wind erosion, and some may indeed be formed in this way. Others, however, are coated with scales that indicate that hydration plays a large part in their formation, and wind is important mainly in removing the weathered products. Some have a hard caprock, but many are in homogeneous rock. Cramer (1963) described pedestal rocks of Wyoming that were formed in bedrock of structural and lithological uniformity. Joint blocks were first rounded by exfoliation then wind action undercut the western face and deflated the debris. The pedestals were undercut most on the eastern side, however, where flaking, caused by hydration and frost, was at a maximum.

Some rock windows and natural arches are formed by the meeting of two flaking concavities that start on opposite sides of a rib of rock.

Tortoise-shell boulders

These boulders, also known as polygonal cracked boulders, elephant hide boulders, and by other names, are case hardened boulders, in which the outer rind has been broken into a polygonal pattern by cracks, often a centimetre or so wide. It has been suggested that some of these may result from a volume increase of the core of a boulder, due to some unknown weathering phenomena, after case hardening. The hard rind is then cracked into fragments.

OTHER FEATURES DUE TO EXPANSION

Fire-cracked boulders

Fire normally produces flakes from boulders, but occasionally cracks them across the middle. This is especially common with boulders fresh from a river bed, as camp fire users often discover, and it is possibly generation of steam in minute cracks that does the damage rather than expansion of the rock itself.

Fragments caused by insolation weathering

The evidence for insolation weathering has been discussed on p. 15. Angular fragments may be produced by blocky disintegration. Cracked boulders may be formed by simple cracking or boulder cleaving. 'Dirt cracking' may also give rise to cracked boulders, and the angular fragments thus produced may be cemented together to form 'dirt crack breccia'.

Features due to ice and frost

The growth of ice is one form of volume change, and creates numerous landforms.

Frost weathering. The mechanism of frost weathering was described on p. 11. It gives rise to sharply angular blocks, which may accumulate to form scree. Ridges may become sharper and form arêtes, with gendarmes projecting on more resistant rock. The formation of glacial cirques possibly involves freezing, though glacial erosion (aided by unloading joints) is of course the dominant process.

Ice heaving. The growth of ice masses in unconsolidated materials gives rise to numerous landforms, but since these are mainly formed by movement of material rather than its breakdown they will be considered only briefly.

Where ice accumulates on porous rocks or unconsolidated material, it is segregated into certain patterns as explained in Chap. II. This pattern gives rise to frost- or ice-wedges and polygonal ground associated with these, or to pingoes, which may be several hundred feet high and are isolated hills formed by the growth of enormous nuclei of ice. Sometimes the ice merely heaves ground into patterns of stones without actually making solid ice wedges, and on material with a wide range of particle size the process forms stone-rings, stone stripes (on steeper grounds), stone garlands and terraces.

Deposits of mixed material due to the slow movement of solifluction deposits under the influence of partial freezing are known as 'head' in England.

Landforms due to expansion of clays

Some soils give rise to a hummocky surface, the mounds and depressions making various patterns and coming in various sizes, the largest being about three feet between depression and mound top. This kind of patterned ground is generally known by the Australian word 'gilgai'.

Many gilgai occur in areas of calcareous soils, and examples have been described by Leeper (1964). The mounds are roughly circular, about three feet across, and separated by flat depressions about five feet across. The surface of the mounds is calcareous and crumbly, and has a very low density. The soil in the depression is not calcareous and is not so cracked as in the mound. Sections through gilgai reveal that the subsoil in the depression is similar to the soil throughout the profile in the mounds, and it seems likely that the mound is formed from the calcareous subsoil. A possible process is as follows: when the soil dries and cracks, pieces of the topsoil fall down the cracks. When the soil becomes wet it swells, and since more material is present than when it shrank, it now exerts a pressure sideways, which is relieved by movement upwards. The final pattern depends on the spacing and pattern of the points or lines where rises occur.

Swelling clays (montmorillonite clay minerals) must be present for this process to operate, but carbonate is not always present. Gilgai have been reported on acid soils, and also from deserts where the desert gilgai were associated with soils having a gypsum layer in the profile (Ollier, 1966).

Simple desiccation can give rise to mud cracks, and these may be related to the giant desiccation features—polygonal ground with open fissures several hundred yards apart—described from Nevada by Willden and Mabey (1961). They seem to have formed during the last few decades.

DIFFERENTIAL WEATHERING

Differential weathering of rocks gives rise to numerous features. Some differential weathering is due to structural or lithological differences in rock, as when a soft layer is hollowed out. Other features are due to rock differences caused by earlier weathering, such as joint cementation or case hardening. Sometimes irregular weathering is found on rocks that are more or less homogeneous. Random patches of vegetation on limestone or granite, for instance, may give rise to solution pits around their roots. Commonly such patches are not exactly random and pick out joints or other structures that are not otherwise apparent.

Differential weathering can occur on all scales. At the upper end of the scale differential weathering and erosion give rise to valleys and hills, and such large-scale phenomena are treated elsewhere (p. 194). The smaller scale features are dealt with here.

Weathering pits

We may use weathering pit as a general term for the many kinds of hollows produced by weathering, which have come to be known by many different names. They may occur on horizontal, sloping or vertical surfaces, and although the terminology does not always discriminate, we shall start with those pits on level surfaces.

Weathering pits may be called *weathering pans* when on a level surface. They are very variable in size, from about one foot across to large holes 60 feet across and 10 feet deep. They have been reported most often from granite (e.g. Smith, 1941), but also from other rocks such as the Hawkesbury sandstone of the Sydney area (Dury, 1966) and metamorphic rocks of Ruwenzori, Uganda (Osmaston, personal communication). Amongst many terms that have been used for such hollows, '*gnamma*' or '*gnamma hole*' appears to be gaining in popularity. An account of gnammas is given by Twidale (1963). The ideal gnamma hole is on a horizontal surface. On an inclined ground surface the hollow might be called an *armchair hollow*, and this marks a transition between gnamma and those hollows on vertical faces which might be called *tafoni*. Returning to the pits on level ground, there are still more terms that can be used. Freise (1938) distinguished *Dellen* (compound granite hollows), *Pingen* (granite hollow) and *Oricangas* (water eyes) which are weathering pits of decreasing size. He believed the larger ones were formed by decaying trees in rain forest, while Oricangas were present in dry areas. Bakker (1960) believes all the weathering pits are formed in the same way, and though retaining Oricangas, he suggests the term *Kociolki* to replace Dellen.

Plants—lichens and mosses particularly—are thought to form a special kind of shallow basin reported from granites in Germany. These are called '*Opferkessel*' (sacrificial cauldrons), a fanciful resemblance that is fortified when they are accompanied by '*Blutrinnen*' (blood drains) leading to the edge of the sacrificial stone. Opferkessel did not form beneath a cover of weathering products but developed on the granite surface exposed to the attack of the atmosphere.

Yet another distinction is made by Demek (1965) who separates '*weathering pits*', which are wide open, from '*dew holes*' which have a narrow opening and become wider inside.

Weathering pits seem to be made by several processes. One kind are hollows due to flaking (see p. 186); others are due to localised granular disintegration; still others are due to granular disintegration taking place below a breached case-hardened surface. The weathering pits that flare out below a narrow hole are almost always of this type. The weathering pits that hold water periodically sometimes develop successive case-hardened surfaces which may be repeatedly breached, giving complicated profiles to the pits. Some weathering pits have vegetation growing in them, and it may be that the organic acids around the roots initiate weathering pits, and they may be the sole cause. Periodic flooding to flush out weathered debris seems to be essential in pit formation.

Some weathering pits are known as '*solution ponds*' or '*solution depressions*', but it is doubtful if solution could be responsible for more than a part of their excavation. Insoluble residues have to be washed or blown out periodically.

Ponds may be broad and shallow, and occur on a variety of rocks not normally considered soluble. Reed (1963) describes some from quartzite ridges in Carolina up to 200 feet across and 3 feet deep which he attributes to solution with the aid of organic substances. Similar features have been described from the same State but on diorite by LeGrand (1952). Apart from their greater size they are similar to other weathering pits.

Cavernous weathering

Cavernous weathering may be used as a general term for the many landforms made by weathering on steep slopes. *Careous weathering*, *etching*, *scalloping* and many other names have been used. Cavernous weathering occurs in many kinds of rock, in many climates and on many scales. Some cavernous weathering pits are known as *tafoni*. Dury (1965) claims that the word means 'window' and should only refer to a hole produced when a weathering hollow breaches the roof of its enclosing boulder, but this restricted usage is not likely to be adopted as tafoni is already too well established.

Some cavernous weathering is due to flaking, and the pits are simply flaking hollows. Blackwelder (1929) shows a flaking hollow and labels it tafoni. Other hollows he illustrates are evidently due to granular disintegration, and are termed *desert niches*. Similar forms are known to occur in environments that are not arid, so desert niche is perhaps an unfortunate term, and '*niche*' might be better. '*Alcove*' is another term sometimes used.

Niches or hollows, produced either by flaking or by granular disintegration, frequently reveal a tendency to grow upwards, into the rock. This is especially common if the rock is case hardened, and sometimes a hollow breaches the case hardening to make a 'window'.

Although cavernous hollows extend upwards and backwards into the rock, they do not grow downwards, and in fact activity stops at the ground surface. In this way *ground level platforms* are produced, by slope retreat, in a manner similar to that described on p. 186. If the ground level is changed by slight erosion, multiple ground level platforms may be formed.

The smallest kinds of cavernous weathering need a different term, and perhaps *scallop* is the most suitable. When scalloping becomes big enough it forms niches.

Minor features of differential weathering

When the weathering rock has a pattern of joints or other lineaments which are resistant to further weathering, it commonly happens that different parts of a rock succumb at different rates. Thus iron-filled joints in a sandstone may be more resistant than the rock itself, and so stand out on a weathered face as a 'boxwork' or 'honeycomb'. In such cases it may be that the enrichment in iron of the boxwork may be at the expense of the rock, which loses cement and therefore crumbles more easily, and the differential hardness is due to part of the weathering process. In other cases, as for instance where quartz veins make the pattern, chemical migration during weathering is not part of the process. Many substances can make the ridges in such structures including iron oxides, secondary silica, calcite and manganese oxides. The term honeycomb may be used in a different sense at times. Demek (1965) illustrates a feature that he calls honeycomb (his Fig. 5), but on the usage presented here it would illustrate multiple scallops in granite weathered to degree 3 (p. 127).

In the reverse case, where a pattern of joints or other lineaments are especially prone to weathering, they have the appearance of many intersecting cuts and grooves and this feature has been aptly named '*butcher block*' weathering (Feder, 1964).

On a small scale there may be a host of minor features with a complicated sequence of developmental forms and an involved geomorphic history. Few examples have been studied in detail, but one example is provided by the *crenellations* and *gutters* of the Millstone Grit described by Cunningham (1964). The gutters originate from the enlargement and coalescence of small hollows initiated by the removal of quartz pebbles, and the ridges between gutters, a few inches high, are the crenellations. Some segregation of iron oxides accompanies the process, and the crenellations are eventually consumed and give rise to gentle ripples and hollows.

Raised rims

Many weathering pits are surrounded by a raised rim that stands not only above the pit, but also above the surrounding rock surface.

Some of these are clearly due to resistance of a cemented or indurated ring around the weathering pit during general surface lowering. Thus Wentworth (1939) reports the cementation of the inside of weathering pits on a shore, and Hills (1949) reports the coating of pits by secondary limonite.

Raised rims are not always so easy to explain. Blank (1951) described 'rock doughnuts' on granite. These are raised annular rings around weather pits, a foot or two across, and some even have an annular depression outside. The raised rim does not appear to be indurated in any way, and Blank was not able to offer any satisfactory explanation.

Similar raised rims are found on a wide variety of rock around depressions (often called potholes, but this is not a good term) on shore platforms. Another example is the flared rim often found on limestone surfaces. These have a pit inside a fairly steep rim, which also rises above the general limestone surface. The rim is notched by many radial grooves, both inside and outside (Fig. 102). The rims are of bedrock, indistinguishable from the rock beneath the depressions, and there is no evidence of secondary calcite growth or cementation.

Differential removal of weathered rock can lead to the formation of other unusua

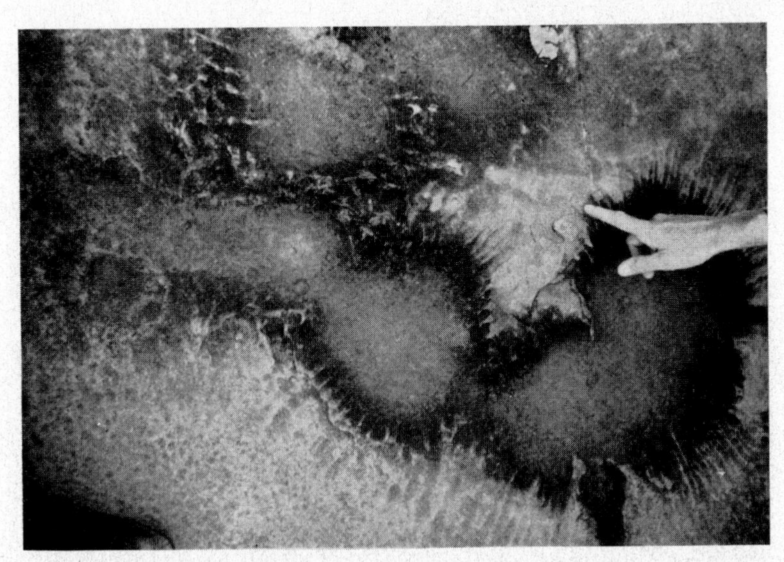

FIG. 102. Flared rims. Solution pits have rims raised above the general limestone surface, with grooves on both inside and outside. Casa Grande National Monument, Arizona.

FIG. 103. Sandstone buns produced by differential weathering. Dakota Rim, Utah.

FIG. 104. Differential weathering, Swifts Creek, Victoria. A dike-like body of granite in schist gives rise to a line of saddles on the spurs. The area is $2\frac{1}{2}$ miles long.

shapes, such as the sandstone spindles reported by Hussey and Tator (1950). These are relatively small, spindle-shaped structures occurring in closely spaced, parallel arrangement on the surface of exposures of an arkose. They result from groundwater solutions attacking the cement to develop zones of weakness along planes of cross-bedding and jointing. There is no detectable difference in composition between the spindles and the enclosing formation except, perhaps, the degree of cementation by gypsum.

Friable sandstones are especially prone to this kind of action, and can produce a wide range of balls, buns and other shapes (see Fig. 103).

Some landforms are due to relatively greater weathering around a protected site.

Butuzova (1962) described *butt-hillocks*, large mounds formed under old trees, where leaching is significantly less than in the sites not protected by the tree trunk.

On a large scale the differential removal of weathered rock can give rise to mushroom rocks, pedestal rocks, rocking stones, and similar bizarre features for which Thornbury suggests the general name *hoodoo rocks*. Differential frost weathering on arêtes may leave upstanding masses of more resistant rock known as gendarmes. On jointed granite tors differential weathering may be responsible for producing some of the battlement effects of castle kopjes.

FIG. 105. Forms due to sculpting of hard granite, Devil's Marbles, Northern Territory, Australia.

Differential weathering and erosion on dykes of igneous rock can give rise to 'trenches' or 'walls', depending on whether the dyke is harder or softer than the surrounding rock (Fig. 104).

Steeply dipping sediments, sandstone dykes and other structures can give rise to similar features. More commonly, instead of a distinct and vertical wall, a resistant band of rock merely gives rise to a ridge with sloping sides. At the upper end of the scale this becomes a landscape of ridges and valleys, and is too large to be considered as a dominantly weathering phenomenon.

Sculpting

Weathering can carve individual boulders into a wide variety of shapes that almost defy classification. The shape, spacing, orientation of depressions and rises seem to be of endless variety. So far only qualitative descriptions of many forms have been given, a situation that will probably persist for a long time. However, even the roughest description is probably better than the application of a well-known term, when it does not exactly apply. Much of geomorphology suffers from a superabundance of terms, but in this respect there is a need for more.

FIG. 106. Sculpting of soft weathered granite (degree 3), Devil's Marbles, Northern Territory, Australia.

Note that the forms taken by granite of different degrees of weathering are quite different even in the same area (Figs. 105, 106).

Sculpting refers to the surface differential weathering only, not subsurface alteration. Thus many corestones become sculpted when they are exposed to surface weathering and erosion.

Forms due to frost action

The main landforms due to frost action are depositional (see p. 215) but it also gives rise to jagged ridges and cliffs, and some arêtes and cirques, though dominantly caused by glacial erosion, are dependent on frost action for plucking and steepening of the back walls. Nivation cirques are also caused by freeze-thaw mechanisms but in this case the debris accumulates below the snow patch that occupies the hollow or cirque. Other features resulting from frost action are described on page 188.

Caves due to differential weathering

Many caves in rocks other than limestone are due to differential weathering of soft bands of rock beneath more resistant overlying beds. Some of these are little more than rock shelters, but others may be several hundred feet long. Water is usually responsible for removing weathered debris, but the caves are not basically erosional features. In

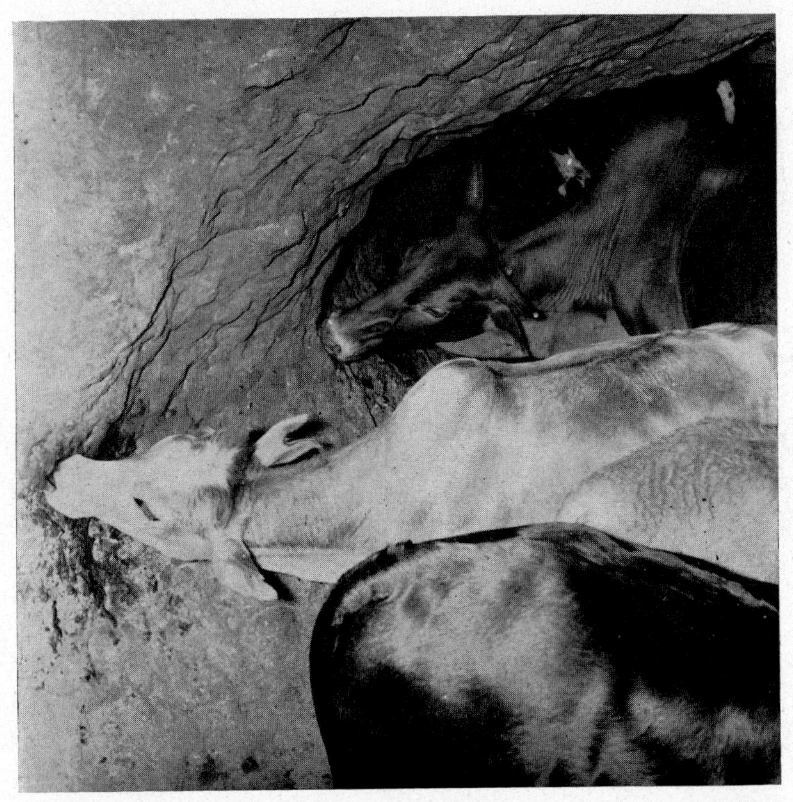

FIG. 107. A rare kind of biotic weathering. Cattle licking mirabilite crystals from the wall of a cave. Mount Elgon, Uganda. (Photograph by J. F. Harrop.)

some instances, especially soluble beds, or beds with a soluble matrix, are the cave-forming layer. One unusual type was described from Mount Elgon (Ollier and Harrop, 1958), where layers of volcanic ash alternate with agglomerate. The ash is more easily washed out, but is also soluble to some extent. The walls of the caves become covered with crystals of mirabilite (sodium sulphate), which is eaten by cattle (Fig. 107). A new crop of crystal grows in a few days after removal.

Mineral boxworks

Limonite boxworks are sometimes formed during the weathering of economic minerals, the limonite following cleavage planes. Since different minerals have different cleavage patterns which can be characteristic, a knowledge of boxworks is useful in prospecting, and an examination of a gossan of pure limonite can indicate the mineralogical composition of the original lode. Some typical mineral boxworks are shown in Fig. 108.

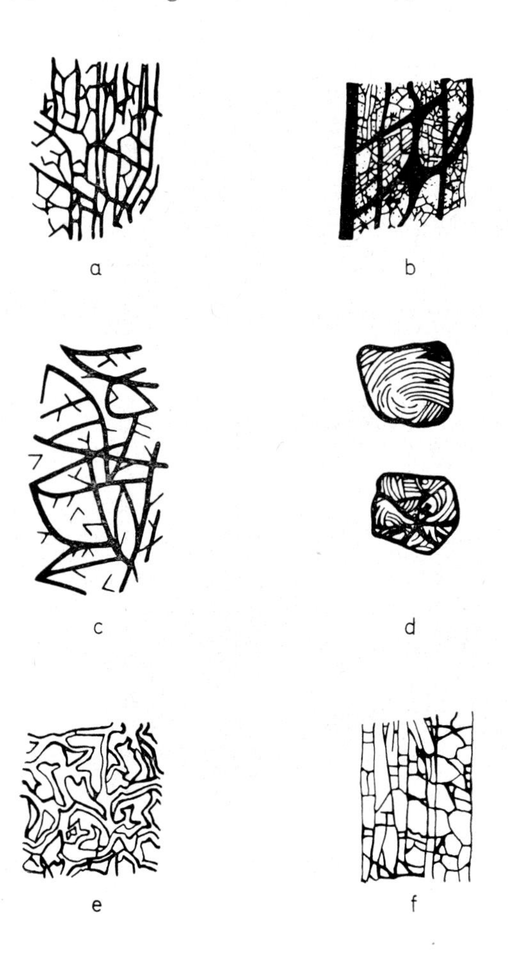

FIG. 108. Mineral boxworks.

a Quadrangular pattern after chalcopyrite (×5). *b* Coarse cellular boxwork after sphalerite (×2). *c* Triangular curved boxwork with eye-shaped cells, after bornite (×5). *d* Foliated boxwork after molybdenite (×20). *e* Contour-like boxwork after tetrahedrite (×5). *f* Rectangular pattern after galena (×3). (Figures from Bateman, 1950.)

LANDFORMS DUE TO STRIPPING

In Chap. X it was shown that the regolith produced by long-continued weathering may be very thick. The weathering front marking the junction between hard rock and weathered rock is sometimes very abrupt, and is usually very irregular. Blocks of unweathered rock may float in the regolith as corestones, as shown in Fig. 109.

Tors and related landforms

During the course of erosion, the basal surface of weathering may be exhumed, and the corestones left behind as the soft material of the regolith is washed away. This stripping gives rise to a number of landforms: it is a kind of differential weathering but on a larger scale than the forms considered under that heading. Some kopjes or castle kopjes are equivalent to tors, others are probably formed by subaerial weathering.

FIG. 109. Corestones of fresh granite in weathering granite that retains the original joint structure. Hong Kong.

Tors are small hills or heaps of boulders, usually about 20 to 60 feet high, rising abruptly from the surrounding gentle ground surface. Some tors may be formed in one-cycle, but many are formed by a two-cycle process of deep weathering followed by exhumation. Linton (1955) first brought the two-cycle origin of tors to prominence, though as often happens in science the idea had been anticipated and the process described by earlier workers such as Falconer (1911) and Handley (1952).

Falconer described two flat-topped hills that rise to a height of 150 ft above the surrounding plain and are composed almost entirely of soft and thoroughly decomposed rock. However, he noted that one of the hills 'although deeply decomposed, still

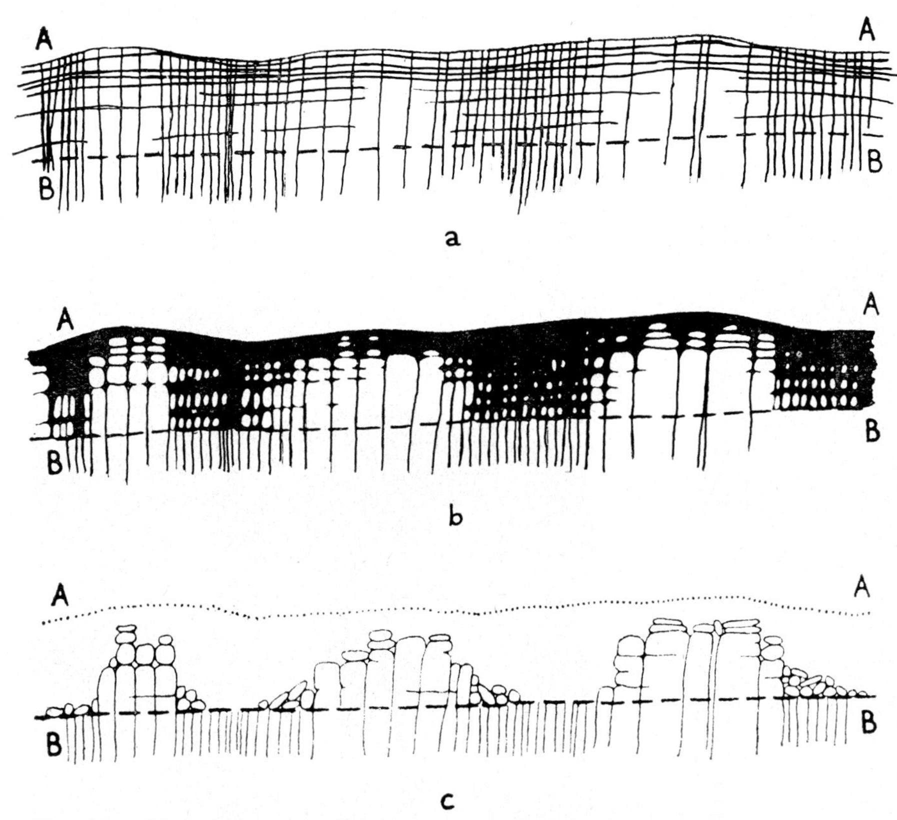

Fig. 110. Linton's (1955) explanation of the evolution of a group of tors, illustrating the importance of joint spacing.

preserves in its lower part detached boulders or cores of unweathered rock. If the subsequent erosion had continued until the weathered material had been entirely removed, the flattened hill would have been replaced by a typical kopje of loose boulders resting upon a smooth and rounded surface of rock below.' This early statement contains the essence of all later ideas of landform creation by stripping. It remains to consider the details.

Linton's (1955) explanation for the tors of Dartmoor, England, is shown in Fig. 110. Weathering proceeds to what he calls a basal platform, and the regolith was then stripped, leaving the platform as the present ground surface, complete with tors. Numerous corestones accumulate on the tors, though the main mass of them is rooted in bedrock. The Dartmoor tors are all made of dense, fresh-looking rock, and the many minor features formed by weathering after stripping seem to be almost absent on Dartmoor.

McCraw (1965) described a very interesting tor assemblage from Otago, New Zealand. Here the bedrock is schist, deeply and irregularly weathered and in places overlain by Tertiary deposits which clearly show that the deep weathering is of pre-

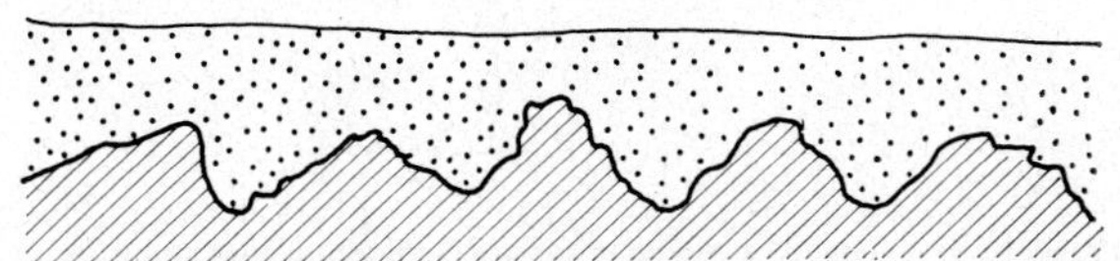

a. Contact between weathered and unweathered schist.

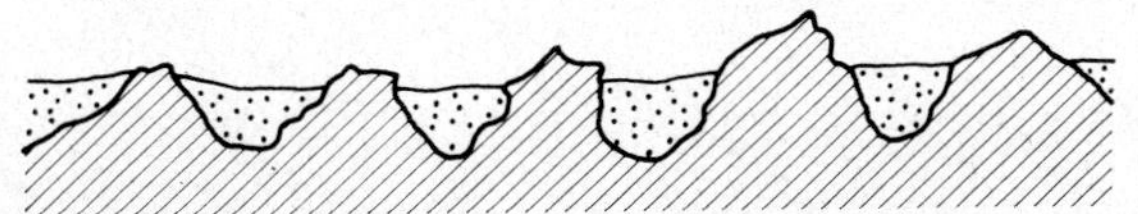

b. Weathered schist partly stripped-'pseudo-tor landscape'

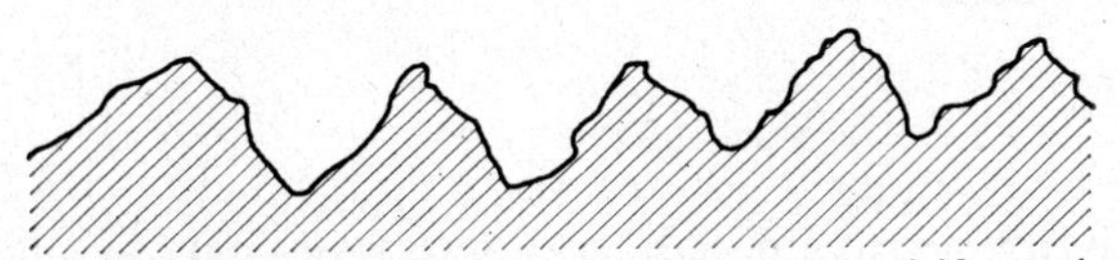

c. Weathered schist completely stripped-'fretted landscape'

FIG. 111. Genesis of a fretted landscape, Otago, New Zealand (from McCraw, 1965).

Tertiary age. Two widely different types of terrain are developed, depending on the attitude of the schist. Where the weathering took place on moderately or steeply dipping schist a landscape described as 'fretted' is exposed (Fig. 111), but where the dip of the schist was less than about 20° a tor landscape is found (Fig. 112). In some

a) Contact between weathered and unweathered flat-lying schist

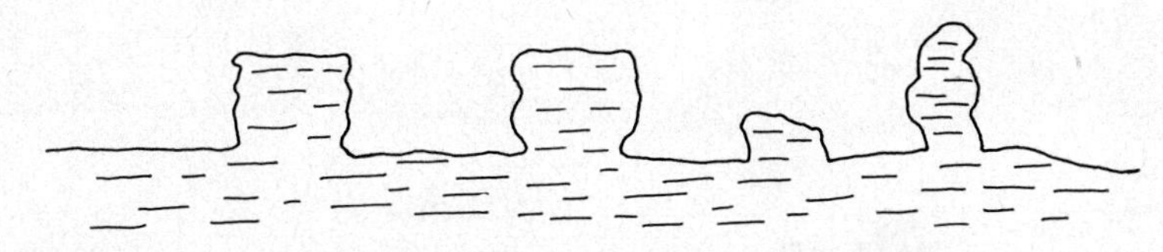

b) Weathered schist completely stripped - 'tor landscape'

FIG. 112. Genesis of a tor landscape, Otago, New Zealand (from McCraw, 1965).

FIG. 113. Penitent rocks, Zambia (from Ackermann, 1962).

places the boundary between the tor and the fretted landscapes is sharp, but in other places is not clearly defined, and both types may be placed in the general class of tors.

As shown in Fig. 111, at an early stage in stripping, schist salients protrude through the mantle of weathered schist, and even the fretted landscape has a resemblance to a tor landscape. In fact, at this stage it could properly be called a tor landscape. Where schistosity is parallel to the ground surface it is clearly possible to have a weathering front more nearly parallel also, and in this case the weathering front really is a basal

FIG. 114. Penitent rocks, Stonehenge, Southern Rhodesia (from Ackermann, 1962).

platform, exposed by stripping as in Fig. 110, and corresponding very closely to Linton's concept.

Further details of the Otago tors are given in Raeside (1949) and Ward (1952).

Another variety of tor formed on rock with a dipping foliation, joint pattern schistosity or bedding is the *penitent rock*, *monk rock* or *Bussersteine*. Penitent rocks from Rhodesia have been described in great detail by Ackermann (1962), and Figs 113 and

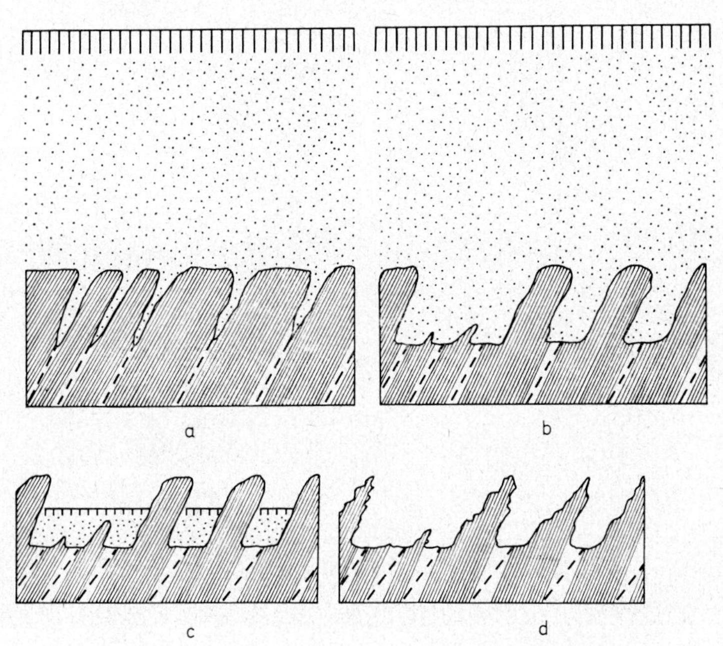

FIG. 115. The origin of penitent rocks (after Ackermann, 1962).

114 are taken from his paper. His explanation of their origin follows the now familiar pattern, and is shown in Fig. 115. It will be noticed that at stage (*c*) the penitents are already present, and it is not necessary to have total stripping as in stage (*d*). Over whole landscapes, such as that shown in Fig. 114, it is most unlikely that stripping will be complete, and pockets of regolith will be present, though with very irregular distribution.

It is not necessary to remove all the regolith to produce tors, and Ollier (1965) has described landscapes in Australia in which the ground surface is largely cut across regolith, giving gently rounded landscapes, with only occasional tors projecting abruptly through the surrounding sea of regolith (Fig. 116). This aspect is dealt with further in the section on inselbergs (p. 206).

Fig. 117, which is similar to Linton's, Fig. 110, shows the simplest relationship between tor (or in this case, inselberg) and exhumation and further weathering. Deep weathering produces the tor, exhumation exposes it, and neither the exposed or still buried part suffers appreciable further weathering during exhumation.

As Linton (1955) says 'This possibility that the weathering implied by the upland tors is late Tertiary is attractive because of the known warmth of our Pliocene climate. To admit such a possibility, however, emphasises the distinction between the two phases of tor formation; exhumation may be much later than the decomposition.'

However, the two phases of tor formation, though no doubt clearly separate in some instances, are not necessarily distinct.

FIG. 116. Gently rounded landscape with occasional tors. Cobaw granite, Victoria.

It is also possible that weathering rates may be more or less equal to the rate of exposure, and two possible cases may be envisaged; the weathering of the exhumed tor may be faster than decomposition of the still buried portion, or the weathering of the buried portion may be greater than that of the exposed part. Both cases have been invoked to account for actual landforms.

Ackerman (1962) illustrates surface weathering attacking the penitents as they emerge, so that the tops have suffered more attack than the bases, and if stripping takes place in phases there will be several phases of surface weathering to correspond.

The evidence of Ackerman from the penitent stones (Fig. 115) shows that so long as the penitent remains buried it suffers little alteration, but once exposed at the surface it decays rapidly. Ollier (1960) also pointed out that once exposed at the surface, tors (or inselbergs) tended to decay by other weathering processes.

This is in contrast to the opinion of Thomas (1965, p. 75). '. . . once exposed, tors decompose only very slowly, while the underlying jointed rock is brought within the vadose zone of intense weathering, and therefore decays rapidly'. He suggests this is

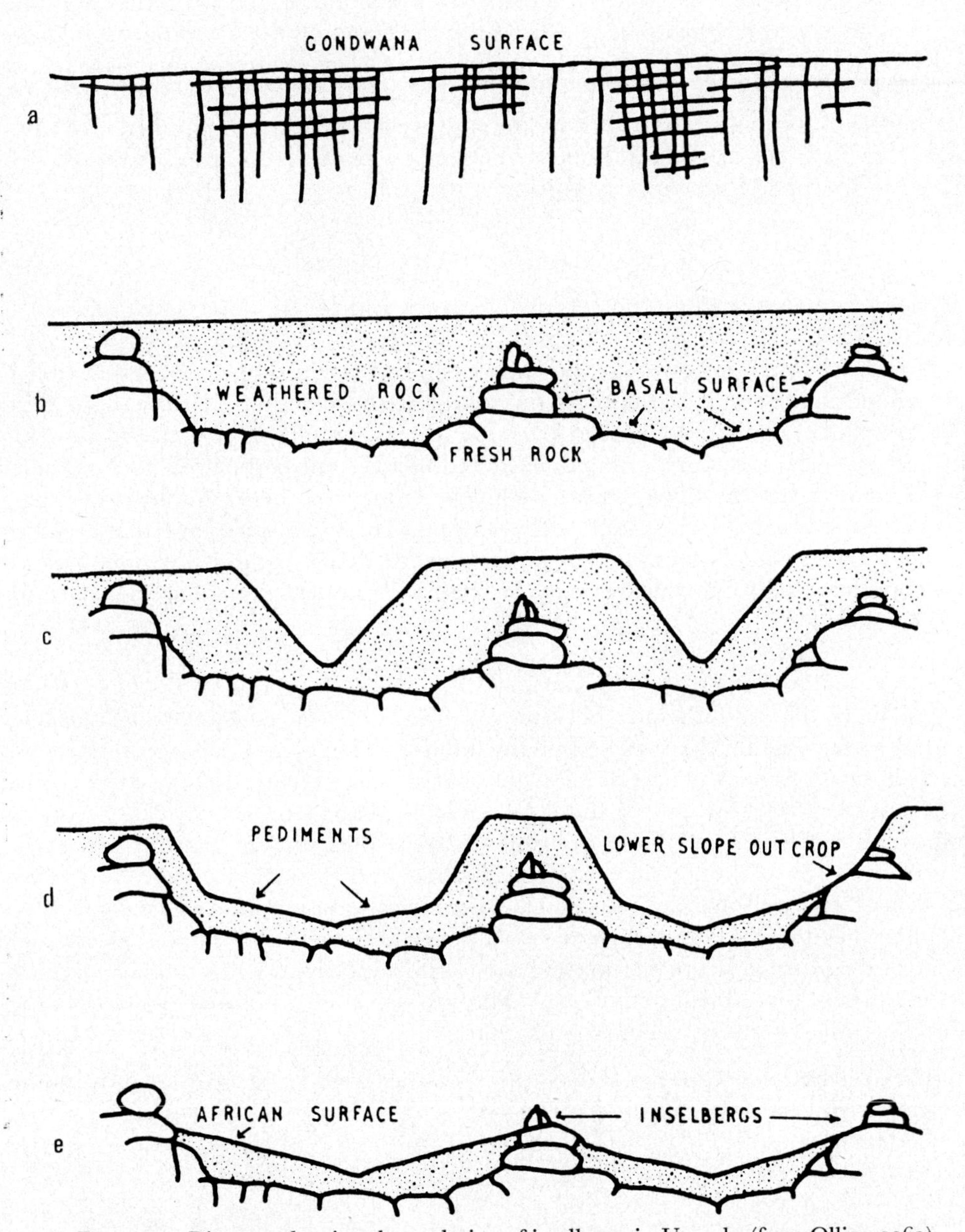

FIG. 117. Diagram showing the evolution of inselbergs in Uganda (from Ollier, 1960).

the mechanism for producing tors of sound rock underlain by very rotted rock with only a few corestones. It could be that both these situations occur in different places.

Thomas (1965) has said that tors are much less common than bornhardts in Nigeria, because they are destroyed relatively quickly in tropical environments. Clearly their abundance is partly controlled by the rate at which they are destroyed by surface weathering, but it is also controlled by the rate at which they are exhumed. In suitable areas they are very abundant, even in the tropics, such as the area of the Buwekula catena in Uganda (Radwanski and Ollier, 1959).

Ruwares

Ruwares are bare rock outcrops of low elevation and gentle slope, low domes of bare bedrock.

Mabbutt (1952), noting that they occur beyond larger domes, suggested that they were the ultimate result of the reduction of domes by sheeting. King (1948) believes they are part of pediments, formed by slope retreat. Büdel appears to be referring to ruwares when he describes 'isolated wash pediments', but he believes they are formed by exhumation, not by slope retreat. Thomas (1965) also believes that many ruwares are formed by exhumation, and believes many may be emergent dome summits, though it seems equally feasible that initially gentle domes could be exhumed too. He also considers that some ruwares may be the final result of the collapse of once much larger domes.

Caves formed by stripping

Occasionally the regolith may be removed from between corestones well below the ground surface, while the surface remains intact. This gives rise to granite caves, an example being Labertouche Cave, Victoria, Australia (Ollier, 1965). Here a surface stream worked its way underground by removing saprolite from between corestones and occupying lower and lower positions. At the present time the upper part of the stream flows into a blind valley, down a sinkhole and through the cave, and comes to the surface at a well-marked resurgence where basal sapping has caused some steepening of the headwall. Points of former resurgence are marked by minor caverns a short distance downstream from the present one, and similarly, former sinkholes are found downstream of the present active one. The cave is about 200 yards long and has a fall of about 50 feet between sinkhole and resurgence.

The concept of stripping can be applied on a larger scale, so that it accounts for the major features of landscapes.

Inselbergs (Bornhardts).

Ollier (1960) suggested that the inselbergs of Uganda were formed in a manner very like the formation of tors according to Linton's theory, but with the preservation of wide areas of regolith between rock outcrops (Fig. 117). To make inselbergs on this scale, weathering would have to be not tens of feet deep but hundreds (which seemed a lot at the time) and it was suggested that some other process may be required to account for those large inselbergs that stand 1000 feet or more above the plains. With our modern ideas of the depth of weathering (reviewed in Chap. X) it now seems feasible that even the largest inselbergs could be formed in this simple way.

Most authors who have considered large inselbergs have thought of more complicated mechanisms, whereby the inselbergs could grow by increments rather than be weathered out in one period and simple exposed by erosion. Falconer (1911) and Willis (1936) proposed that in each of many cycles of erosion and weathering, the inselbergs, being little jointed, suffered little weathering while the surroundings were weathered to a

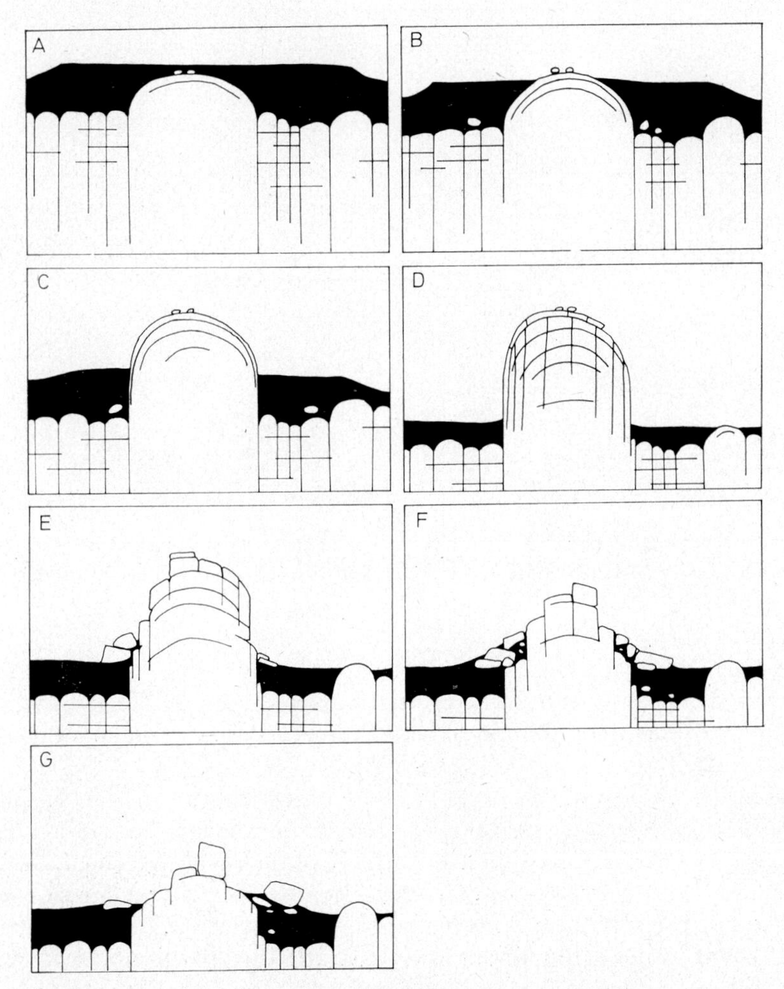

FIG. 118. The evolution of inselbergs (Bornhardts) according to Thomas (1965). Each phase of stripping is accompanied by further lowering of the basal surface of weathering around the bornhardt. In this way, the dome may eventually exceed in height the original depth of the weathering profile.

depth of many feet. In the next cycle the saprolite was eroded, thus increasing the relative height of the inselberg, and weathering again attacked the base. In this manner the inselberg gradually became higher and higher above the plain.

This suggestion has gained some support from later observations. Clayton (1956) described rings of deep weathering around inselbergs and suggested that the bare rock of the inselberg sheds water rapidly so weathers little, whereas its immediate surroundings are flushed with water and so weather with greater speed. Slight erosion of this zone can produce linear or annular depressions around the inselbergs. Twidale (1962) has also suggested that the retention of water in debris at the foot of an inselberg may cause extra weathering.

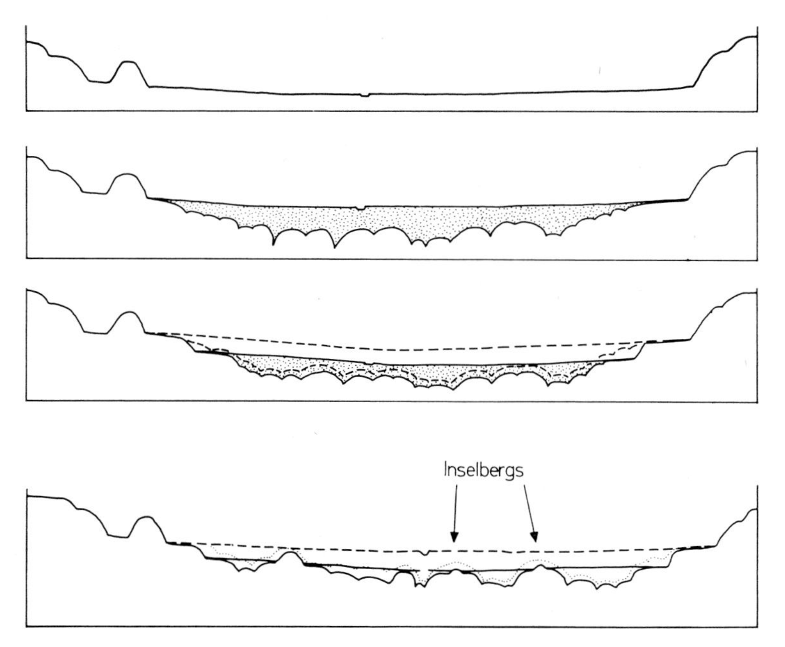

Fig. 119. Tropical deep weathering and stripping according to Büdel (1957).

Some such mechanism of rotting at the base of inselbergs has been supported more recently by Thomas (1965) who says that 'small bornhardts having a relative relief of less than 200 feet could arise from the stripping of the regolith formed during a single prolonged phase of deep weathering. Larger domes which may commonly exceed 500 feet in elevation clearly require a more elaborate explanation.' He believes that deep weathering accompanies exhumation, and illustrates the growth of inselbergs as shown in Fig. 118.

Büdel is another author who invokes deep weathering to form inselbergs, in a rather different manner shown in Fig. 119. The larger inselbergs in front of mountain rims he calls 'outlying inselbergs' (Ausliegerinselberge) and does not attribute them to stripping. Those on the plains, the only part of the landscape where he invokes deep weathering, he called 'shield inselbergs' or 'isolated wash pediments', but from his accounts these are little more than ruwares (see p. 206).

Etchplains

Tors and inselbergs are really minor features on most landscapes; the plains between outcrops cover a much greater area. Some plains are largely created by deep weathering.

Wayland (1933) suggested that plains may be formed by alternate deep weathering and stripping, and termed such features 'etched plains'. He envisaged that a plain would weather 'deeply' (about 10 feet), and in the next erosive cycle the regolith would be stripped to reveal a new bedrock plain, approximately parallel to the original. This etched plain would then be deeply weathered and stripped in turn, and the process might be repeated many times. Minor irregularities in the depth of etching could give rise to tors. If the irregularities persisted through many cycles of etching and erosion they might give rise to inselbergs as in the Willis conception described above.

Wayland's concept of an etched plain (now usually called an etchplain) differs from later ideas on the depth of etching. Ten feet would now be considered very slight, and modern ideas of etching invoke hundreds or even thousands of feet of weathering.

Büdel's 'double surface of erosion' is a kind of etching. He envisages a plain weathering to depths of 30 to 60 metres between episodes of stripping, and so giving rise to lower plains, dotted with ruwares or 'shield inselbergs'. His idea is illustrated in Fig. 119.

Ollier (1965) envisaged irregular deep weathering rather than vast sheets parallel to the surface. The regolith is thought to be in deep pockets separated by underground ribs of unweathered rock. When a stream cuts down in a period of erosion it is slowed when it reaches hard bedrock and lateral erosion cuts minor plains across the regolith behind. Thus the base level of the main streams controls the removal of the regolith, not all of which is removed (Fig. 120). In some areas, indeed, very little regolith is removed and the landscape is carved out of soft regolith, and consists of rolling country with only a few outcrops of bare rock. It is in fact a clay landscape rather than a 'typical' granite landscape of craggy rock outcrops.

Deeply weathered granite may give rise to areas of negative relief if the surrounding rocks do not weather so deeply. The hornfels rim of metamorphism around a plutonic granite often stands up as a ridge around a granite plain, emphasising the difference in susceptibility to deep weathering. Fig. 121 shows a beautiful granite of negative relief. The metamorphic aureole stands several hundred feet above the granite, which at the outlet to the basin is covered by several hundred feet of alluvium, and is rotted to even greater depths.

The Murmungee Basin is an exceptionally large area of negative relief, but smaller plains cut across regolith are common, and are sometimes regarded as puzzling features. Why is the regolith not completely removed by erosion since it is soft? The usual

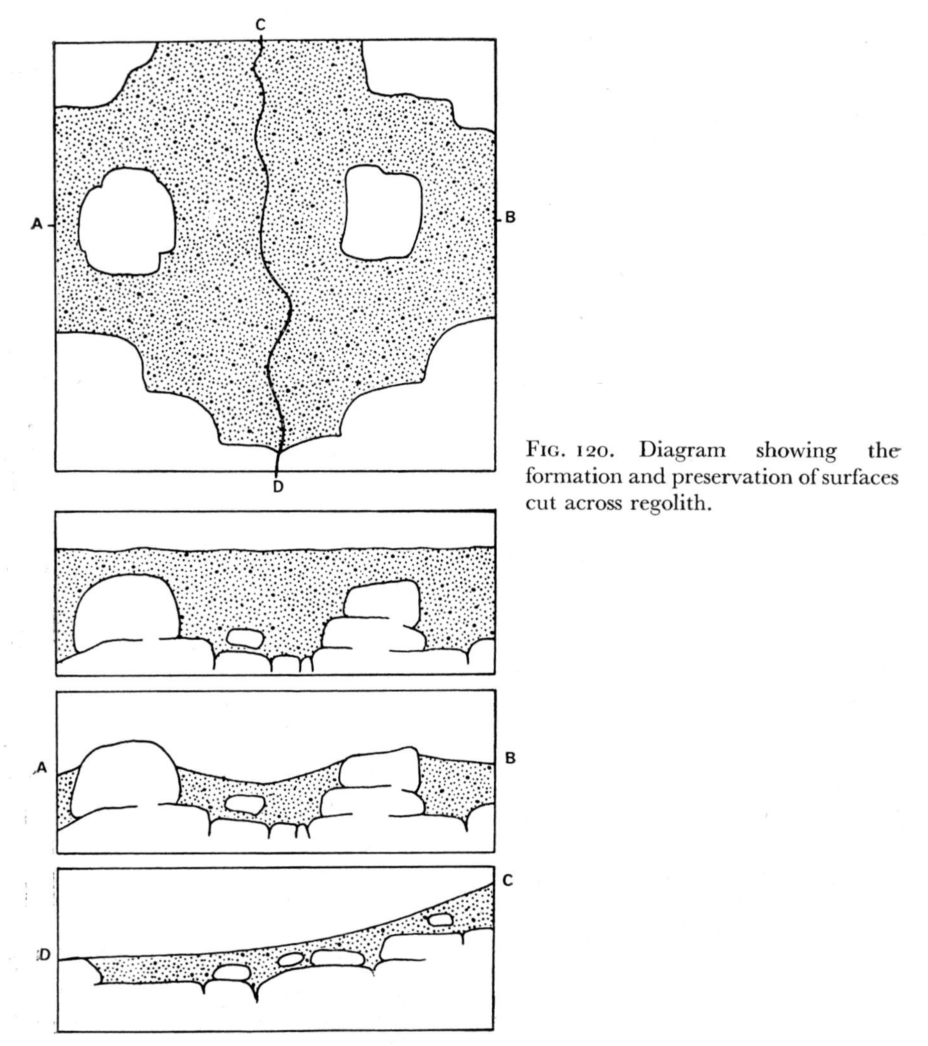

FIG. 120. Diagram showing the formation and preservation of surfaces cut across regolith.

answer is that a downstream bar of hard rock protects the upstream regolith, as shown on Fig. 120.

The rates of deep weathering and landscape formation are variable, and must be considered in the same way that they were considered in relation to inselberg formation. Does deep weathering precede stripping and then virtually stop, or do stripping and deep weathering go on together at similar rates?

It will be noticed that so far we have considered examples of both types. Fig. 117 (p. 205) (Uganda) and Fig. 110 (p. 200) (England) are examples of well-separated phases of weathering and exhumation. Fig. 119 (p. 208) (Tanzania) and Fig. 118 (p. 207) (Nigeria) are examples of weathering continuing during exhumation. Unfortunately Thomas (1966) has begged the question in his definition of etchplain

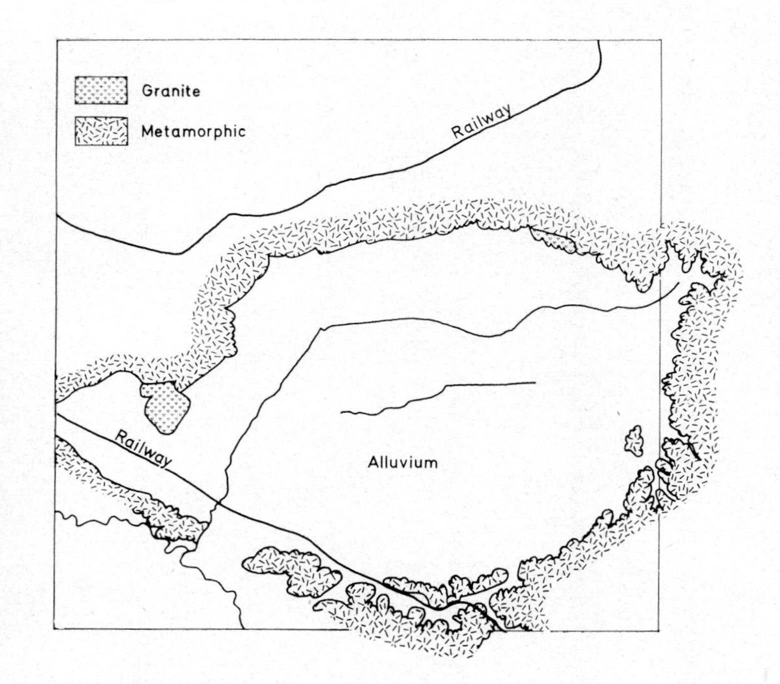

FIG. 121. *Left*: Air photograph of the Murmungee Basin. (Reproduced by courtesy of the National Mapping Authority of Australia from mosaic Photo 792 Zone 7, Beechworth.)

Right: Geological map of the Murmungee Basin shown in the photograph. The basin is in deeply weathered granite and is surrounded by a rim of metamorphosed sedimentary rocks.

below which stipulates the simultaneous lowering of the weathering front and the land surface.

In Wayland's original concept of the etched plain it was assumed that the weathered material was completely stripped in the following erosive phase, just as grooves are eaten away in the process of etching used by artists. Can we use etched plain, or etchplain for those plains where the regolith is only partially removed? I believe not, and suggest the following usage:

1. Deeply weathered plain—no other term needed.
2. Partial etchplain—a plain with partially stripped regolith, as in Fig. 117, p. 205.
3. Etchplain—a plain from which regolith is virtually all stripped, as in Wayland's concept of etched plain (e.g. Fig. 115d).
4. Complex etchplain—a plain on which regolith production and stripping are contemporaneous and at comparable rates.

The usage is different from that suggested by Thomas (1965) which was:

1. Etchplain (*sensu stricto*)—plains formed by the simultaneous reduction of the weathering front and the landsurface.
2. Dissected etchplains—deeply weathered plains dissected by streams which expose little of the weathering front.
3. Partially stripped etchplains—stripping of regolith from upstanding portions of the basal surface, hills and inselbergs emerge, but basins of weathering may become still deeper.
4. Dominantly stripped etchplains—even the lower portions of the basal surface become exposed except for a few deep pockets.
5. Incised etchplains—etchplains in which the solid rock profiles become widely affected by stream erosion.

On Fig. 117 the valleys are placed conveniently above the deepest part of the regolith, but there is no reason in reality why this should happen, and indeed it could only happen in special circumstances. The rivers flowing on the surface of a deeply weathered plain follow a consistent pattern of downhill slopes, but as shown on p. 205 the basal surface of weathering is very irregular. As a river cuts down it becomes superimposed on the underlying material. If it meets a hard mass of rock with a sloping surface, the river may swing laterally, as in uniclinal shifting, to a position where the regolith is deeper (Fig. 122). Alternatively it may simply become superimposed on to the hard rock.

FRAGMENTED ROCKS AND RESIDUAL DEPOSITS

The particles and deposits of particles produced by weathering are themselves features of landforms and landscapes just as much as the sculptured rock faces from which they are derived. The mode of formation of these fragments and deposits has already been discussed in various places in the book, and here they are merely brought together to emphasise similarities and differences.

Insoluble residues

These result from solution, especially of limestone. Commonly they consist of clay, but some limestones have quite a lot of sand.

Granular disintegration sand

Granular disintegration is especially prevalent on coarse-grained rocks, most notably on granite. Both mechanical and chemical weathering may be responsible for its

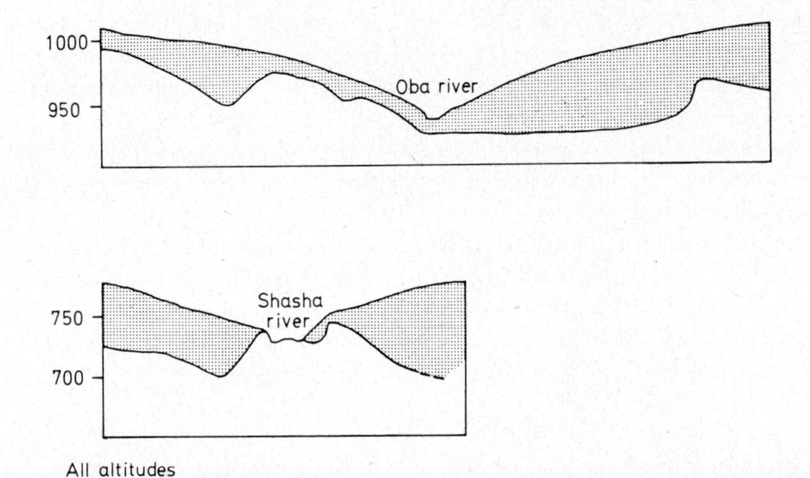

FIG. 122. Cross sections of the Oba River and Shasha River, Western Nigeria. Regolith (dotted) over Precambrian gneiss (from Thomas, 1966).

production. Essentially, the individual mineral grains, though often little weathered themselves, become only loosely attached to other grains, and can be rubbed off the weathering rock by hand. The grains accumulate around weathering outcrops as granular disintegration sand.

Weathering of granite may produce a gritty or sandy weathered saprolite. This has not accumulated from fallen fragments like granular disintegration sand, but is *in situ*. Such a deposit could be called *grus*, though the term is sometimes applied to granular disintegration sand or redistributed regolith.

Loess

Loess is a deposit made up of fragments, mainly quartz, derived by weathering. Some authorities believe that frost weathering or glacial erosion is important in producing fragments of the size present (see p. 24), and others consider loess to be entirely formed by weathering, and discount the aeolian hypothesis completely (Berg, 1964).

Rock fragments

Rock fragments of many kinds are produced by weathering. Some may be distinguished genetically, such as frost riven fragments which are very angular; fire-cracked boulders, which are broken rounded pebbles with one or more straight faces; flakes produced by the various flaking processes described on p. 186; blocks formed by the disintegration of unloading sheets; fragments formed by insolation weathering, which are usually angular, though they may be remnants of originally rounded boulders; Kernsprünge are large boulders, usually of granite that have been split into a small number of large fragments; corestones; also known as 'niggerheads' or 'woolsacks', are rock fragments produced by subsurface weathering.

Rock fragments may also be classified by shape, though the rounded fragments have either been abraded, as in the case of beach pebbles or river gravels, or produced by spheroidal weathering or flaking.

Most angular rock fragments are produced by splitting of larger fragments, though windkanter are produced by wind abrasion. The shape of angular fragments depends on the planes of fissility in the parent rock to some extent. Joint blocks, platy fragments, or even extreme forms such as 'pencil shales' may be produced from suitable rocks.

Corestones

Often associated with tors are occasional loose corestones known as 'woolsacks', 'niggerheads' or by other names. They are frequently perched in improbable positions, and may give rise to rocking stones. The 'Devil's Marbles' of Central Australia are well-rounded perched corestones left by stripping, and occur in association with tors (Fig. 99).

Some landforms are made by the accumulation of weathered blocks, boulders and fragments.

Boulderfields (Felsenmeere) are areas of accumulation of corestones, left behind after sheet erosion carries away the fine-grained and friable decomposition products from an area of deep weathering with corestones.

Boulderstreams are similar features formed on steeper slopes. These may be formed rapidly, and Fezer (quoted in Wilhelmy, 1958) stated that new block streams were produced in the northern Black Forest after 1945 as a result of heavy wartime timber cutting on steep slopes. Of course the formation of the waste mantle with its included corestones would have taken very much longer.

Blockfields are accumulations of angular blocks due to frost shattering, and quite distinct (in theory at least) from boulderfields of rounded corestones. Blockfields occur in high latitudes and in mountainous regions. Blockstreams (stone rivers), also have sharp edged and angular blocks, and occur in the same areas as blockfields.

Another form of accumulation of blocks is the *rock-glacier* (Blockgletscher) known from many glaciated areas. This may have been originally a debris cover over a glacier, but the blocks may also move by freeze-thaw or other mechanisms.

Complicated situations can occur when both corestone formation and frost action occur in the same area. Matsumoto (1964) described landforms of accumulated boulders in the Abukuma and Kitakami mountains of Japan where such a situation is present. 'Blockfields' of hard, fresh, but rounded boulders 1 to 3 m in diameter, were present on slopes of 5° to 15°. They were originally corestones due to preglacial deep weathering, but reached their present position by periglacial action.

In Europe there is considerable debate as to what extent the Felsenmeere indicate glacial or periglacial conditions, and to what extent they are corestones indicating warmer conditions.

Masses of corestones may occasionally resemble moraine material especially the block moraine or bear-den type of moraine (Geschiebewälle) and can occasionally cause confusion. Some alleged moraines in Victoria, Australia, were later shown by Beavis (1959) to be corestone remains, which fortunately had been sufficiently undisturbed to reveal joint and other structural patterns consistent with that of neighbouring rock, and included a weathered dyke.

Talus and scree slopes
(shingle slips in New Zealand)

Talus and scree are virtually interchangeable, though talus originally meant the slope rather than the material, and there is a tendency to use scree more specifically for debris derived by frost action and talus for waste produced by any means so long as it rests at the foot of a cliff or steep slope.

The surface of scree slopes is at the angle of repose for the fragments that make up the scree; the larger the fragments the steeper the slope. Scree slopes are often almost straight. Some scree is in more or less continuous sheets along valleysides; some is in the form of cones or fans, derived from funnel-like gullies called chimneys or couloirs. Scree and talus surfaces are generally mobile, and there is no soil formation or colonisation by vegetation while weathering continues unabated.

In many areas screes are not active but are relic formations resulting from periglacial conditions in past glacial or periglacial periods. For instance in Wales, Ball (1966) says that many screes are now covered by soil and vegetation, though others have been exposed by erosion and are now subject to resorting and accretion.

A deposit similar to layered scree may be deposited in nivation cirques. Watson (1966) described nivation cirques from Wales, with a *drift platform* up to 50 ft thick, consisting of beds of stony layers and solifluction deposits with a clay matrix. It is believed that the debris is produced by freeze-thaw and moves underneath the snow to the lower part of the snow patch.

Screes can sometimes form from rocks that are more commonly attacked chemically, and also in arid regions. Hey (1963) reports two limestone screes from Libya formed by frost action. The older screes are recemented (presumably indicating a slight change in climate allowing solution and reprecipitation of carbonate), but later screes remain uncemented.

Periglacial deposits formed by weathering and mass movement may sometimes be difficult to distinguish from true glacial deposits. Tricart and Cailleux (1962) maintain that snow patch deposits may be differentiated from true moraines by the character of the deposit, which consists of only coarse material without much clay, and has generally more angular blocks. Watson and Watson (1967) used the orientation of stones to distinguish glacial from periglacial deposits: in ground moraine the preferred stone orientation is parallel to the direction of glacier flow; in solifluction deposits the preferred stone orientation is parallel to the slope on which movement took place.

RESIDUAL DEPOSITS OF ECONOMIC IMPORTANCE

A number of weathering residues are of economic importance as mineral deposits, and have therefore been studied in greater detail than most residues, providing much useful information on processes of weathering.

Residual clay deposits

Clay formations result from normal weathering processes on a variety of rocks, but the chief source-rocks of economic clays are massive igneous rocks, especially the more silica-rich ones. Clay deposits have roughly the same form as the original rock: many of the kaolin deposits of the southern U.S.A. are derived from pegmatite dykes, and occur as dykelike deposits up to 300 feet wide and 120 feet deep. Residual clay deposits are rarer than sedimentary clays, and are generally absent from areas that have suffered glaciation.

Residual iron ore deposits

Laterites provide most of the iron ores formed by weathering. Extensive deposits occur in Cuba, where they rest on plateaus cut across serpentine, and are probably of Tertiary age. Similar deposits are found in Colombia, Venezuela and the Guianas. Nonlateritic enrichment in iron gives rise to the iron ores of Bilbao, Spain, where a blanket of iron oxides has been derived from the weathering of a limestone rich in iron carbonate. The brown ores of northern Texas are weathering residues, partly in greensand, on Eocene beds. Other brown ores are found in Virginia, usually as scattered lumps in clay, but occasionally in masses several hundred feet deep.

Bauxite

Bauxite is formed in tropical areas where the temperature is over 25°C most of the time, and where abundant water leaches porous rock rich in aluminium. It is formed on both igneous and metamorphic rocks, and on limestones and clays rich in aluminium, and is associated with land of low relief, but where groundwater movement is able to leach silica away. The optimum conditions for the formation of bauxite apparently existed between Middle Cretaceous and Eocene, when most of the world's commercial deposits were formed.

Bauxite occurs as blanket deposits, always associated with flat erosion surfaces, as in British Guiana, Ghana and northern Queensland, or can occur as irregular pockets or masses, especially on limestone or dolomite. Half of the world's aluminium production comes from deposits associated with limestone. Bauxite can be eroded and redeposited as detrital bauxite, but most of the so-called interstratified bauxite deposits are old blanket deposits that have been buried beneath later sediments, and invariably occupy unconformities (Bateman, 1950).

Bauxite can be produced from the desilicification of feldspars or other rock minerals, or from kaolin. Numerous examples are given by Harder (1949) of fresh rock altering directly to bauxite with a knife-edge contact (weathering front) and no intermediate stage of kaolinite formation is detected.

The great difference between bauxite and laterite is that silica is removed in the case of bauxite and retained in laterite, so that bauxite is characterised by gibbsite (or other aluminium oxides) and laterite has kaolinite clay minerals. Silica retention is aided by the presence of organic matter, by strong acids, by drought, and by groundwater conditions that prevent removal of silica. These conditions must be absent for bauxite formation. A bauxite can also be re-silicified to kaolinite by silica introduced at a later time.

Gibbsite is probably the first formed Al mineral in a bauxite deposit. This ages into boehmite, and eventually into diaspore. All stages are found in bauxites, but it is interesting that dominantly diaspore rocks are confined to Palaeozoic rocks, dominantly boehmite deposits to Mesozoic rocks, and Tertiary deposits are principally made of gibbsite. This change represents long-term ageing, which may be regarded as either weathering or diagenesis, and does not represent a change in weathering style through time.

Residual nickel deposits

New Caledonia was once the world's foremost producer of nickel, which is found there in lateritic profiles on peridotite and serpentine. Nickel is evenly distributed in the parent rock, and has been concentrated by weathering up to thirty times its original content. The upper part of the laterite profile has iron oxides, and the nickel is concentrated in a zone 1 to 20 feet below the surface; exceptionally the cover reaches a thickness of 100 feet. The best ores are found where weathering has been most effective, and are concentrated near the top of the weathered rock and below the laterite. The laterite in New Caledonia does not appear to have formed on such a perfect plain as is usual, and the best nickel concentrations are on gentle slopes, saddles and spurs and not beneath the flattest areas.

Fig. 123 shows the zones of weathering in the New Caledonian nickel-bearing profiles, which are very like the normal weathering profiles described in Chap. X.

Other economic residual deposits

The formation of residual manganese deposits is very similar to that of iron deposits, but there is much more solution and re-deposition of the manganese minerals. Limestone and schist enriched in manganese are the commonest source rocks, and a warm

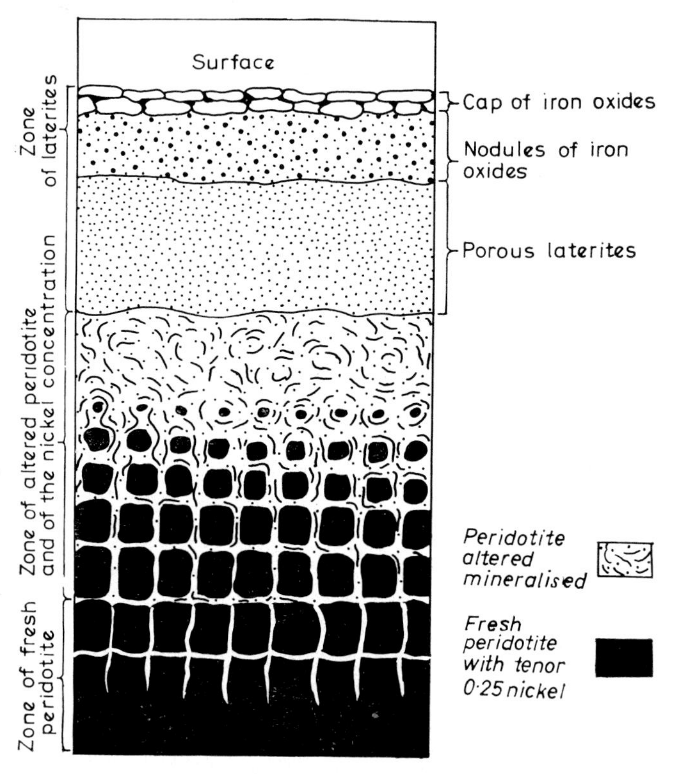

Fig. 123. Typical section through nickeliferous laterite, New Caledonia.

though not necessarily tropical, climate seems to be required. Residual manganese deposits are found in many non-glaciated areas, and large deposits are known in India, Ghana, Brazil and Morocco.

Ochre and other pigments are produced by weathering, and consist mainly of limonite, hematite, manganese oxides and clay.

Black oxides of cobalt are found in residual accumulations over some copper ore deposits in Katanga.

In Virginia and Tennessee mineable masses of zinc ore have accumulated in residual clay over a shaly dolomite. The ores occur next to the limestone at the bottom of the clay (Fig. 124).

Much barite in the U.S.A. is derived from residual deposits.

LIMESTONE AND SOLUTIONAL LANDFORMS

Solution gives rise to many landforms both large and small, especially in limestone country where the term 'karst' is applied to the special features of relief and drainage. Unfortunately the terminology applied to these forms is vast and complicated, because almost every country or region with limestone scenery evolves its own landform names, and many words from many languages have been used as technical terms. What follows is only a brief summary of limestone (and other solutional) landforms.

Solution tends to make originally plane surface irregular by producing pits and grooves. When limestone dissolves very irregularly, *spongeworks* or *boxworks* may be

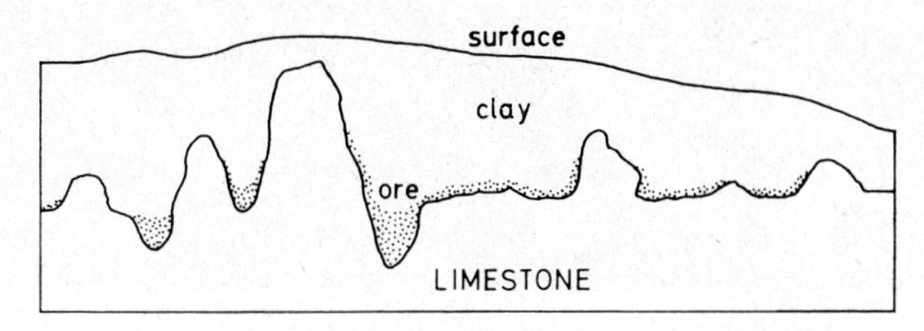

FIG. 124. Section in open cut at Bertha zinc mines, Virginia.

produced by differential weathering, at the surface or in caves. Trickling films of water can give rise to small concave hollows an inch or so across, producing a '*cockled*' surface. Frequently the hollows are arranged in fairly distinct horizontal rows to give *ripple patterns*.

Where a stream flows over limestone, and especially in active caves, *asymmetrical scallops* may be produced. These are from about 1 inch to 20 inches in length, aligned in the direction of water flow, and concave with the steepest side upstream (Fig. 125). In some caves, possibly where water is more stagnant, *symmetrical scallops* are made by solution. These are spherical hollow surfaces, a foot or so across.

Karren (German) or *lapiés* (French) are irregular to elongate hollows formed by solution. Bögli (1960) has classified Karren into many varieties, of which the most significant are probably Rundkarren and Rillenkarren. The former have rounded divides between the grooves, and some are possibly formed beneath a cover of soil, peat or snow; the latter have sharp divides and are due to running water (Figs. 126 and 127). Rillenkarren are especially common on steep slopes, but in English the term fluting is often used for the solution grooves found on vertical or very steep limestone surfaces. Other types of Karren in Bögli's classification are listed below.

Kluftkarren—grikes or large solutional slots;
Rinnenkarren—(drain grooves) solution grooves that widen at depth;
Regenrinnenkarren—(rain furrows) broader and straighter kind of Rinnenkarren;
Trittkarren—flat-bottomed, step-like karren;

Fig. 125. Asymmetrical scallops (from Moore and Nicholas, 1964).

Fig. 126. Rundkarren; Mole Creek, Tasmania. Soil and litter being removed by recent deforestation and grazing. (Photograph by J. N. Jennings.)

FIG. 127. Rillenkarren and swamp undercut. (Photograph by G. E. Wilford.)

FIG. 128. Pseudokarren. Solution grooves on granodiorite at Granite Rock, Victoria. (Photograph by A. A. Baker.)

Maanderkarren—solution grooves with meanders;
Wandkarren—solution grooves on vertical walls;
Hohlkarren—karren in caves;
Deckenkarren—anastomosing channels on cave roofs;
Spitzkarren—steep grooves between needle-like spires;
Schichtfugenkarren—ribs of rock corresponding to particular beds.

Solution grooves on granite and occasionally on other rocks, not generally regarded as soluble, are known as *pseudokarren* (Fig. 128).

Bare, horizontal limestone surfaces are called limestone pavements. Solution frequently dissolves out the joint pattern on pavements, and the resulting grooves or trenches, which are commonly about 4 to 8 feet deep and a foot or so across, are called *grikes*. The flat areas between grikes are called *clints*.

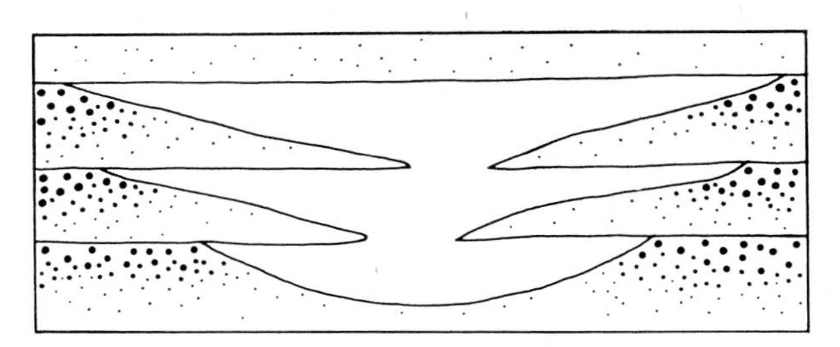

FIG. 129. Solution shelves in cave passage, County Clare, Ireland.

In limestone country various depressions in the ground surface are due to solution. A surface stream may enter a cave through a *swallow hole* or *swallet*. Active swallets commonly have former swallets further along the line of the valley. These, and other scattered depressions that lead to caves, are called *sink holes*. The term *doline* is often used for small depressions. Larger depressions, sometimes formed by the coalescence of several dolines, may be called *uvalas*. Very large depressions, several miles across, are called *poljes*. These are flat bottomed and alluviated, but often completely surrounded by steep cliffs, and are the most impressive testimony to the power of solution.

Marked lateral solution of limestone is often seen at the ground surface, where a plain or swamp meets a limestone cliff. Solution at the edges of swamps seems to be especially common in tropical areas, where *swamp undercuts* are produced (Fig. 127), often at several heights indicating changes in level of the swamp surface.

When water flows underground, in cave passages, it can dissolve the rock on all wetted surfaces. In a completely filled passage in uniform rock solution may be the same in all directions, and a circular section develops to the passage, which becomes a tube.

Commonly, however, the limestone is not equally soluble in all directions, and lithological or structural differences are etched out by differential solution. In County Clare, Ireland, some limestone beds have coarser-grained limestone at the base grading into finer limestone at the top. The coarser limestone is more easily dissolved, and cave passages develop *solution shelves*, which may be many feet wide and only a few inches thick (Fig. 129).

Fig. 130 shows a splendid example of differential weathering in two sorts of limestone. The Senonian limestone has well marked fissility and variation in solubility, and produced an *angular cave passage*: the homogeneous Urgonian limestone develops a *tube* form, in this case half a tube.

A *half tube* is a technical term for another feature which has a horizontal base and arched roof. Half tubes vary in size from a fraction of an inch to several feet across. To make the arcuate roof of a half tube the passage needs to be full of water, giving equal solution in all directions, except of course downwards. It is thought that insoluble residues accumulate on the floor of the tube and prevent attack on the underlying limestone.

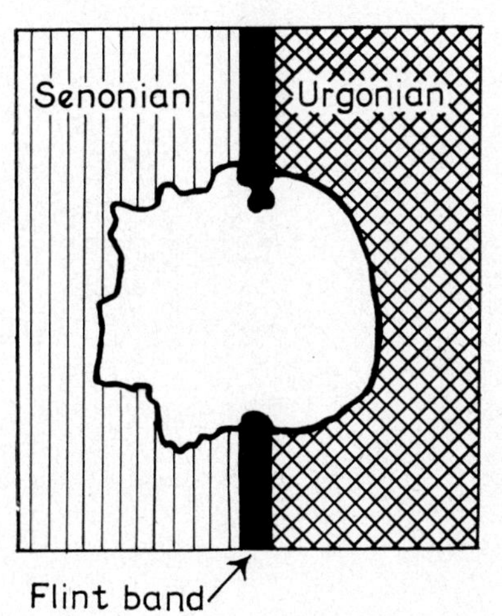

FIG. 130. Cross section of cave passage in "Trou qui souffle", at the junction of bedded Senonian and compact Urgonian limestone (from Gèze, 1965).

When water first seeps along a bedding plane it commonly dissolves out an anastomosing pattern of small half tubes an inch or so across. In caves with bedding plane roofs these are seen as *bedding plane anastomoses*.

As such half tubes enlarge, the limestone between is removed until all that are left are pointed or blade shaped projections on the roof of the cave, known as *pendants*.

Free flowing streams in caves (*vadose streams*), even when carrying considerable debris, probably accomplish most of their excavation of caves by solution. This is sometimes indicated by bridges of insoluble rock such as chert that completely traverse a cave passage. It would be impossible for a stream to corrode its passage without cutting through the insoluble band, but solution can go around such obstacles.

Caves themselves are solutional features, and their shapes reveal much about the nature of limestone solution. Most caves have a marked horizontal development, even in steeply dipping or folded limestone. This suggests that their position, and solution of limestone, follows the water table in limestone country. In plan many caves have a clearly joint-controlled pattern (Fig. 131) showing that joints control the course of solution in the horizontal direction.

Some caves, or parts of caves, have been very largely excavated in the vadose zone, by streams with free air surface. These have plans more like normal river systems, with consistent downhill gradients, and their passages are marked by meanders, asymmetrical scallops, and a general lack of phreatic features. Other caves were formed in the phreatic zone below the water table. These may be distinguished by lack of regular gradients, tubes, spongeworks, blind passages and symmetrical scallops. Many caves

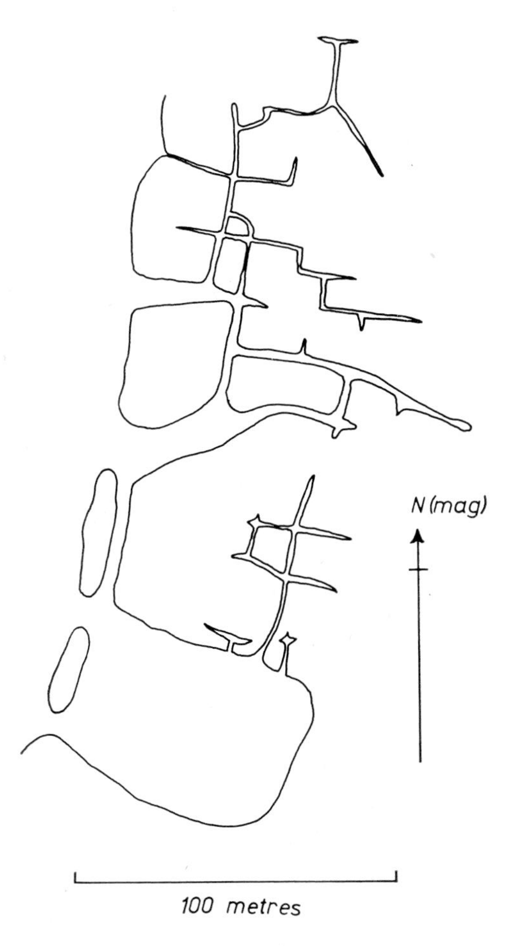

FIG. 131. Plan of Matupi cave, Mont Hoyo, Eastern Congo (from Ollier and Harrop, 1964).

have phreatic features in detail, but their general near horizontal development and sometimes regular gradients suggest they were formed just below the water table as shallow phreatic caves.

As well as near level passages there are vertical features in caves.

One notable vertical feature is the *domepit*. Domepits are vertical shafts, usually between 5 and 25 feet in diameter and a hundred or more feet high. They have fluted walls due to solution by trickling water. Domepits are generally situated below valley heads or sink holes on the surface, but have no constant relationship to nearby cave passages. They often extend high above cave passages, and far below the floor, and sometimes only a narrow opening connects a domepit to a passage it almost missed. Domepits are thought to be enlarged fissures or joint intersections down which water moves directly from the surface to the water table, regardless of the presence of higher cave passages.

Another vertical feature for which no very satisfactory explanation has yet been given is the *bell hole* (Wilford, 1966). Bell holes grow in cave roofs, sometimes in groups; they are circular in cross-section and vertically grooved, a few feet high, and terminate quite definitely in hard, unfissured bedrock. They clearly have no direct connection with water from the surface, and are possibly phreatic features caused by turbulence at the cave roof.

Solution of limestone underground may undermine the overlying strata, leading to collapse. Some dolines are *collapse dolines* rather than *solution dolines*. Collapses into caves tend to be circular or elliptical in plan. *Cenotes* are round steep-sided lakes in limestone country, those of Yucatan being the most famous. Some cenotes are roughly cylindrical, but others appear to flare outwards at depth. The openings are irregularly scattered, and as they may be found on hilltops, mid-slopes or in valleys, they are thought to be caused by collapse into large underground caverns.

Solution of limestone produces an insoluble residue that may be retained on the ground surface. This does not seem to impede further solution, indeed if the residue becomes acid by biotic activity it may well enhance solution. The base of soil on limestone is usually very irregular, and the limestone surface has a series of depressions and ridges like the Karren seen on bare limestone surfaces.

Besides Karren, more spectacular depressions may form, including *solution pipes*. These are commonly circular in cross-section and have vertical sides, and when excavated they closely resemble man-made wells. Some appear to be formed at joint intersections, others are apparently in solid rock with no guiding fissures. Some are occupied by long tree roots, but whether these cause, or merely follow, the solution pipe is not clear. Solution pipes are normally full of soil; often the lower part is insoluble residue from the limestone, the rest is largely soil that has fallen in from above. Solution pipes may be found on hard limestone, but the best ones seem to form on soft, porous limestones.

Fig. 132 shows another feature that superficially resembles a solution pipe, but is called a *soil tongue* or *solution tongue* (Yehle, 1954). Solution tongues are only found in calcareous gravels. They are made by solution dissolving the carbonate, and the space being filled by insoluble residue and some material from above. In many gravels there are distinct pebble lines and where these cross the tongue they sag. Some solution tongues are circular in cross-section, and virtually the same as solution pipes, but others are wall-like, and the walls make a polygonal pattern like that associated with ice wedges. The patterned ground is in fact indistinguishable from that of ice wedges, but the more rounded base of the tongue, the sagging lines of pebbles (if present), and the lack of pressure features in neighbouring material may be used to distinguish the two features.

Limestone landforms are very much affected by lithological control, and most distinctive limestone landforms are developed best in hard limestone of low porosity but with well developed joints. Soft limestone such as chalk and calcarenite dissolve very evenly and give rise to gently rounded landforms.

Climate is another marked control on limestone landforms, though Jennings and

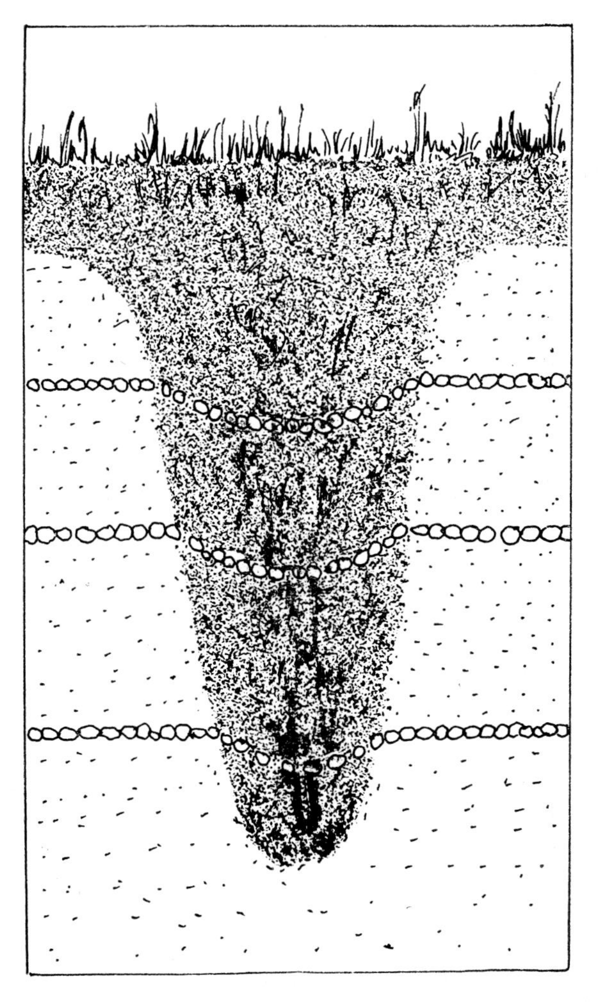

FIG. 132. Solution tongue. Note marker bands of gravel can be traced across the tongue.

Bik (1962) have shown that some of the landforms commonly attributed to different climates can occur in close proximity.

In general, tropical areas produce tower karst, with steep limestone hills rising above surrounding flat plains—*karst border plains*—or separated from each other by steep-sided, flat-bottomed *corridors* like giant grikes. In temperate regions doline karst is dominant, where depressions of various kinds are sunk below the general level of the limestone surface.

SLOPES

Landscapes are very largely made up of hillslopes—the slopes joining summits to drainage lines. The form of slope depends on the processes of erosion, mass movement and weathering, and often it is not easy to draw a line between these three factors. Because of this, although our concern here is with the effect and importance of weathering on slopes, some general features of slope development will have to be discussed.

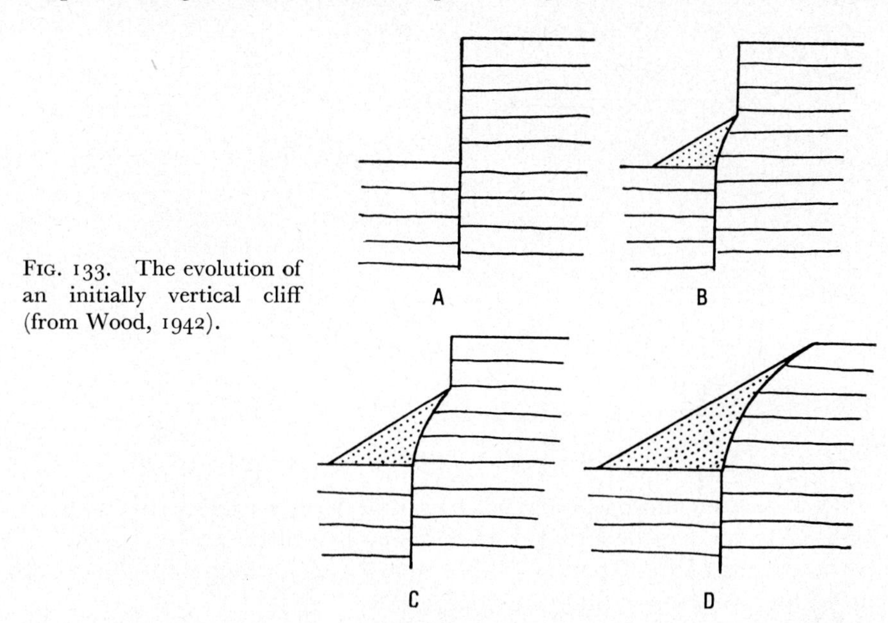

FIG. 133. The evolution of an initially vertical cliff (from Wood, 1942).

One of the most significant analyses of slope development was presented by Wood (1942). He considered the evolution of an initially vertical cliff, which was assumed to be homogeneous, and which produced debris by free fall of fragments from the cliff face. It was assumed also that the talus heap was not weathered and had the same volume as the parent rock. When scree forms it protects the base of the rock face from weathering, and as the face above retreats, adding to the scree as it goes, a curved rock face is produced beneath the scree, and finally the buried face becomes tangential to the scree (Fig. 133).

Such slopes are seldom seen in nature, but the concept of the interrelationship between weathering, protection and slope development proved most valuable.

SLOPE ELEMENTS

Wood proposed a nomenclature for the different parts of a slope which has since been built on and modified by many writers, including King (1957) who proposed the four basic units shown in Fig. 134.

Different processes operate on different parts of slopes, and it will be convenient to discuss slope units or elements in turn.

Waxing slope

The waxing slope is the convex crest of a hill or scarp, and is perhaps the most difficult of all slope elements to account for satisfactorily. Most authorities believe creep is the dominant agent in its formation. Birot (1949) has suggested that debris formed at the

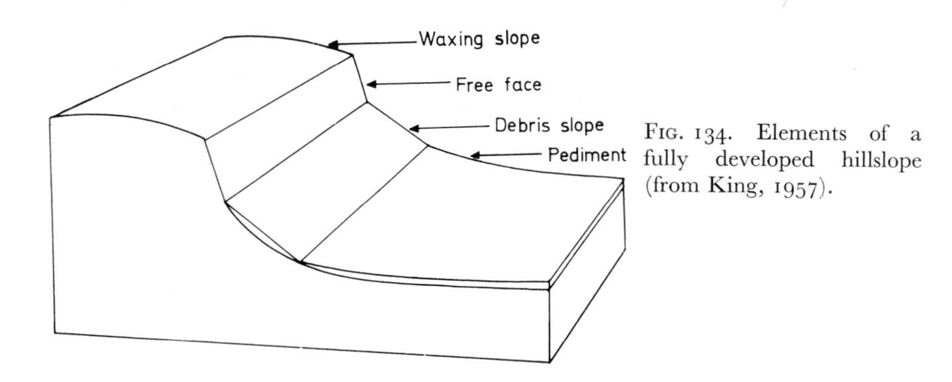

FIG. 134. Elements of a fully developed hillslope (from King, 1957).

top of a slope moves downhill, where it protects underlying rock from further weathering and erosion. Meanwhile the exposed upper slope is subjected to further weathering, and a process of constant re-exposure of fresh rock is a factor in the rounding of hilltops, as the top will weather faster than the slope below.

Sparks (1960) has pointed out some limitations to thinking of slopes in two dimensions. Many slopes are either convex in plan, as at the end of a spur, or concave as at the head of a valley. Material moving down a slope convex in plan would tend to spread, and each point of the slope would be passed by less debris than it would on a slope straight in plan, and the profile should become less convex, which does not happen in nature. The opposite would be true for a slope concave in plan, where converging material should cause greater convexity. However, spreading material might expose the underlying rock to greater weathering and offset the decrease in slope. Also, where material converges the thicker cover of debris could protect the bedrock from weathering and thereby decrease the slope (Fig. 135).

The free face

The free face is the outcrop of bare bedrock exposed below the waxing slope. It is the most active element in backwearing of the slope as a whole. Weathered products are rapidly removed, which encourages further weathering, and so on. Weathering is rapid, but weathering products do not accumulate.

The debris slope

The debris slope is covered with detritus derived from the free face above. Debris on the upper part of the debris slope is frequently coarse, and rests at its angle of repose, if only temporarily, for it is a mobile slope and debris moves down the slope. The debris frequently undergoes further weathering, which comminutes it and may increase its clay content, which in turn reduces its permeability, giving rise to numerous complications in further slope evolution. On this kind of slope there is usually a reduction in particle size with distance downslope.

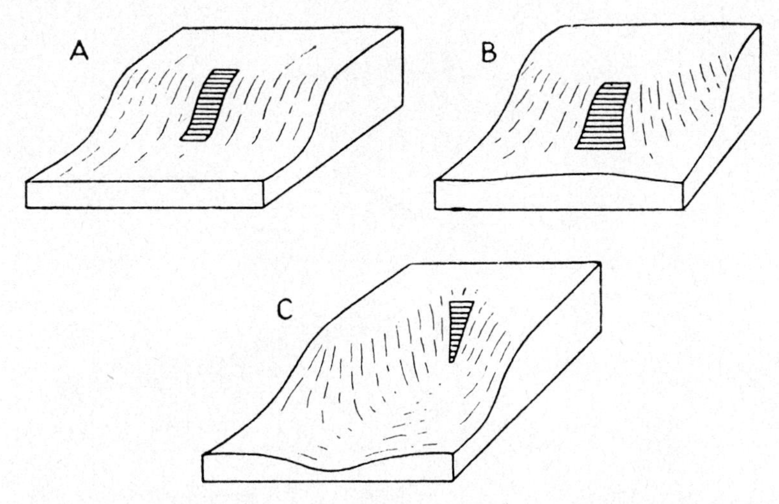

FIG. 135. Tracks available for movement of waste (from Sparks, 1960). *a* Slope straight in plan. *b* Slope convex in plan. *c* Slope concave in plan.

Where comminution is of little importance, as in scree slopes, a bedrock slope may be formed at such an angle that scree can just traverse the slope, that is at about the angle of rest of the scree. Accumulation of scree at the top of the slope increases the angle above the angle of rest, and so scree moves down. This only happens if scree does not accumulate at the bottom. Such a slope has been known as 'Richter's slope of denudation' (Richter, 1901; Scheidegger, 1961).

To cite a particular example, Twidale (1967) describes hill slopes of the Flinders Ranges, Australia, and writes, 'Where debris is efficiently evacuated, and there is therefore no significant accumulation, debris slopes on any given rock type have remarkably uniform inclination. This is the inclination just sufficient to permit the available debris to maintain progress downslope, and is an equilibrium angle.'

Bakker and Le Heux (1952) enunciated a 'geomorphological law' of such slopes, as follows:

> 'In every type of weathering-removal recession of steep walls, a denudation slope with a rectilinear cross-profile and a constant slope angle equalling the

scree's angle in nature will be formed, provided no, or hardly any, scree is deposited on the terrace at the foot of the initial wall.'

The piedmont angle

Sometimes, especially in arid areas, there is a distinct break of slope at the upper end of the pediment slope, and the origin of this piedmont angle is controversial. Twidale (1967) believes that weathering followed by erosion of weathered material is an important factor. His hypothesis is shown diagrammatically in Fig. 136. Runoff from the often bare hillslopes percolates into the rock where it reaches the plain, causing intense weathering. The subsurface weathering permits scarp foot erosion and stream incision, the lower slopes are dissected and isolated, the new plain level extends headwards, and the piedmont angle sharpened.

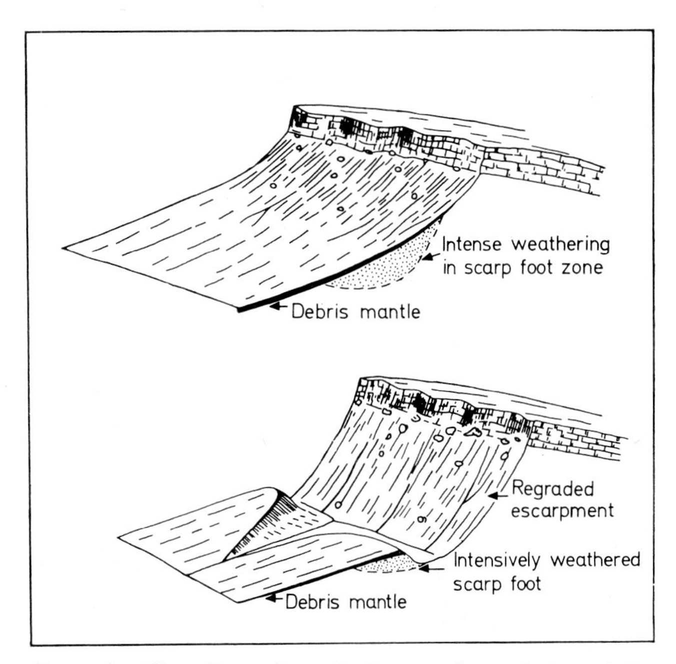

FIG. 136. The effect of weathering on the evolution of the piedmont angle, according to Twidale, 1967.

The pediment

The pediment is a very controversial landform in many ways, and the term is used here simply as the basal slope element, in the sense of King (1957). The French restrict the use of pediment to slopes on hard rock (using glacis for similar slopes on soft rock), and pediments are also frequently regarded as confined to the arid zone, but such limitations are not implied here.

Very commonly the pediments of arid regions are made of hard, fresh rock, where weathering is slight. However, in other climates analogous pediments may be deeply weathered.

An interesting aspect of weathering on pediments has been described by Mabbutt (1966) from Central Australia. Granitic pediments are mainly sub-alluvial, but bedrock is roughly levelled although there is no lateral erosion by streams. Such pediments are levelled by subsoil weathering beneath the mantle. In this way the levelness of the alluvium is transmitted to the bedrock floor. The process is said to be most active near the hill foot, where alluvial mantling and stripping alternate frequently.

SLOPE DEVELOPMENT

Many models of slope development have been proposed to account for the different form of slopes, and to explain slope evolution by parallel retreat, slope decline, or some more complicated method.

As an example of a non-mathematical model, we can consider Baulig's (1940) theory of slope grading, summarised by Cotton (1952b). Weathering products from the upper slopes increase in thickness and become finer towards the foot of the slope. The production of waste increases or diminishes as the layer protecting bedrock thins or thickens. The physical processes of weathering are more active where the layer is thin or coarse, and where it is thin it is generally also coarse. Chemical weathering is most effective where the layer is thicker and finer and thus holds more water. Chemical weathering is much less important on the convex slope than on the lower concave slopes, where increasing fineness and impermeability hold water better. The actual form of the slope depends therefore on various interdependent facts, and the rate at which debris is produced by weathering is controlled by the thickness and the fineness of the layer of material covering and protecting the underlying rock.

Baulig's model may have some application, but it is essentially theoretical, and not derived from specific instances of carefully studied slopes. Some of the theoretical ideas are not valid in all instances. For instance, in the example of pediments in central Australia studied by Mabbutt and described above, the debris cover, far from protecting the underlying bedrock, caused an increase in weathering.

Many mathematical models of slope development have been devised to investigate the evolution of slopes under specific conditions (e.g. Lehmann, 1933; Scheidegger, 1961) which necessarily accept certain assumptions. It may be assumed that the intensity of weathering action is a function of the declivity of the slope; that weathering acts normally to the slope; that weathering proceeds at an equal rate on all exposed points of a slope. The validity of these assumptions has hardly been tested so far, but they are certainly not self-evident. Increase in steepness, for example, may speed up weathering because removal of debris is easier; on the other hand, flatter sites retain water longer than steep sites, and so have greater chemical weathering. Where the balance actually lies on any particular slope can only be determined by observation of slope form, weathering products and operative processes: it cannot be deduced from theory.

INVESTIGATIONS OF WEATHERING ON SLOPES

There are still too few descriptions of slopes in relation to weathering and superficial deposits to derive many firm conclusions, but there are enough to point out the limitations of the theoretical approach.

Young (1963) studied slopes and regoliths in three areas of Britain, which all provided similar information. The regolith remained relatively uniform, both in thickness

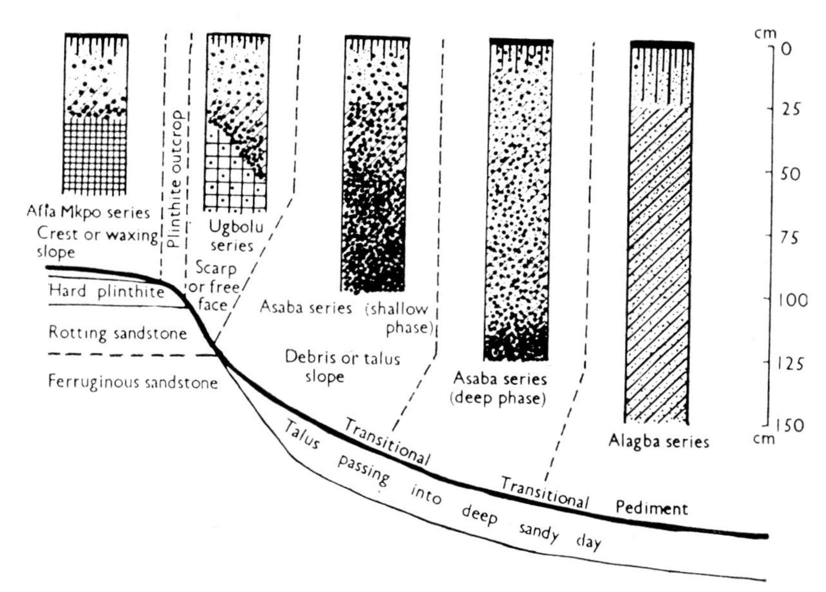

FIG. 137. Soil land slope relationships in South West Nigeria (from Moss, 1965).

and degree of reduction, over the long and gently sloping convex sectors with a gradual curvature. On steeper slopes it tended to thicken, and always tended to be greatest in thickness on the steepest segments. On concavities there was a further increase in thickness combined with a large downslope increase in weathering.

In contrast, the slopes studied by Ollier and Thomasson (1957) in the Chiltern Hills of England revealed that the deposits on the upper, convex slope became progressively thinner downslope, that there was no accumulation on the free face, and there was a progressive increase in thickness of the mixed debris on lower slopes (Fig. 87, p. 168).

A detailed account of a tropical slope from S.W. Nigeria is provided by Moss (1965) and is shown in Fig. 137. The waxing slope is generally associated with a relict soil that thins towards the free face. The free face is marked by an outcrop of ironstone (plinthite) and a narrow band of truncated soil immediately below it. The debris slope is characterised by soils with detrital fragments of ironstone, thickening downslope and passing imperceptibly into the pediment with extensive development of non-mottled ferrallitic material.

This example described by Moss is uncomplicated by concurrent precipitation of iron oxides, but the very thick ironstone outcrops found commonly around the edge of laterite-capped plateaus are apparently due to the issue of laterally moving iron-bearing water at the free face. In these cases the ironstone (plinthite) thins rapidly with distance from the face of the cliff and may be absent at some distance into the plateau.

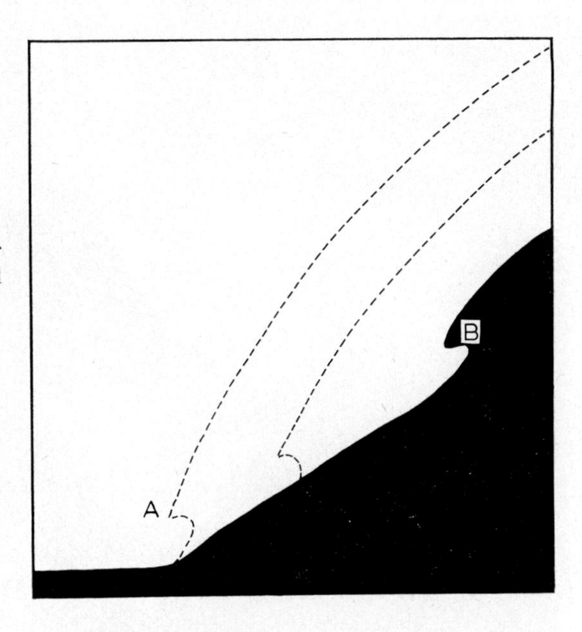

Fig. 138. The evolution of lower slopes at Ayers Rock, Central Australia.

Some of the escarpments of the Colorado plateau were once thought to have been stable for long periods because there was little accumulation of debris at their foot. However, Schumm and Chorley (1966) have pointed out that the scarps are in fact actively eroding, but the debris is rapidly destroyed, partly by mechanical shattering in the original rock falls, and partly by frost action. In general, debris production is slower than debris destruction by weathering.

Sometimes weathering and slope formation give rise to exceptional landforms that are not to be classified among the usual slope elements, as in the following example from Central Australia (Ollier and Tuddenham, 1962).

Around the base of Ayers Rock, where the entire surface is of bedrock, weathering by flaking occurs on all surfaces but nevertheless distinctive slope elements develop.

Flaking in shallow concavities (negative spheroidal weathering) exaggerates their curvature and gives rise to caves. These flake on all surfaces, but especially at the top and back, so that the cave works upwards and backwards, generating a new slope beneath as shown in Fig. 138.

The slope above the cave must also be retreating by back weathering, for it does not overhang the origin of the lower level.

Weathering on this upper surface is also by flaking, but of a different intensity from that on the lower slope. In fact, the lower slope has hardly been altered since it was formed, but the upper slope has retreated from A to B in Fig. 138. The different intensity is also marked by greater differential weathering, and the upper slope is very ribbed while the lower slope is comparatively smooth, as can be seen on Fig. 139.

FIG. 139. The lower slopes of Ayers Rock, Central Australia. The ribbing that picks out steep bedding on the upper slope is absent on the lower slope. The two slope elements meet in an abrupt angle, often marked by a cave.

COASTAL LANDFORMS

Many coastal landforms are undoubtedly caused by marine erosion, but others are due to various forms of weathering. The origin of coastal landforms is often in dispute, and it is unfortunate that some landforms have a presumed origin incorporated in their name, as in 'abrasion platform' and 'solution notch'.

Those landforms attributed to weathering will be described here, together with some evidence of their mode of origin. The other side of the picture—the arguments in support of erosional origin of these or similar landforms—will not be pursued, and no assessment of the relative importance of weathering and erosion in coastal morphogenesis will be attempted. The reader must therefore remain aware that what follows is a somewhat biased and one-sided account.

A COASTAL TRAVERSE

On cliffed coasts a series of features may be formed between land and sea, as shown in Fig. 140. Of course it would be very unlikely that all the features would ever be present together, and commonly one or more features are missing from any real section. The features will be described in turn.

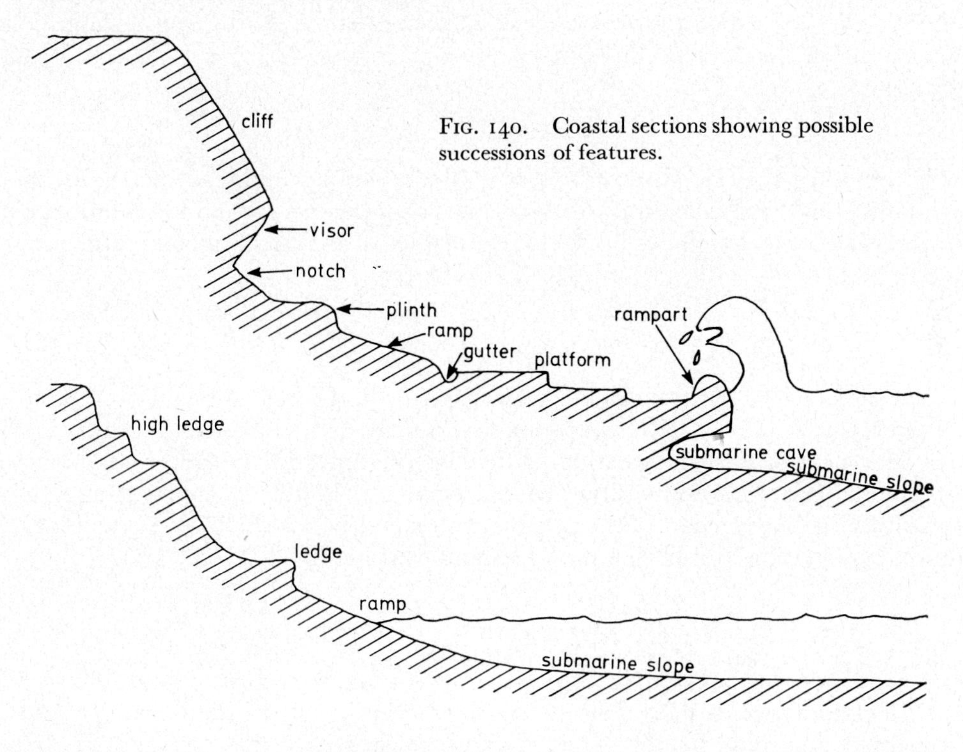

Fig. 140. Coastal sections showing possible successions of features.

The cliff

Marine cliffs differ from normal slopes (some of which may come to be washed by the sea, as on drowned coasts) in that the cliff surface is formed by marine undercutting, and modified by the wash of sea spray. Salt action, solution, and especially wetting and drying, may be important minor processes on cliffs and give rise to lapiés, honeycombs, caves and other minor features.

The visor

If the base of a cliff is undercut by a notch, the angular projection so formed is called a visor. The visor is not simply due to collapse of material over the notch, but is largely caused by solution. On limestones and calcareous aeolianites the underside of the visor is indurated, and in all cases the undersurface is pitted with very irregular solution cavities.

The notch

At the base of cliffs and rock stacks that rise above the platform a notch (wave notch or solution notch) overhung by a visor is commonly present. Hodgkin (1964) has shown that the notch develops very rapidly, and certainly by solution. Maximum solution occurs at about the middle of the notch, which corresponds roughly to mean high sea-level. Notches are especially well-developed on limestone coasts, and do not appear to be present on basalt or granite coasts.

The plinth

On the coasts of Victoria, Australia, there is often a slight prominence below the notch, which Hills (1968) has called the plinth. This is convex outwards and rises immediately from the platform. It follows all the irregularities of the notch in plan and must be formed concurrently with it, presumably by solution.

The ramp

This is a smooth rock surface sloping seawards at 5° to 8° from the cliff base, from the notch, or from the plinth. Ramps probably result largely from abrasion due to waves of translation and swash, and weathering may be insignificant. However, on calcareous coasts the ramps are generally pitted with circular, flat-bottomed, intersecting hollows, especially at the lower edge. The pits do not extend below the level of the platform. Solution and wetting-and-drying may be significant here.

The gutter

Between the wave ramp and the platform there may be a gutter (also 'channel' or 'moat') incised a foot or more below the landward edge of the platform. It contains sand, pebbles and small boulders, and is formed by abrasion.

The platform

The platform (shore platform) consists of a number of adjoining flat areas at slightly different elevations. They are essentially horizontal, with irregularities only about an inch in height, and steps that can be somewhat larger between platforms. They are bare in patches, but are mostly covered by algal growth and mussel colonies in the intertidal zone.

Platforms have been reported on many different rocks, including limestone, sandstone and basalt, but they do not appear to be formed on granite, except when weathered. Where beds of different strata are exposed on a beach, lithological differences give rise to platforms at different heights, the differences ranging from a few inches to several feet, and in extreme cases even 8 feet (Hills, 1968). Platforms on sandstones are higher than those on mudstone.

Platforms are believed to be produced by water-layer weathering, which in turn is caused by wetting-and-drying (see p. 21). Rocks projecting above the platform are reduced by wetting-and-drying, but when the rock reaches a level at which it remains permanently wet (aided perhaps by a mat of algae and other organisms) the process stops. As explained on p. 22 the process works around the edge of pools which extend

Fig. 141. Blocks of arkose resting on a mudstone shore platform, giving rise to pedestals, Cape Patterson, Victoria. (Photograph by A. A. Baker.)

laterally and coalesce, and pools at slightly different levels will produce platforms at slightly different levels.

Hills (1968) describes a very interesting instance of a fallen block lying on a platform which has been lowered some two or three inches all around, leaving the block on a pedestal that matches the shape of the block exactly. The fallen block would protect the underlying rock from rain, protect it from drying out, and also kill the organisms underneath; all of these effects would reduce weathering. Similar blocks are shown in Fig. 141. The same author also described large calcareous concretions that now stand on similar pedestals above a water-layer lowered platform surface.

High platforms

Most platforms are in the intertidal range, but some platforms caused by water-layer weathering may be found at heights considerably above normal sea-level. Hills (1968) reports platforms 16 feet above normal tide-level, and water-layer effects extending up to 20 feet above sea-level on cliffs and stacks exposed to violent wave action.

The rampart

On the outer edge of shore platforms there is sometimes a rampart—a ridge of rock rising some feet higher than the platform. The rampart is preserved because it is wetted more continuously, by breakers, than the rocks of the platform behind it. So long as the shore platform continues to be lowered, the rampart can persist, even if eroded on its outer edge: if the platform is reduced to water level the rampart cannot be regenerated if it is once destroyed by erosion.

The low tide cliff

At the outer edge of shore platforms there is a low cliff, often about 15 feet high, separating the platform from the submarine slope. This cliff is due to submarine erosion, and the features of the cliff in no way resemble the shoreline cliff with its visor and notch.

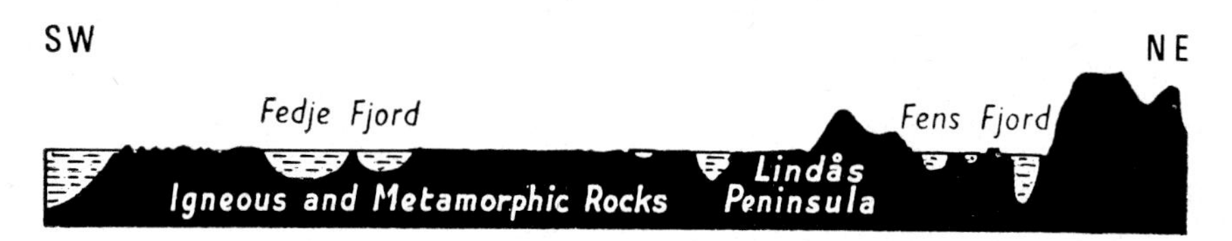

FIG. 142. Section across the strandflat north of Bergen, Norway. Length of section 32 miles (from Holmes, 1965).

Submarine caverns

On soft limestone coasts the low tide cliff may be undercut, producing caverns and channels that penetrate several yards under the platform. The platform surface in this case is indurated by secondary carbonate, while the limestone beneath is unaltered, and differential erosion rather than weathering causes the undercutting.

The submarine slope

This is due to marine erosion. When a shore platform is not present the submarine slope is continuous with the ramp.

FEATURES NOT INCLUDED IN THE TRAVERSE DESCRIBED ABOVE

The traverse described above included features of simple coasts in areas of warm climate with no complications of lithology, process or history. Where such complications occur, certain other landforms may be present.

Strandflat

The strandflat is an exceptionally wide marine platform on the Norwegian coast, varying in width from a few kilometres to 60 km, and ending abruptly on its landward side against mountains (Fig. 142). It is now partly above sea-level due to postglacial isostatic uplift. Strandflats are also found in Greenland, Spitzbergen and on other fiord coasts. Nansen (1922) considered that ice-foot weathering (see p. 12) was partly responsible for the formation of the strandflat.

Slope-over-wall cliffs

In the last glacial phase of low sea-level, marine cliffs, no longer subject to wave attack, were degraded by frost weathering and accumulated mantles of frost shattered debris.

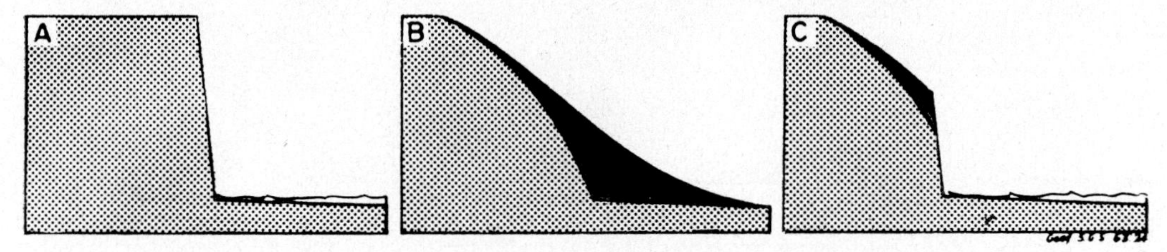

FIG. 143. The evolution of slope-over-wall cliffs (from Bird, 1964).

Since the postglacial transgression such slopes have been rejuvenated and steepened again. Those cliffs that retain part of their frost-made form have slope-over-wall profiles (Fig. 143).

Submarine limestone caves

The limestone islands of Bermuda (and probably many others) are riddled with caves, apparently of phreatic origin (see p. 224) and undoubtedly formed by fresh water, but which are now filled by salt water. Bretz (1961) has invoked the Ghyben-Herzberg hypothesis to account for this situation. According to this hypothesis an island of adequate size can maintain a lens of fresh water, but smaller islands cannot. In glacial times there was a lowering of sea-level, so Bermuda was much bigger than now and had a large lens of fresh water, enabling phreatic development of caves to take place. The postglacial rise in sea-level destroyed these conditions and the caves became filled with sea water as at present.

MINOR FEATURES

Coastal solution features

Coastal landforms caused by solution include honeycombs, solution pits and grooves, and a variety of lapiés or Karren, similar to those formed on inland limestones but often more jagged.

FIG. 144. Elevated pits with limonitic lining on an arkose shore platform at Cape Patterson, Victoria. (Photograph by A. A. Baker.)

FIG. 145. Imprisoned boulders, Cape Patterson, Victoria. (Photograph by A. A. Baker.)

Elevated pits

Small pits are filled by the tide or by rainwater and this keeps the immediately surrounding rock wet, while the platform around is reduced in height by wetting-and-drying. Eventually the pit comes to be like a miniature volcano with the crater filled with water and the sides keep damp, rising above a miniature plain which occasionally dries out. Such pits are found on limestone, basalt, arkose and other rocks. The 'crater' may be lined with limonite, as on Fig. 144.

Imprisoned boulders

Heaps of boulders that are just too large to be transported but small enough to be oscillated by water movement rub against each other and eventually come to form an interlocking array of facetted boulders. Baker (1959) has called these 'imprisoned boulders' (Fig. 145).

XIII WEATHERING RATES

Weathering may be sufficiently fast to yield results even in short-term laboratory experiments. An early and fairly crude experiment by Hilger (1897) is an example. Samples of limestone, sandstone and mica schist were broken into uniform particles 1 to 2 cm across, and these were exposed to the atmosphere for a total of 17 years. At intervals during this period the specimens were examined to determine how much fine earth (particles less than 0·5 mm) had been produced, and at the end of the 17 years the specimens were analysed chemically. All the rocks were attacked both physically and chemically; the limestone suffered most chemical attack, the sandstone exhibited most physical disintegration and the mica schist was intermediate in both cases.

To study weathering over longer periods it is necessary to find methods other than direct experiment. One way is to have speeded-up experiments, such as those of Griggs (1936), described in Chap. II. He had an apparatus which produced many cycles of heating and cooling in a short time, and he believed he could simulate the weathering of diurnal temperature changes for a period of 244 years. However, such experiments are always open to question, and perhaps of more value are those observations on natural weathering of specimens that can be dated.

One sort of well-dated surface that offers opportunity to study weathering rates is the tombstone. Geikie (1880) made a survey of Edinburgh churchyards and found that a good-quality sandstone weathered very little in 200 years, a slate of 90 years age had clear engraving, but a marble stone had partly crumbled into sand. The 'tombstone method' was refined by Goodchild (1890). He studied some very old tombstones and calculated the weathering rates for different sorts of limestone shown in Table 19.

Table 19

Type of limestone	No. of years to produce 1 inch of weathering
Kirkby Stephen	500
Tailbrig	300
Penrith	250
Askrigg	240

Dated buildings offer another opportunity for study of weathering rates. Hirchwald (1908) examined numerous buildings of different date made from rocks from known quarries, and was able to give a qualitative idea of weathering rates, as shown in Table 20.

Table 20

Rock	Age	Condition
Brochterbeck sandstone	100	Good
	550	Slight weathering
	770	Strongly weathered in parts
	900	Most of it strongly weathered
Rothenburg sandstone	55	Significant traces of weathering
	80	Rather strongly weathered
	100	Very strongly weathered
Nahetal porphyry	150	No significant weathering
	400	Distinct surface weathering, no change in interior

Use of very old structures was made by Barton (1916) in his study of granite weathering in Egypt. South of Aswan, surfaces of granite blocks that have been continuously above water for 4000 years are still quite fresh. Facings at Giza, 1400 years older, show exfoliation and granular flaking to depths of 0·5 cm to 0·8 cm on exposed surfaces. Below soil level, alteration is greater still, perhaps 5 to 10 mm per 5000 years. These observations suggest that in generally dry areas the presence of moisture is of major importance in weathering rates.

More information from the same area, but for limestone, is provided by Emery (1960), who studied the weathering of the Great Pyramid of Giza. This was originally faced with well-fitting limestone blocks that protected the rocks of the core from weathering until the facing was removed about 1000 years ago. Since then there has been differential weathering of the four main rocks of the core.

The most durable is a hard, grey, dense limestone, which is so little affected that it retains the quarry tool marks of 2800 B.C. The quality of this stone was recognised by the builders, who used it mainly on corners and other vulnerable parts. More common, is a soft grey limestone, characterised on weathered surfaces by many small pits 1 to 2 cm deep. The rocks that weather fast were, wisely, little used. One is grey shaly limestone which has developed niches with soft, scaly walls, many exceeding 20 cm in depth. The other is a friable yellow limy, shaly sandstone with many close joints, which turns into rubble.

During the process of weathering, most of the debris remained as talus on individual tiers, and around the base of the pyramid. Emery calculates that there was a total of 50,000 cubic metres of debris before parts of the pyramid were cleared. Since the rock has been exposed for about 1000 years, 50 cubic metres is the average yearly supply. Allowing a porosity of the talus of 50%, this corresponds to an average loss of 0·2 mm per year over the whole surface of the pyramid. This represents only about 0·01% of the total volume, and it is gratifying to calculate that, at this rate, the pyramids will survive for 10,000,000 years to come.

Akimtzev (1932) compared weathering of a limestone building stone with weathering of nearby soil also on similar limestone. The building concerned was the Kamenetz fortress in the Ukraine, which had been accumulating soil since its destruction 230 years before the time of the observations. This had an average thickness of 12 inches, which was, in fact, more than that of the surrounding country,despite its much longer period of weathering! In many properties the historical and the natural soils were very similar, as shown in Table 21. A few hundred years' weathering is evidently quite sufficient to account for the natural soil of the area.

Table 21

	Soil in fortress	Natural soil
Depth of soil (cm)	10–40	8
Humus %	3·5	3·8
Clay %	50–56	53
$CaCO_3$ %	5	5
Exchangeable Ca	0·85	0·89
pH	7·7	7·67

Besides buildings there are other artificial structures that can be dated, and give indications of weathering rates. Spoil heaps from mining operations have been utilised in this way.

Mitchell (1959) studied the ecology of tin mine spoil heaps in Malaya, and found that even after 25 years the fertility of the soil on the spoil heaps was only one-fifth or less of that of unmined areas. He believed that this was because the spoil heaps lacked clay and organic matter, and therefore water-holding capacity. If some clay were present to retain water then weathering would be much faster.

In a study of soil formation on spoil heaps of the Northamptonshire ironfield, Bridges (1961) found that aspect was as important as time. The overburden had been left in ridges, south-facing slopes had developed impervious to freely drained calcareous soils, but there was only very slight soil formation on north-facing slopes. This very local variation indicates the need for caution in generalising too freely about weathering rates.

Rather similar to spoil heaps are the middens or shell mounds that have sometimes been left in coastal areas. Dupuis *et al.* (1965) examined historical rendzinas developed on shell mounds. Those of Saint Michel en l'Herm, France, which were dated about 1066, show rapid pedogenesis.

In Holland the reclaimed land of the polders provides plenty of opportunity to examine the rate of weathering of new-formed land. Hissink (1938) has provided evidence on the leaching of calcium carbonate, which was originally about 10%. After 300 years it had completely disappeared. The data are shown in Fig. 146.

Another study of leaching utilised a drained lake floor. Tamm (1920) studied the podzolisation of the sediments of the former Lake Ragunda in Sweden, which was drained in 1796. The sandy parent material contained 0·5% calcium carbonate, which was leached out to a depth of 10 inches under pine-heath, and to 25 inches under a mixed forest rich in mosses.

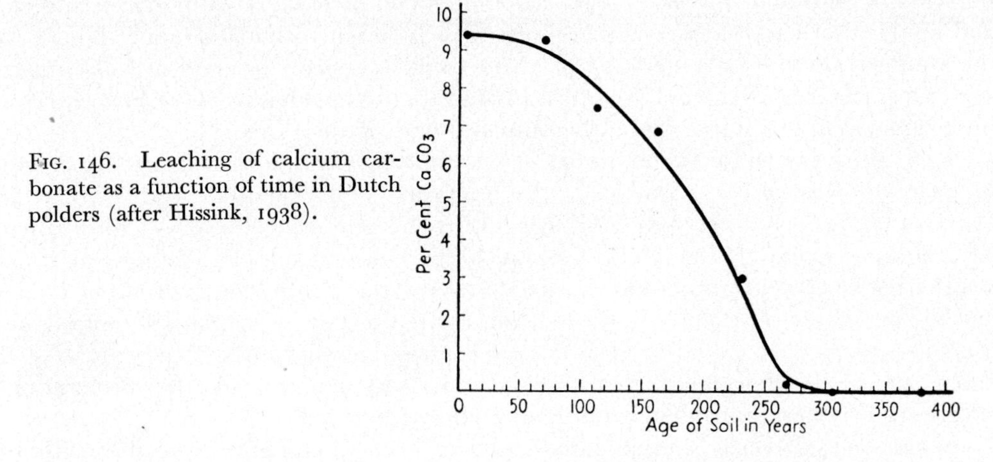

FIG. 146. Leaching of calcium carbonate as a function of time in Dutch polders (after Hissink, 1938).

A very elegant study of the time factor was made by Salisbury (1925) on a series of sand dunes at Southport, England. The ridges could be dated from descriptions, from old maps, and by tree-ring analysis of the willows that established themselves very

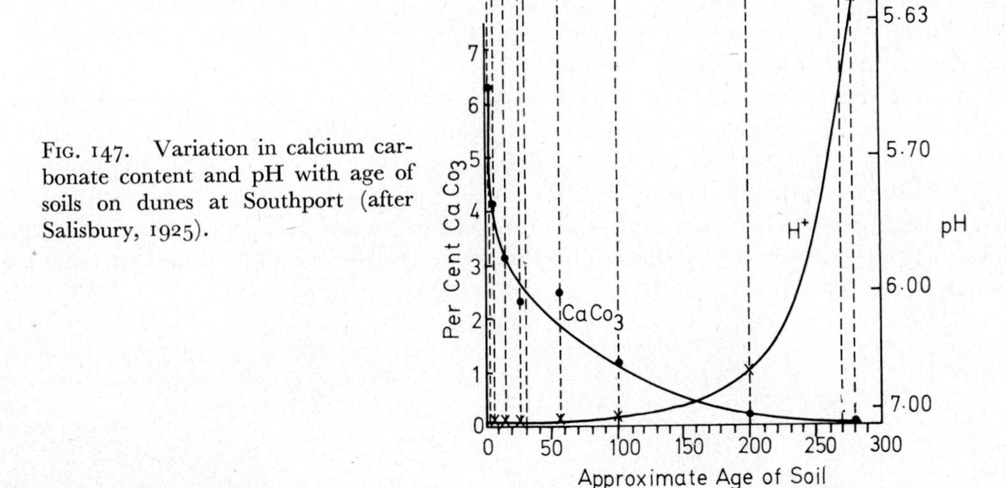

FIG. 147. Variation in calcium carbonate content and pH with age of soils on dunes at Southport (after Salisbury, 1925).

quickly in the swales between the ridges. The curves of pH and $CaCO_3$ are shown in Fig. 147. It is seen that the carbonate falls rapidly at first and then slows down, while pH has the opposite trend, changing slowly at first and then more rapidly after 200 years.

Yet another time sequence is provided by river terraces. In Victoria (Australia) a constantly recurring sequence of terraces have deeply weathered material and krasnozem soils on the older terrace, and chernozem-like soils on the younger terraces. Boulders of granite in the old terrace have often completely rotted to gravel and clay, and boulders of basalt have weathered extensively, often with exfoliation. Boulders in the younger alluvium are very fresh. Mineralogy can also be used to show relative age of terraces, and Neiheisel (1963) has found that older Pleistocene river terraces are impoverished in less stable minerals compared with Recent ones.

Retreating glaciers leave moraines of different ages which provide an opportunity to examine weathering sequences. In Victoria Valley, Antarctica, Calkin and Cailleux (1962) examined four moraines left by the retreating Lower Victoria glacier. The moraines contained boulders of granite and gneiss which had undergone cavernous weathering since deposition, and the size of cavity, the number of cavities per boulder and the percentage of boulders affected, was related to age. Incidence of hollows went from 86 % to 57 %, 28 % and 12 % on successively younger moraines. Unfortunately, the actual age of the moraines is not known, and no explanation for the mechanism of cavernous weathering in these conditions is suggested.

Schreckenthal (1935) studied dated moraines of the Tyrol and found that within 80 years, although the climate was seemingly unfavourable, there was significant alteration. Soil acidity increased, silt became relatively abundant, and some clay was formed.

Hansel and White (1960) studied the effect of time in the genesis of soils on tills. The Wisconsin glaciation produced till in several substages, and in the Mid-West of the United States the soils of the Tazwell substage may be about 6400 years older than the soils of the Cary substage. The fertility of the soils has declined with time as the K-supplying power of the soils has decreased, presumably due to the formation of illite.

Volcanic eruptions produce fresh parent material that provides excellent material for time studies of weathering.

Van Baren (1931) gave an account of soil formation on pyroclastics in Indonesia derived from the eruption of Krakatoa in 1883. The inadequate profile description merely lists a surface soil (35 mm), a middle layer and the parent rock, but the analytical data (Table 22) indicate considerable weathering in a 45-year period. Silica is leached from the surface, where alumina and iron oxides accumulate; there is very marked leaching of potassium and sodium, and there is a marked increase in fine particles and water in the topsoil.

Table 22

Constituents	Rock %	Middle layer	Surface soil
		%	%
SiO_2	67·55	65·87	61·13
Al_2O_3	15·19	16·31	17·24
Fe_2O_3	1·52	1·74	2·56
FeO	2·15	2·05	2·59
CaO	2·89	3·07	3·61

On a longer time-scale, Hay (1960) described the rate of clay formation and mineral alteration in a 4000-year-old volcanic ash soil on St Vincent, West Indies. A clayey soil, 6 feet thick, has formed on deposits that were originally andesitic ash. Glass has altered to halloysite, allophane and hydrated iron oxides. Anorthite crystals are etched in the soil and frosted in the underlying ash. Many crystals of hypersthene and a few of augite and olivine are etched. Labradorite, hornblende and magnetite remain unaltered. It is calculated that clayey soil formed from ash at the rate of 1½ to 2 ft/1000 years, and glass decomposed at the rate of 15 g/cm^2/100 years. Evidently most of the weathering that gave rise to the soil was weathering of volcanic glass, and in fact the stability of some of the minerals present is surprisingly high.

Similarly, in Papua on The Hydrographers strato-volcano, weathering rates can be calculated as the age of the volcano has been obtained by potassium argon dating as 650,000 years (Ruxton, 1966). The weathering profiles on sites with little or no erosion have an upper zone 1·5 to 7·5 m (5 to 25 ft) of silty clay over a lower zone of 15 to 30 m (50 to 100 ft) of clayey silt with rock structure preserved. Taking the larger figures, this works out as only 58 mm in 1000 years, which is rather low for such weatherable material as volcanic ash. Presumably the weathering rate has slowed down because of the accumulation of weathering products.

In limestone terrain where solution is paramount, a great deal of work has been done on rates of solution.

One approach to the rate of solution is chemical analysis of drainage waters. Drainage waters are analysed to determine their calcium content, and if a reasonable average figure can be obtained, and the total run-off from a region can be determined, then the amount of calcium lost can be calculated, and this in turn can be expressed as amount of lowering per 1000 years (or some such figure). Corbel has provided a formula to calculate the number of millimetres of limestone per 1000 years removed by solution in a limestone area:

$$\text{Erosion in } m^3/yr/km^2 = \frac{4ET}{100}$$

where E = run-off of water in decimetres,
T = amount of $CaCO_3$ in mg/litre.

An alternative formula is proposed by Groom and Williams (1965):

$$\text{Solution in } m^3/yr/km^2 = \frac{M}{Sa \times 10^6}$$

where M is the mass of limestone removed in one year in grams,
S is the specific gravity of the limestone,
a is the catchment area on km^2.

Fortunately it is easy to convert the amount of limestone dissolved as calculated by Corbel's formula and that of Groom and Williams into a rate, for with suitable units the figure remains the same, thus:

16·0 $m^3/km^2/yr$ total quantity removed is equivalent to 16·0 mm/1000 yr surface lowering.

Groom and Williams found by their own formula a rate of lowering of 15·77 mm/1000 years, for an area in South Wales; using the same data but Corbel's formula they found an average of 40 mm/1000 years. They believe the considerable discrepancy is due to the impossibility of assessing the average calcium content of stream waters in an area of wide variations of stream flow, as well as some local geological factors.

M. M. Sweeting (1966) reported work on solution rates in parts of northern England.

Erratic blocks occur scattered on limestone pavements and stand on pedestals 30 to 50 cm above the surrounding pavement. The height of the pedestals indicates the amount of general surface lowering since the erratics were lodged, about 12,000 years. In the same area when glacial drift is freshly cleared, the limestone surface beneath is found to be fresh, and glacial striae are preserved. Sweeting cleared a 2 square-metre area in 1947. By 1957 only a few striae were left; by 1960 all had disappeared and the limestone surface lowered by 3 to 5 cm. In another experiment runnels up to 15 cm deep were formed in 13 years. Both experiments were under very favourable conditions for solution, and the rate cannot be extrapolated to determine the age of the pedestals.

Applying Corbel's formula to data from northern England, Sweeting (1966) obtained the following results:

0·04 mm per year by surface solution,
0·043 mm per year by underground solution,
0·083 mm per year total lowering.

The surface rate corresponds to 41 mm lowering in 1000 years. In the 12,000 years available since the erratics were thought to be lodged, there would be 12 × 41 mm, or about 49 cm of surface lowering. This result is quite close to the observed height of the pedestals (see above). Goodchild's estimate of 1 inch in 500 years (see p. 242) gives a result of 2 ft or 60 cm in 12,000 years, which is a comparable result.

F. Bauer (1963) calculated the rate of solution in the Austrian Alps during post-glacial time as 9-12·5 mm/1000 years on bare limestone sealed off from outside influences. Under soil and vegetation (dwarf pines) the rate was much higher—28 mm/1000 years. Similar results have been obtained from the Swiss Alps by Bögli (1961).

One conclusion can definitely be drawn from work on limestone solution rates: there is great variation. Ingle Smith (1965) has provided data for a number of British limestone areas, and Corbel (1959) provides data for many other areas. Corbel attributes most of the variation to temperature, but this is much too simple a notion. Soil carbon dioxide variation is a potent factor (as Bauer showed) and the lithology, and especially the porosity of the rock, allowing greater contact between water and mineral particles, is also of importance. The effect of other ions can affect the solubility of calcium carbonate, and Cigna (1963) has shown that under laboratory conditions saline solutions dissolved up to 10% more calcium carbonate than did equivalent volumes of fresh water under the same conditions.

Chemical methods of determining weathering rates have rarely been applied to rocks other than limestone because of the many complications, but one such study is that of Leneuf and Aubert (1960), who attempted to measure the rate of ferrallitisation of a two-mica granite of known chemical composition in the forest zone of the Ivory Coast. They worked out the amount of water drained from the soil each year, and the quantity of Si, Ca, Mg, K and Na removed each year. This data enabled them to calculate the time taken for complete ferrallitisation of granite under prevailing conditions to be 22,000 to 77,000 years per metre.

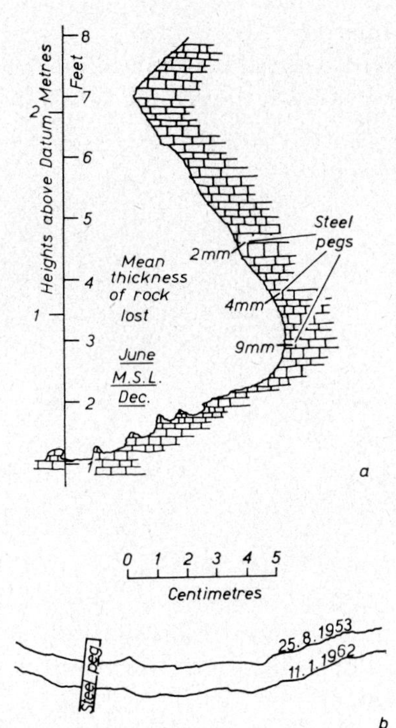

FIG. 148. Weathering at Point Peron, West Australia (from Hodgkin, 1964).

a Profile of rock face to show position of steel pegs and amount of rock lost 1953 to 1962. *b* Portion of tracings from plaster casts.

Rates of weathering of rocks exposed to the sea are possibly complicated by the effects of mechanical erosion, but a number of studies in selected localities have yielded data that give an approximate idea of weathering rate, especially solution of limestone.

On Norfolk Island, the steps of a jetty built of limestone between 1939 and 1955 have been considerably weathered, and Hodgkin (1964) was able to draw the following conclusions:

1. At the lowest level of the intertidal zone the rate of erosion (and presumably of weathering) is between 0·5 mm and 1·0 mm a year.
2. Rate of erosion decreases with height and decreased frequency of wetting.
3. Weathering is most rapid in depressions where water is retained.
4. Nevertheless, weathering of a vertical surface is little slower than that of a horizontal surface.

At Point Peron, Western Australia, the same writer used a technique of direct measurement to derive the rate of erosion. Stainless-steel rods were driven into the hard limestone of a coastal notch at various heights above sea-level. Plaster casts were taken of the notch over a period of years, and the amount of retreat of the rock relative to the end of the pegs could be measured, as shown in Fig. 148. It was found that the whole of the rock face was eroded, and most attack was on the lower level, where the rate was 1 mm per year.

The different rates of erosion at different heights accounts for the formation of the notch on many coasts, though the actual mechanism is not explained, and it is not clear

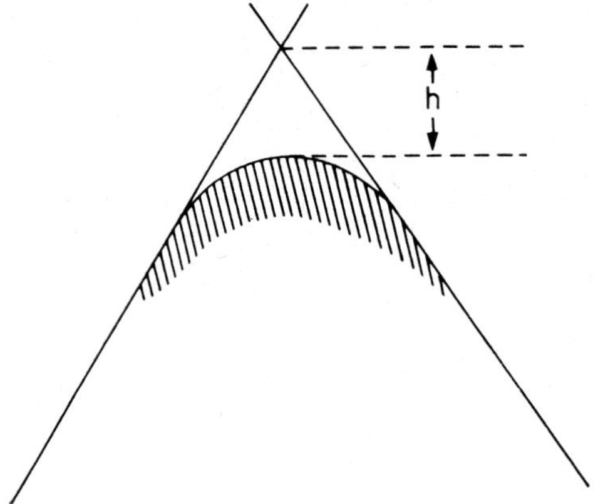

FIG. 149. The blunting of an originally sharp blade of rock.

why erosion apparently ceases abruptly or becomes very much slower at about low-water level. It seems probable, however, that frequency of wetting and drying is of major importance.

The rates calculated by Hodgkin are similar to those obtained by other workers. Emery (1941) used the weathering of dated inscriptions at La Jolla, California, to derive a rate of 0·5 mm a year. Kaye (1959) examined intertidal nips developed in limestone in Puerto Rico which was placed intertidally by earth movements 155 years previously. He calculated a rate of weathering of about 1 mm a year.

If weathering rates were sufficiently well known it might be possible from an examination of a specimen or profile to calculate how long it had been weathering. Two possible ways of doing this have been suggested by Černohouz and Šolc (1966). Originally sharp edges on blocks of sandstone become rounded with weathering, and the more they weather the more rounded they become. The simplest measure of the blunting is the distance from the present edge to the projected original edge (Fig. 149). A formula can be derived using this distance, h, and constants that are found experimentally, and the method was said to be usable in the time range 200 to 100,000 years. The second method concerns the formation of weathering crust on basalt, the thickness of which, d, is related to time in the following manner:

$$d = A \log (1+Bt)$$

where d is thickness in mm, time t is in millennia, and A and B are constants. For basalt from Bohemia the constants were as follows:

$$A = 4{\cdot}64 \pm 0{\cdot}05.$$
$$B = 0{\cdot}010 \pm 0{\cdot}001.$$

The authors claim that the methods are accurate to within 10 to 20%.

The overall rates of chemical denudation for the continents have been calculated by Livingstone (1963), who provides the figures in Table 23.

Table 23

Continent	Long tons/sq. mile	Metric tons/sq. km.
Australia	6	2·3
Africa	63	24
South America	73	28
Asia	83	32
North America	85	33
Europe	110	42

The rivers of the world deliver 3,900,000,000 tons of dissolved material to the sea each year.

The average concentration of the important constituents in parts per million is shown in Table 24.

Table 24

Bicarbonate	58·4	Sodium	6·3
Calcium	15·0	Magnesium	4·1
Silica	13·1	Potassium	2·3
Sulphate	11·2	Nitrate	1·0
Chloride	7·8	Iron	0·67

The total average is 120 ppm of dissolved solids. These ten constituents account for most of the dissolved material, though all but 37 of the naturally occurring elements have been detected in lake or river water.

Some weathering rates, such as those of coastal limestone, may remain almost constant for long periods, because conditions remain virtually the same and there are no complications due to accumulation of weathered material. Other weathering rates vary with time for a number of reasons. Some are markedly affected by accumulation of weathered products, which slow down access of water or air and hence the overall rate; others slow down because there is only a limited supply of weatherable material. The results discussed from the Southport sand dunes and the drained polders show marked changes in rate of leaching (Figs 146 and 147), and, while on the dunes the rate is rapid at first and then slows down, on the polders it is slow to start but then gets faster. In the first case the material is freely drained, so leaching is effective from the beginning. In the second case the heavy clay is at first badly drained, and only when weathering and 'ripening' have improved the soil structure does through drainage and leaching of carbonate becomes significant. Of course, when most of the carbonate has been removed the rate will fall again.

If weathering rates approach asymptotically to zero there will come a time when profiles remain virtually unchanged. Such weathering profiles would be analogous to soil profiles termed 'mature'. Without discussing the validity of this concept further, the point can be made that very often weathering rates are most rapid at the start and become increasingly slow. Thus, for example, a weathering rind on basalt thickens at a decreasing rate, and the formula relating thickness to time is logarithmic (p. 250).

Laboratory investigations of weathering rates usually concern the start of weathering, so rates will be higher than over the entire course of weathering. Furthermore, experiments tend to be performed in favourable conditions for alteration on very suitable material, so that measurable alteration will occur, while on the earth's surface there will be large areas where conditions are not so suitable, thus reducing the average rate. In general, therefore, experimental determinations of weathering rates tend to be too high, and it is unwise to extrapolate such rates to natural weathering of large areas over long periods.

This chapter has been concerned with fairly rapid weathering for the most part, but slow weathering must not be ignored. At the ground surface weathering products may be present that have formed recently, even in the past few days or weeks. In the zone of hydrolysis and reduction, weathering may have been going on for a time that is measured in millions of years or geological periods. Even deep oxidation is relatively slow, and time in the order of a geological period may be required to form economic oxide deposits. Many of the oxidised zones of the copper deposits of the United States were formed during Tertiary times.

XIV WEATHERING THROUGH GEOLOGICAL TIME

THE style and intensity of weathering has changed through geological time in all parts of the earth. In what are now cool temperate lands there is abundant evidence of glacial and periglacial weathering during the Quaternary, and in many places it is possible to elucidate an alternation of mild and severe climates—glacial and interglacial; pluvial and arid—with considerable precision.

Weathering features preserved in rocks show that climate and weathering have changed not only in Quaternary times, but through all geological time. Thus in Ireland there are Tertiary laterite weathering profiles preserved beneath lavas, yet it would be impossible for laterites to form under the present conditions in Ireland. In Triassic time England was largely a desert, as was Scotland in Torridonian (Precambrian) time. In contrast, South Africa, India and Australia had glacial climates in Permo-Carboniferous time.

Some changes in climate and weathering, such as those of the Quaternary, may be accounted for by changes in solar radiation, though the exact cause of ice ages and the mechanism whereby changes are brought about remain controversial. Some of the older changes, when studied in detail, seem incapable of explanation by any simple climatic change, but make sense if continental drift is invoked. Irving (1958) has summarised some of the evidence for former climates around the world, and shown that though the geographical distribution of past climates is at first sight haphazard, if the continents are reassembled, the former climates fall into simple latitudinal zones. Furthermore, the latitudes indicated by weathering and other climatic indicators are those indicated by palaeomagnetic determination of latitude.

Whether or not continental drift is thought necessary or relevant, the examples of weathering changes discussed above are caused by climatic change, and might be in accord with the Principle of Uniformitarianism. However, there are also progressive changes in the history of weathering.

When the earth first formed, its primitive atmosphere and hydrosphere would have been very different from those of today, and our ideas of the composition of the primitive atmosphere depend to some extent on how we think the earth originated. If it started as hot incandescent material, the first atmosphere would have consisted of light elements, which would have been rapidly lost to space by diffusion. As the earth cooled, the rate of loss of light elements would decrease, and below 2000°C H_2O and CO_2 are stable and could have been formed if hydrogen, oxygen and carbon were available. Nitrogen if present, would remain uncombined. With further fall in temperature, water would liquefy, and the volume of water, by dissolving carbon dioxide and nitrogen, in proportion to their partial pressures, would control the composition of the atmosphere.

If the earth formed by accretion of planitesimals, the atmosphere would have been formed of gases released during accretion. Some evidence in favour of this hypothesis is presented by Mason (1966).

It is also possible that the earth's atmosphere results almost entirely from emanations of the lithosphere, in particular from volcanic eruptions. The main constituents of present day volcanic gases are water and carbon dioxide, though carbon dioxide makes up only a few hundredths of one per cent of the present atmosphere.

On any hypothesis, the primitive atmosphere contained large amounts of carbon dioxide which had to be split up to provide the oxygen of the present atmosphere. Photosynthesis is by far the likeliest cause of this splitting. However, oxygen was apparently available in much of Precambrian time, and the state of oxidation (measured by the $FeO : Fe_2O_3$ ratio) is much the same in Precambrian and later rocks. This implies that photosynthesising plants originated well back in the Precambrian. Rasool and McGovern (1966) have calculated that the primitive atmosphere remained reducing for 10^9 years.

Carbon dioxide is not only the original source of the oxygen in the atmosphere, but also of much of the carbon in the biosphere and in carbonate rocks. It has been calculated that if carbon dioxide were the only, or greatly predominant source of carbon, there would be overproduction of oxygen, but Cotton (1952) has pointed out that oxidation of juvenile hydrogen would dispose of this excess.

It seems most probable that the atmosphere originated from volcanic activity to a large extent, early in the earth's history. Throughout geological time there have been gains and losses of certain components (summarised in Mason, 1966) but the composition has remained much the same since Cambrian times, and even earlier.

There is considerable controversy about precisely when the atmosphere changed from reducing to oxidising, and much of the evidence comes from Precambrian ores. Thus the banket ores of South Africa contain gold and uraninite in a conglomerate, a combination not found in any younger deposits and which, if genuinely detrital, suggests a reducing environment. The banded iron ores of the Lake Superior type occur on all the continents within a certain range of Precambrian time, and not in any younger deposits. Their formation implies a non-oxidising CO_2 rich atmosphere under which large quantities of iron could be carried to the sites of deposition in the soluble ferrous state. Cloud (1965) has suggested that this ferrous iron may have served as a vast reservoir for oxygen, retarding the evolution of free oxygen to the atmosphere. Davidson (1964) has argued in favour of a more uniformitarian view, and appears to believe that an oxidising atmosphere has been present for over 3000 million years.

Davidson also describes other mineral deposits that have a bearing on past weathering conditions.

Two deposits of supposed laterite in the northern Onega district of the U.S.S.R. are of Upper Devonian and Lower Carboniferous age, a date that might correspond with the first invasion of the land by a forest flora, suggesting that organic processes may be essential in lateritisation. However aluminium-rich deposits are known from even earlier periods, possibly Lower Cambrian in the Buryat Republic, and also from the Lower Devonian in Siberia and the Devonian of the Urals. These unusual deposits, interbedded with dolomite, have generally been regarded as marine precipitates similar to oolitic iron ores, the alumina being derived from weathering of nearby land under warm humid conditions, but the Ural deposits have recently been interpreted as subaerial bauxites formed on the karst surface of Lower Devonian limestones.

Minor changes in atmospheric composition have been suggested as possible causes of changing climates in the past, especially of ice ages. Carbon dioxide in particular, being a selective absorber of radiation, has been thought responsible for periods of refrigeration. However, water vapour has the same property, and is much more likely in fact to vary in quantity in the atmosphere. The amount of carbon dioxide in the atmosphere is controlled by the amount in the oceans, which hold a much greater quantity (the CO_2 content of sea water is 20 g/cm^2 of the earth's surface; of the atmosphere 0·4 g/cm^2). Thus to double the CO_2 content of the atmosphere would require not just the addition of the same amount of CO_2 to the air, but a very much greater amount added to the sea.

There is considerable disagreement about the origin and evolution of the oceans. The amount of water and salt in the primitive ocean is unknown, but the amount was probably small in proportion to the present hydrosphere. Throughout geological history the oceans have grown by the addition of primary magmatic water, especially from volcanic eruptions.

Mason (1958) states that the uniformitarian view, that the ocean has been similar to that of the present day as far back as the Archaean, is borne out by available geochemical data. The evidence of sedimentary rocks and fossils indicates that the salt content has probably changed little over the same period.

The actual increase in salt at present is very small; the total annual contribution from the land is only an infinitesimal fraction, $5{\cdot}5 \times 10^{-8}$, of the quantity of dissolved solids in the ocean.

The cations of the salts dissolved in sea water have most probably been derived by weathering of the rocks of the lithosphere, but the main anions (Cl, SO_4, Br, etc.) cannot be simply the product of rock weathering since there are more in the ocean than in the earth's crust as a whole (Rubey, 1951). The anions are derived primarily from volcanic exhalations, where they are found to have a strikingly similar quantitative ratio. If it is accepted that the atmosphere and ocean have grown through geological time largely from volcanic exhalations, it is important to realise that growth may well have been considerably more rapid in early geological times, when the crust could have been thinner, and a greater production of radioactive heat from short-lived isotopes resulted in more frequent volcanic activity. However, even though volcanic exhalations probably produced most of the volatile materials now at or near the earth's surface the contribution of weathering cannot be neglected. Quantitative estimates of the contributions have been made by Rubey (1951, Table 25).

The hydrolysis of silicate rocks through geological time has had the most profound effects, as indicated by Keller (1957). Since the time when chemical weathering began

Table 25. Estimated quantities (in units of 10^{20} grammes) of volatile materials now at or near the earth's surface, compared with quantities of these materials that have been supplied by the weathering of crystalline rock (data from Rubey, 1951).

	H_2O	Total C as CO_3	Cl	N	S	H, B, Br, A, F, etc.
In present atmosphere, hydrosphere and biosphere	14,600	1·5	276	39	13	1·7
Buried in ancient sedimentary rocks	2,100	920	30	4·0	15	15
TOTAL	16,700	921	306	43	28	16·7
Supplied by weathering of crystalline rocks	130	11	5	0·6	6	3·5
'Excess' volatiles not accounted for by rock weathering	16,600	910	300	42	22	13

a great deal of water has been dissociated into H and OH ions. The OH ions have gone with metal cations to the ocean, making it alkaline. The H ions have combined with alumino-silicate anions to form clay minerals. The world's surface thus represents one big chemical reaction: the land is the acid part, and the ocean is the alkaline part. However, the constituents dissolved in the ocean are in a steady state balance, though streams supply very varied constituents (Mackenzie and Garrels, 1966). The mass balance between stream-borne dissolved constituents and the steady state ocean system demonstrates the importance of alumino-silicate reactions in the oceans.

Disposal of the dissolved solids requires synthesis of clay minerals from degraded alumino-silicates by a reaction which may be generalised as

$$\text{Amorphous Al silicate} + SiO_2 + HCO + \text{cations} = \text{cation clay mineral} + CO_2 + H_2O$$

Such reactions imply control of the major ion ratio in sea water, and of the CO_2 pressure of the atmosphere, by equilibria involving alumino-silicates. It is these 'reverse weathering' reactions which, by removing HCO_3 and alkali metals from the oceanic system, prevent the ocean from attaining the composition of a soda lake. The clay minerals most likely to be synthesised in these conditions are illite and chlorite, which is in accord with our knowledge of the abundance of clay mineral species through time.

In its very early days the earth was undoubtedly very different from the earth of today, but by 4000 million years ago, according to Donn, Donn and Valentine (1965):

siallic rocks existed
erosion by running water had started
the surface temperature was above freezing point over much of the earth.

Given this model of the earth, how would weathering proceed?

Donn *et al.* adopt the simple view that rates of weathering and erosion vary together: they write, 'The rate of weathering of bedrock and resulting erosion vary directly with temperature, atmospheric moisture content, and organic acids from vegetation. Since these all increased with geologic time it is suggested that the rate of bedrock erosion, which would have been slow initially, must have accelerated.'

This conclusion is open to a number of criticisms and qualifications.

1. Weathering rate varies with the factors mentioned (temperature, atmospheric moisture content, organic acids) but we cannot be sure that erosion does. Weathering rates and erosion rates may not vary together in the same way with respect to other factors, and in fact it is possible to have little erosion alongside very rapid weathering.

2. It cannot be assumed that temperature increased with geologic time. Geologica evidence suggests rather that it fluctuated, with glacial periods in the Quaternary, Permo-Carboniferous and at several times in the Precambrian, and with periods of mild climate in between.

3. We have little or no direct evidence of atmospheric moisture content from past Periods, and can only infer it from the evidence of past weathering, erosion and deposition. It is debateable whether geological evidence indicates a progressive increase in atmospheric moisture.

There appears to be a consistent change in abundance of clay minerals through time.

Montmorillonite is absent in sediments older than the Mesozoic though common in younger sediments. Presumably the main cause of its absence in old rocks is destruction or conversion to other minerals. Diagenesis or low-grade metamorphism could easily accomplish this, the mineral losing water and picking up cations, especially K and Mg, and so becoming converted to chlorite or mica.

Another possibility is that conditions for initial formation of montmorillonite may not have been as favourable in the past as at present. However, this would require a lack of volcanic activity and peculiar weathering conditions on a grand scale, which seems extremely unlikely.

Grim (1953) suggested that kaolinite may show a similar trend, being rarer in pre-Devonian rocks than younger ones, due to progressive metamorphism. Kaolinite would, of course, persist much longer than montmorillonite.

Since both montmorillonite and kaolinite tend to disappear with age, the commonest clay minerals in old rocks are illite and chlorite. This is mainly due to alteration of other minerals, and not to changing conditions at the earth's surface, although changes in condition may have occurred. In particular, illite is probably the clay mineral most stable in marine conditions, and of course many sedimentary rocks are old marine deposits in which whatever clays were deposited have undergone diagenesis in marine conditions.

In Precambrian times, even when the ocean and atmosphere had attained almost their present form, weathering was quite different from that of today. There was no plant life on land, and so soil as we know it could not be formed. Transpiration was unknown. Ther was no root respiration to provide CO_2 and no cation exchange by roots to provide H ions. Rainfall would be direct, with no vegetation to protect underlying rock, and there would be no matt of roots to hold water and weathering products. No humic acids would accumulate, and cheluviation could not speed up translocation of ions.

In these conditions, of absence of biotic weathering and reduction of chemical weathering assisted by biological products, the overall weathering rate might be expected to be smaller. However, physical weathering could be more intense in such circumstances, and insolation, if it is indeed effective, would have operated upon uninterrupted bare rock surfaces.

With no plants or soil to hold water, run-off would have been intense, and any weathering products would have probably been removed very rapidly. This removal of weathering products would, of course, encourage further weathering, both physical and chemical.

Thus there are some factors reducing weathering (lack of organisms) and some factors speeding it up (rapid removal of weathered products), and one would need to know the balance between these two in order to assess the total effect on weathering.

The appearance of land plants (summarised in Fig. 150) would have had a marked effect on weathering.

The first land plants appeared in the Silurian, and by the Middle Devonian vascular plants were widely established in many of the land areas of the world. Ferns, horsetails and tree ferns were common, and some of the taller lycopods were 40 feet high. Extensive coal remains of Carboniferous age testify to the abundant vegetation of the time, and beneath coal seams are 'seat earths'—fossil soils, often recognisable as hydromorphic soils. By this stage weathering and soil formation had taken on something like their modern form. Grasses spread in the Jurassic, though turf-making grasses, which have a marked effect on erosion rates, came only after the Miocene, and trees appeared in the Cretaceous. The increase in biotic activity brought about an increase in weathering, largely due to the retention of water, so that solution, hydration and hydrolysis were probably increased, and also due to more direct biological effects. At the same time there would be a reduction in erosion due to vegetation protecting rock and soil, and the retention of weathering products would have caused a slowing down of some types of weathering. The effects on the overall rates of weathering would have depended on the balance between the weathering-increasing factors, and the weathering-reducing factors.

In connection with the early evolution of plants and soils, an interesting fact was pointed out by Lewis and Ensminger (1948). Low-order plants have a much greater capacity than higher plants to extract the K ion from minerals that are relatively insoluble. They suggest that the soil in early periods in the earth's history was unlike modern, thoroughly weathered soils, and that the lower non-seed plants of early times had no source of freely available potassium. 'The fact that the older plants are more efficient than the newer ones in obtaining their ions from what might be called unavailable sources, indicates that their direct ancestors lived in an environment where frugality and slow growth were a necessity.'

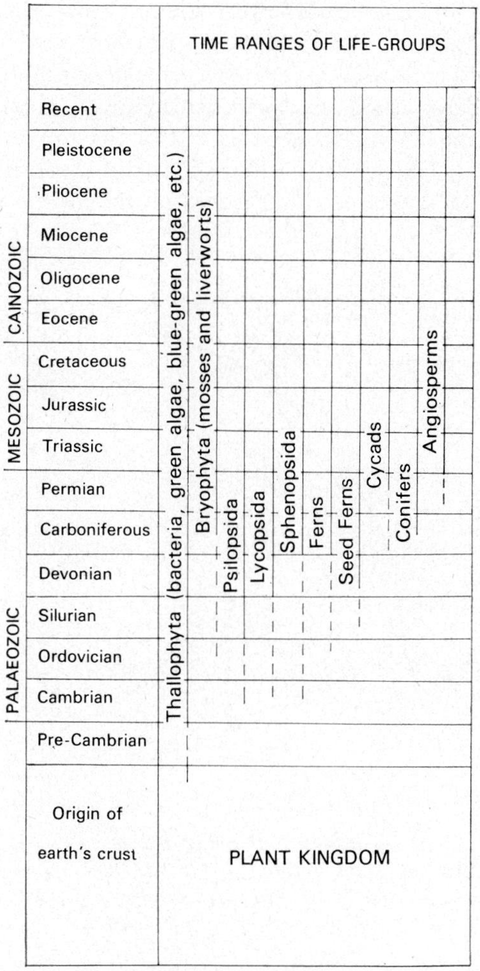

TIME RANGES OF LIFE-GROUPS

CAINOZOIC: Recent, Pleistocene, Pliocene, Miocene, Oligocene, Eocene
MESOZOIC: Cretaceous, Jurassic, Triassic
PALAEOZOIC: Permian, Carboniferous, Devonian, Silurian, Ordovician, Cambrian
Pre-Cambrian
Origin of earth's crust

Thallophyta (bacteria, green algae, blue-green algae, etc.)
Bryophyta (mosses and liverworts)
Psilopsida
Lycopsida
Sphenopsida
Ferns
Seed Ferns
Cycads
Conifers
Angiosperms

PLANT KINGDOM

FIG. 150. Time ranges of the plant kingdom.

If we assume that weathering and erosion are in proportion to sedimentation, then one way to estimate weathering and erosion rates through geological time would be to examine changes in sedimentation rates. Holmes (1957) has shown that the average rate of sediment accumulation has increased since Cambrian time, but he believes this is probably due to an increase in earth movements rather than an increase in weathering rates. Furthermore, Hudson (1964) has questioned the validity of increasing sedimentation rates, and concluded that the evidence for it is unsatisfactory.

Since the Upper Precambrian we can adopt a fairly uniformitarian attitude to weathering, in that water and the atmosphere were present, more or less as today. Only the steady increase in the biotic factor is non-uniformitarian. As it turns out, the greatest differences in weathering during the last 600 million years have been caused by climatic fluctuations rather than biological changes. There are many indications of former climates besides the evidence of weathering, that lead to the following conclusions which are drawn mainly from Schwarzbach (1963).

In Eocambrian times (the uppermost Precambrian) an ice age occurred in many parts of the world, and presumably physical weathering would have been intense.

In the Lower Palaeozoic, all of the northern hemisphere, including land in high latitudes, had a warm climate, as did Australia, and the climate was also somewhat arid. At the same time, traces of glaciation are found in South America and South

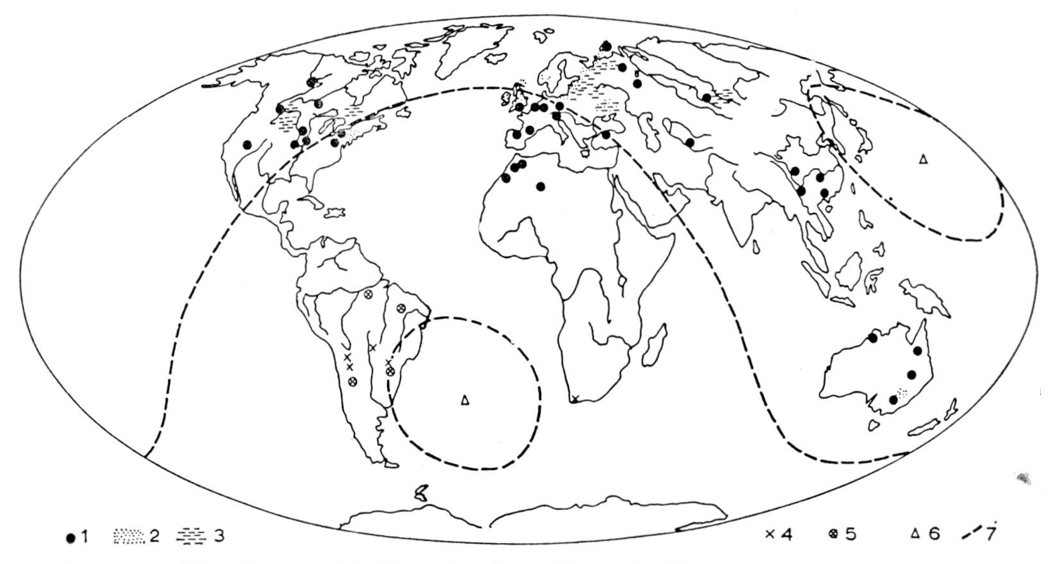

FIG. 151. Climatic map of the Devonian (from Schwarzbach).

1 Coral limestones. 2 Old Red Facies. 3 Evaporites (after F. Lotze). 4 Iapo Formation and Table Mountain Tillite. 5 Upper Devonian Tillite. 6 Paleomagnetic South Pole (after Runcorn) and North Pole. 7 Equator and polar circles.

Africa. Schwarzbach (1963) has shown that the Devonian palaeomagnetic data fits into this picture, with a south pole centred in the South Atlantic (Fig. 151), without invoking continental drift.

In the Lower Carboniferous the whole world seemed to enjoy an equable and warm climate, but in the Upper Carboniferous there was a marked contrast between the northern and southern hemispheres. The southern continents were extensively glaciated, while the northern ones had warm moist climates, becoming arid in the Permian.

An important fact in Palaeozoic palaeogeography is that almost all of Africa south of the Tropic of Cancer, peninsular India and the Brazilian shield were land during almost the entire Palaeozoic era.

In Mesozoic times the whole world experienced a warm phase, and glaciation was completely absent. World climates in the Jurassic were particularly uniform, but in the Upper Jurassic and Cretaceous climatic variations once again became important. In Australia the Mesozoic was a time of humid rather than arid conditions.

At the start of the Tertiary the world was still considerably warmer than it is now. There were no ice caps, trees grew in polar regions, and the climate was more uniform over the earth. In the Upper Tertiary, cooling set in (Fig. 152) and conditions became almost like those of today. Glaciers may have appeared, but some authorities would say that, by definition, glaciation marks the beginning of the Quaternary. The first glaciations started over three million years ago.

The Quaternary is mainly an ice age, with several glacial periods when temperatures were considerably below those of the present day, separated by interglacial periods when temperatures were possibly slightly higher than today. There has perhaps been a

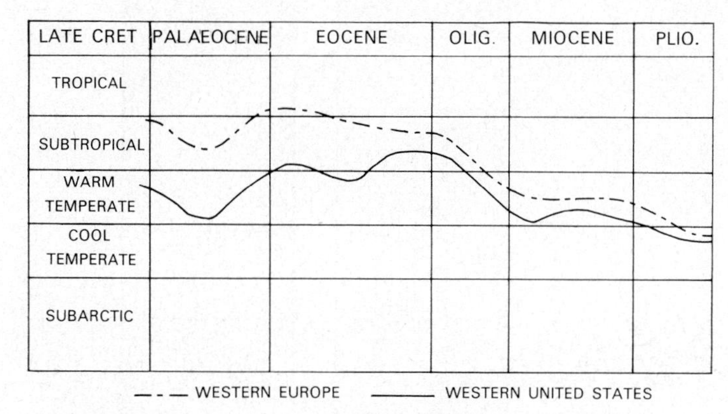

FIG. 152. Curves illustrating the inferred climatic regimes of Western Europe and Western United States since Late Cretaceous times (from Dorf, 1964).

progressive drop in interglacial temperatures from the start of the Pleistocene to the present day. The Last Glaciation ended about 10,000 years ago, and there was a climatic optimum around 5000 to 3000 B.C. when temperatures, in Europe at least, were 2° to 3° higher than now.

Fluctuations of climate during the Quaternary did not only affect glaciated areas, but caused movements of all climatic belts (Fig. 153). Very few areas suffered no changes in climate, and many areas had quite drastic changes.

This very brief review of the world's changing climate is inadequate for any thorough consideration of the history of the world's weathering, but it can act as a background. It is apparent that some of the major climates of the world lasted for Periods or even Eras of geological time, whereas in the Quaternary there have been rapid alternations.

Weathering phenomena indicate specific environment, climates and histories, and the time element cannot be overlooked in weathering studies. We cannot simply apply present-day processes to the past, but must take into account stratigraphy and changes in weathering through geological time. Extensive frost-weathering is attributable to Quaternary times; warm climate weathering may possibly be attributed to interglacials; very deep weathering probably requires warm and moist climates to persist for long periods, and so is perhaps attributable to the Tertiary, or even Mesozoic, times when the climate was suitable.

To take one detailed example: It was once thought that boulderfields in Germany were being formed at the present time. Later they were attributed to periglacial action, and therefore regarded as evidence of cold Quaternary climate. Eventually

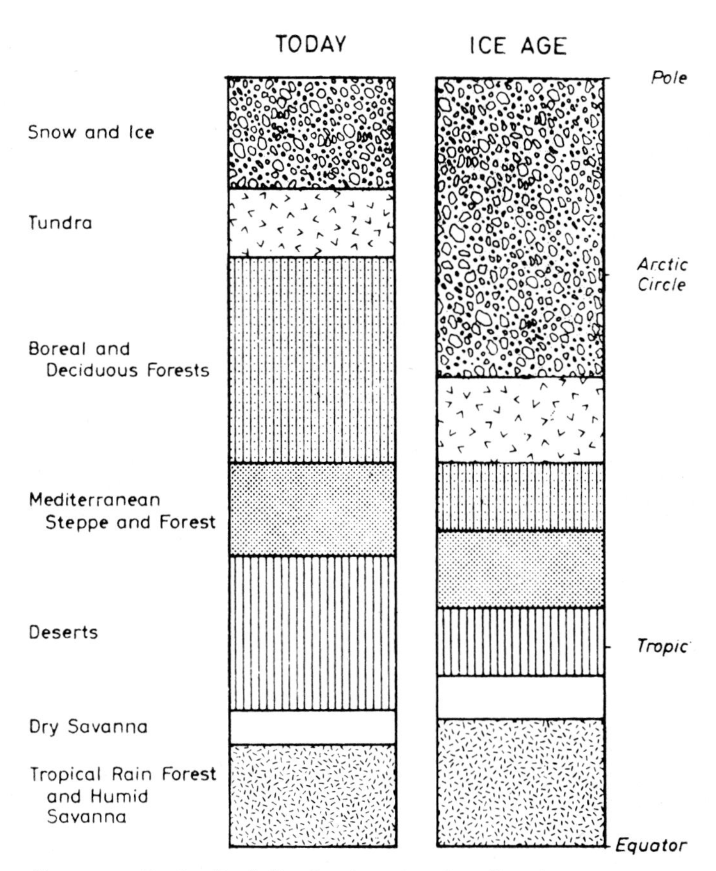

FIG. 153. Latitudinal distribution of major climatic zones in the northern hemisphere today and during each glacial period (after Büdel, 1951).

it was realised that present processes are exposing corestones from a regolith formed in an earlier period of chemical weathering: since mechanical weathering is associated with glacial periods, the corestone weathering was attributed to interglacials when chemical weathering could be dominant. However, it is doubtful whether the intensity of weathering associated with the corestones could be produced in such a short time as an interglacial, and the weathering may be better attributed to some time in the Tertiary. If weathering reached extreme depths, such as described in Chap. X, whole eras may be required to produce the regoliths.

The only sound way to work out weathering histories at present is to study those examples where contiguous rocks allow stratigraphic principles to be applied and detailed geomorphic histories to be elucidated. Unweathered material overlying weathered rock provides a minimum age of the weathering of the latter. A maximum date is harder to obtain, but an estimate can be obtained from the oldest strata that has not suffered the weathering in question.

In the case of the German boulderfields, for example, boulderfields were found beneath loess, and so were older than the loess. By looking at oldest strata affected, Hovermann (1949) came to the conclusion that the weathering was Mio-Pliocene.

Deep weathering in Scotland has been described by Fitzpatrick (1963), who notes that Pliocene gravels contain flints that have suffered little weathering. The deep weathering in Scotland is thought to be of Pliocene or Pre-Pliocene age.

In north-western New South Wales and the adjoining area of Queensland there is a deep weathering profile with duricrust developed on a formerly continuous pediplain. This has been cut by a volcanic dyke, and the freshness of the dolerite indicates that it was emplaced later than the deep weathering. A potassium/argon date of 22·7 million years was obtained from the dolerite, which can be regarded as a minimum age. Assuming the Miocene-Oligocene boundary at about 26 million years, the age of the deep weathering and duricrust is early Miocene at latest, with the strong possibility of a greater age (Langford-Smith, Dury and McDougall, 1966).

The chronology of deep weathering of granites in the Monaro area of New South Wales has been worked out by W. R. Browne (1964). Numerous tors are overlain and surrounded by 'grey billy', a silicified sand and gravel deposit formed beneath an Oligocene basalt cover. There is no doubt that the tors were in existence as such before the Oligocene basalt was poured out, and before the sub-basaltic sediments were laid down. Making use of other evidence besides that of the immediate area, Browne concludes that during a long period of Cretaceous peneplanation the granite was decomposed to a considerable depth. About the close of Eocene time moderate uplift revived erosion, and partial stripping produced a tor topography. This was covered by Oligocene basalt, which preserved it for a long time. Exhumation probably began in Pliocene time and has continued to the present day.

Browne's conclusion may possibly be extended to other areas, not only in Australia but on a world-wide scale. The Cretaceous was a period of little orogeny and extensive planation, and, as we have seen, the climate over much of the world was suitable for extensive chemical weathering.

In time it may be found that much weathering falls into a large scale and simple history. In Africa, Australia and South America, and possibly even on other continents, the history of the plainlands, elucidated and described by Lester King (1962), could possibly be correlated with a history of weathering. The Gondwana surface, of Mesozoic age, had a continuous land history of scores of millions of years, and for long periods suffered negligible erosion. Perhaps many of the examples of extreme deep-weathering can be attributed to this time. Other, and possibly lesser periods of extensive weathering would have occurred through the Tertiary, and even in some areas in the Quaternary.

XV TECHNIQUES

INTRODUCTION

WEATHERING studies use techniques borrowed from many disciplines including geology, geomorphology, pedology, mineralogy, sedimentary petrology and chemistry. It is clearly impossible to summarise all techniques in one chapter, and the present account is meant only as an introduction and a guide to further literature. I have assumed that the methods used in field geology and geomorphology are known to the reader, and the summaries of techniques that follow are designed to make the relevant literature in other fields intelligible to geomorphologists.

SOIL AND WEATHERING PROFILE DESCRIPTION

In describing weathering profiles many of the techniques used by pedologists in soil profile description can be applied, and of course many weathering profiles are little more than soil profiles. The techniques are described in various books, notably the *Soil Survey Manual* published by the U.S. Department of Agriculture (1951). Only the major aspects of soil description will be given briefly here.

HORIZON DETERMINATION

The first step in a profile description is to divide it into various horizons. These may be given designations as in soil science, such as A1, B3, etc., but this is not very important. The depths and thicknesses of each horizon are recorded and the horizon boundaries are described. Horizon boundaries vary in distinctness and in surface topography. The distinctness depends on the thickness of the transitional area between horizons, and may be described as: (1) *abrupt*, if less than 1 in thick; (2) *clear*, if about 1 to 2½ in thick; (3) *gradual*, if 2½ to 5 in thick; and (4) *diffuse*, if more than 5 in wide. The topography of the horizon boundary is described as: (1) *smooth*, if nearly a plane; (2) *wavy* or undulating if pockets are wider than deep; (3) *irregular*, if pockets are deeper than wide; and (4) *broken*, if parts of the horizon are unconnected with other parts.

When the horizons have been determined, the material of each horizon is described. The description may consist of a list of many properties, such as pH, friability, plasticity, presence of roots or concretions, etc., but the properties most generally given are colour, texture and structure.

COLOUR

Colour of soil (or other material) is determined by comparing a sample with the colours in the *Munsell Soil Color Charts* (published by the Munsell Color Company, Inc., Baltimore, U.S.A.). Colour is reduced to three properties, which are:

1. *Hue*

This is the dominant spectral colour, mainly the redness or yellowness of the specimen. Different pages in the Munsell book have different hues, with names like 'ten red' (10 R) or 'two point five yellow red' (2·5 YR).

2. *Value*

This is the lightness or darkness of the colour, and varies vertically on the charts.

3. *Chroma*

Chroma is the intensity of the basic spectral colour, and varies across the page.

The colour of a specimen is given by the Hue, Value and Chroma in that order, e.g. 2·5 YR 3/4 or 5R 6/6.

Colour varies with the wetness of the specimen, and either both wet and dry colours should be given, or it should be stated whether the colour refers to wet or dry specimen.

TEXTURE

By texture soil scientists mean the particle size distribution of soil material less than 2 mm in diameter. Not only is coarser material excluded from the texture, but a completely different method is used to express the distribution from that used by sedimentologists (see p. 268). First the amounts of sand, silt and clay are determined (or in the field estimated). On the International Classification the size limits of these fractions are:

Sand (coarse)	2·0–0·2 mm
(fine)	0·2–0·02
Silt	0·02–0·002
Clay	Less than 0·002

The International scale has the advantage of consistency about the figure 2, but there is an alternative scale, the American Classification, which has the advantage that the silt (and other) fractions can be determined by rubbing the soil between the fingers. The size limits of the fractions are:

Very coarse sand	2–1 mm
Coarse sand	1–0·5
Medium sand	0·5–0·25
Fine sand	0·25–0·1
Very fine sand	0·1–0·05
Silt	0·05–0·002
Clay	Less than 0·002

Ignoring the subdivisions of the sand class, the quantities of sand, silt and clay enable the particle size distribution to be plotted on a triangular diagram like Fig. 154.

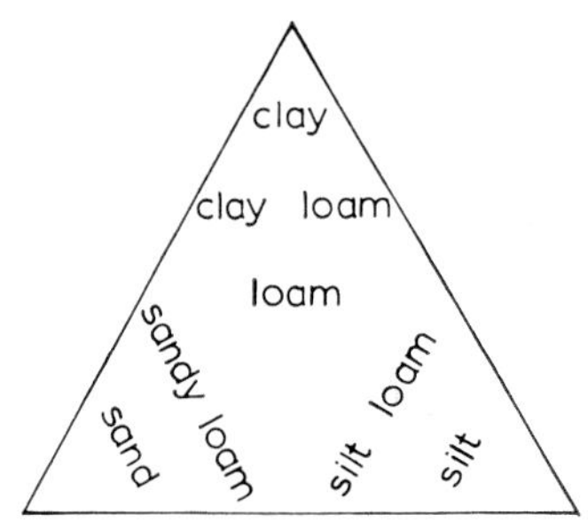

FIG. 154. Triangular diagram showing simple relationships of soil textural classes.

Basically a uniform mixture of sand, silt and clay is called a loam, and occupies the central part of the triangle, and is surrounded by various other textural classes as shown in Fig. 154. The divisions on the American diagram (Fig. 155)are based on the same logic. Experienced soil scientists are rarely more than one class out when using the American divisions, and are frequently correct. This system is therefore useful in the field, which is the one great advantage of soil texture methods over the more elaborate methods of particle size distribution study described later.

STRUCTURE

Soils and weathered rock break into aggregates of particular size and shape, called peds. The basic types of soil structure are platy, prismatic, columnar (like prismatic but rounded at the top), angular blocky, subangular blocky, granular and crumb. The approximate size of the peds should also be given. If the soil consists of large peds that break into smaller peds, both should be described. Structureless is the term used when there is no observable aggregation, and the soil may be termed massive if coherent, or single grain if not.

In weathering profiles it is also important to describe remnants of original rock structures when present, but of course the word structure has quite a different meaning in this sense. Fig. 156 shows some examples of soil structures.

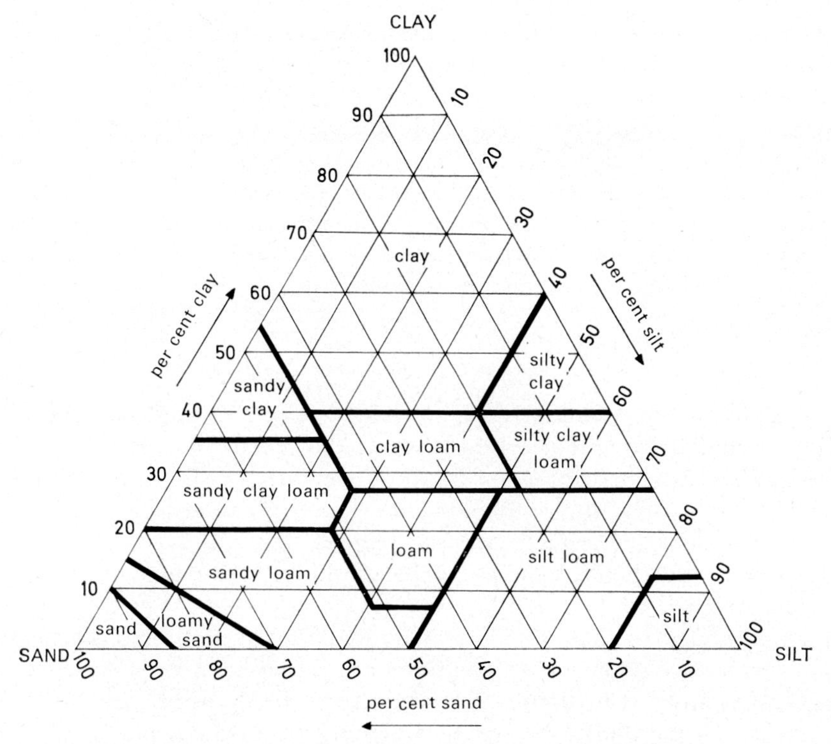

FIG. 155. Chart showing the percentages of clay (below 0·002 mm), silt (0·002 to 0·05 mm) and sand (0·05 to 2·0 mm) in the basic soil textural classes.

MECHANICAL ANALYSIS

In many studies of weathering it is necessary to know the particle size distribution of samples of material—not only of soil but of grus, scree, hillwash or other weathered material. Mechanical analysis is the name of the technique by which the particle size distribution is determined. Even in routine description of a single deposit a particle

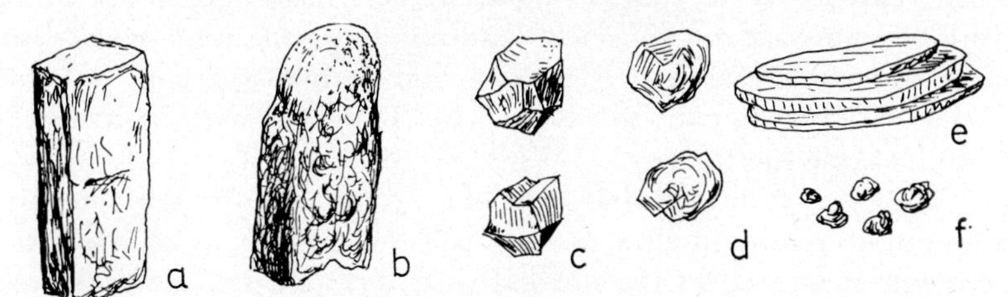

FIG. 156. Soil structure.

a Prismatic. *b* Columnar. *c* Angular blocky (corners are sharp or pointed). *d* Subangular block (faces may meet in a sharp angle, but corners are rounded). *e* Platy. *f* Crumb.

size distribution is helpful, but in some instances it is possible from a study of several analyses to see something of the course of weathering, such as an increase in the amount of silt, breakdown to clay, etc.

There are three operations involved in mechanical analysis. First, it is necessary to obtain the particle size distribution, and a number of techniques are available for this. Secondly, the particle size distribution may be represented by various conventional graphs or diagrams. Thirdly, there are numerous parameters (statistically derived attributes of the particle size distribution) that are used as a kind of shorthand in describing sediments and in comparing one with another.

MECHANICAL ANALYSIS METHODS

For coarse material the sediment may be simply passed through a series of sieves of graduated size, and the weight of sediment retained by each sieve weighed. The weights are then expressed as a percentage of the total weight. With clean and dry material it is possible to 'dry sieve' the sediment, but when fine material is present it may be necessary to 'wet sieve', in which case the various fractions are washed through the sieves with water, and then dried out before weighing. Of course, the 'fines' must be collected and weighed too.

For fine material (less than 2 mm) techniques based on settling of particles through a column of water are used. A weighed amount of sediment is dispersed in a certain amount of water, and left to settle. At certain times readings are taken—by means of a hydrometer, a special plumb bob, or by weighing samples collected at a stated depth by pipette.

Fine particles tend to coagulate, and conventional methods are used to disperse the sediment before analysis. A popular dispersing agent is Calgon (Sodium hexametaphosphate). In some cases other treatment may be necessary, such as removal of organic matter.

To a certain extent mechanical analyses are conventional, and cannot ever be completely precise. Grains are not equidimensional, and lath-shaped grains will go through a hole in a sieve when end on but may be too long to go through sideways. In a given time of shaking a certain proportion will pass through a sieve, but not all. In mechanical analyses by settling, calculations are based on spherical particles of regular density. Grains are not spherical, and not all of the same density, so this is an approximation. For this reason it is usually best to follow a conventional method, which has the advantage that the results can be compared with many other analyses carried out by the same method.

There are, however, certain times when one wants not so much convention as the best approximation to reality in a particular instance. It may be that weathered flakes would disaggregate if treated in the normal way and give too many 'fines'. In this case it is possible to use either thin sections or polished surfaces of the specimen, and measure the areas of a large number of grains under the microscope. Statistical techniques are then available for getting an idea of the actual grain size distribution in the specimen. For fine-grained material this method is unsatisfactory.

PARTICLE SIZE DISTRIBUTION

In most scientific work it is normal to divide the various grain sizes of a sediment on a geometrical scale in millimetres, so that sieves have apertures of 2 mm, 1 mm, 0·5 mm, 0·25 mm, etc. Such a scale is in general use in sedimentology, and is known as the Wentworth scale. The sizes mentioned are also given other numbers on a ϕ scale (the Greek letter 'phi'), shown below, so that particle sizes can be discussed in whole numbers, and to facilitate further work.

Table 26. The Wentworth scale

Aperture of sieve or grain size in mm	ϕ
4·0	−2
2·83	−1·5
2·0	−1
1·41	−0·5
1·0	0·0
0·7	0·5
0·5	1·0
0·35	1·5
0·25	2·0
0·177	2·5
0·125	3·0
0·088	3·5
0·062	4·0

Representation of distribution

Normal conventional diagrams are often used for illustrating particle size distribution. Histograms, bar charts, frequency curves, and especially cumulative frequency curves, can all be used. Fig. 157 shows the same data plotted in several ways. Frequency

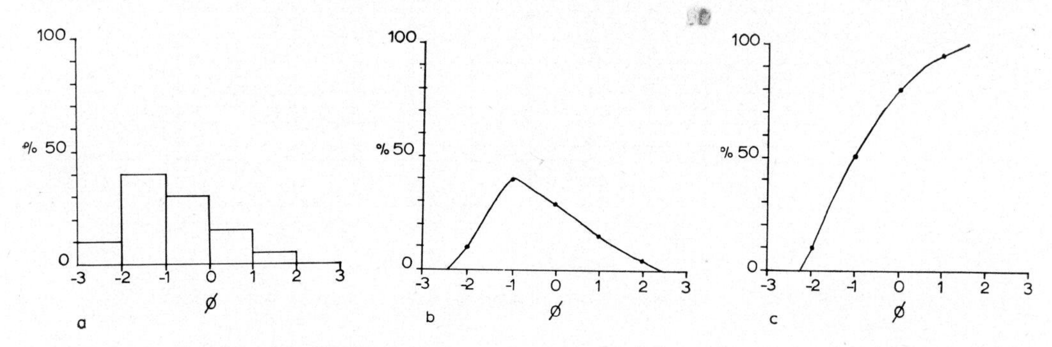

FIG. 157. Particle size distribution of a weathered granite plotted as: *a* Histogram. *b* Frequency curve. *c* Cumulative frequency curve.

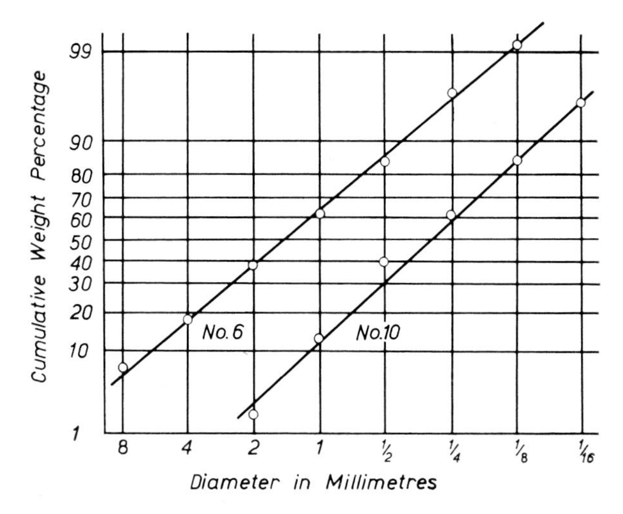

FIG. 158. Particle size distribution of two clastic sediments plotted on probability paper (from Pettijohn, 1957).

curves enable certain properties of the distribution to be seen very easily, such as skewness and kurtosis. Cumulative frequency curves have the great advantage that percentile readings (see p. 274) can be made with ease at any desired point, and this is very useful if parameters are to be calculated.

A well-sorted sediment has what is known as a 'normal distribution' and this gives a symmetrical curve as in Fig. 163, p. 274. There exists a special sort of graph paper called probability paper, on which such a distribution comes out as a straight line (Fig. 158). Deviations from a straight line are then most easily seen.

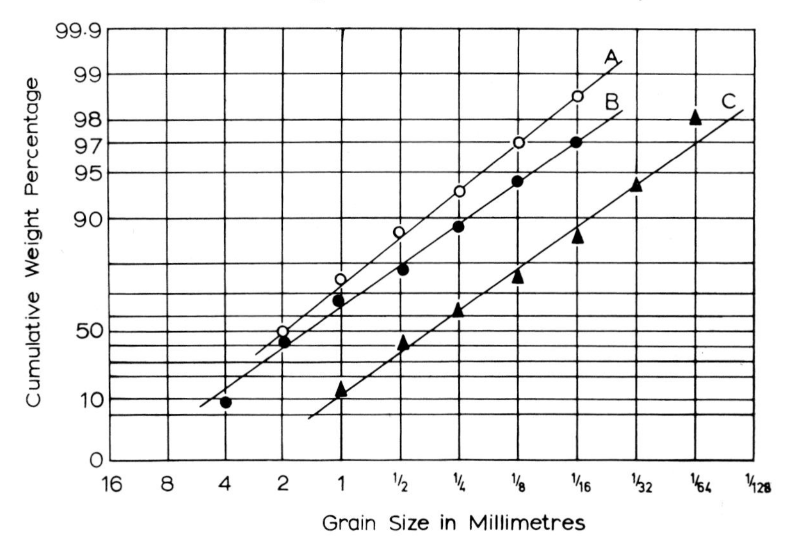

FIG. 159. Particle size distribution plotted on Rosin-Rammler paper (data from Pettijohn, 1957).

A Artificially crushed quartz. *B* Disintegrated igneous rock. *C* Detritus from weathered gneiss.

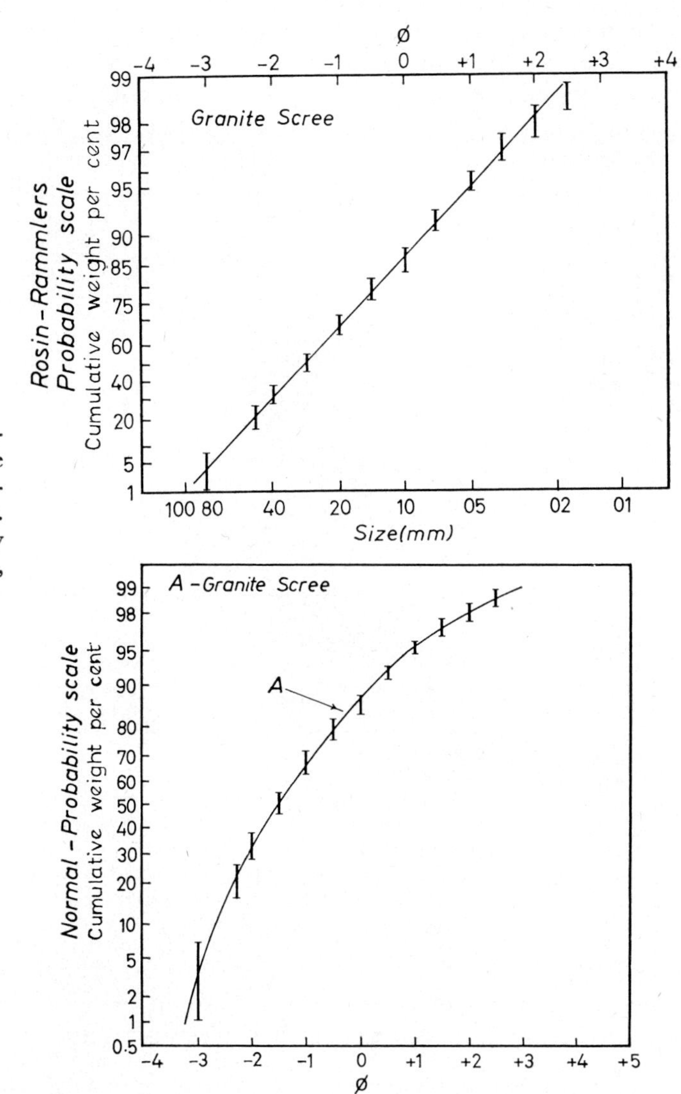

FIG. 160. Particle size distribution of granite scree plotted on: *Above* Rosin-Rammler probability paper. *Below* Normal probability paper (from Kittelman, 1964).

Plotting mechanical analyses on probability paper has become increasingly common in the past few years, but it does have many disadvantages. A 1 % unit around 95 % probability is about four times as great as that around 50 % probability, and the same holds true for the corresponding units at low probabilities. The units are therefore increasingly exaggerated in both directions from the 50 % probability locations, and deviations from a straight line are exaggerated at both ends of the distribution. Now most particle size distribution measurements suffer their greatest inaccuracies at the ends of the distribution, and so plotting on probability paper exaggerates these inaccuracies, and emphasises the least reliable aspects of the particle size analysis.

It is sometimes thought that certain materials do not follow the normal distribution, but another distribution based on the 'law of crushing'. Mechanically broken rock, and crushed coal, etc., are said to follow this distribution, and so is pyroclastic material (volcanic ash, scoria, etc.) and weathered material. A kind of graph paper has been devised on which such distributions come out as straight lines. This is the Rosin and Rammler distribution paper (Fig. 159). If such distributions really were typical of

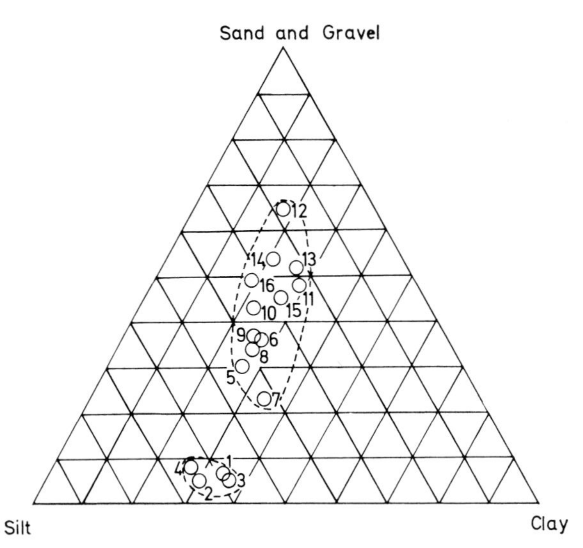

FIG. 161. Mechanical analysis of samples from a profile on till, Illinois. Samples were taken at one-foot intervals starting at six inches and numbered consecutively downwards. The lower part of the profile is weathering till, but the top samples (1-4) appear to represent an addition of loess (from Gravenor, 1954).

weathered rocks this plotting technique would be very useful in weathering studies. However Irani and Callis (1963) have pointed out some major objections. They write:

'The Rosin-Rammler equation is strictly empirical and relies heavily on curve fitting, and has been found to be applicable only in the case of powdered coal having a small range in particle size. Even then a modified log-normal distribution equation can be shown to give just as good a fit. In addition, the Rosin-Rammler distribution cannot be conveniently used to obtain the various statistical measures of the particle size distribution such as weight-mean size.'

Fig. 160 shows an example of size frequency distribution plotted on Rosin-Rammler's paper, and for comparison the same data plotted on probability paper. On the Rosin-Rammler graph the hypothesis of linearity is rejected at the 95% level, but the goodness of fit is qualitatively much better than with the normal-probability paper.

If a distribution can be expressed in amounts of three components then the sediment can be represented by a point on a triangular diagram. This is commonly done with soils where the percentages of sand, silt and clay provide a unique fix on a triangular diagram (Fig. 154, p. 266).

Less common diagrams can be used with other components, such as gravel, sand and clay, or any three components chosen for a particular purpose. Fig. 161 provides an example, and shows clearly that the profile is not due to weathering alone.

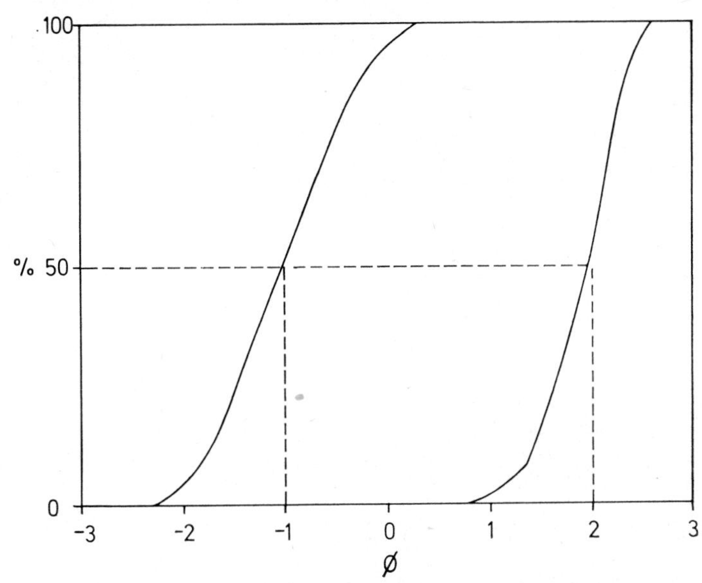

FIG. 162. Cumulative curves of different medial value. Curve *a* on the left, curve *b* on the right.

PARAMETERS

Parameters are expressions of certain aspects of the distribution, and to express various properties of the distribution a number of conventional parameters are generally used.

Average size

In sediment studies the median is the most generally used measure of central tendency, and therefore of average size. If the ϕ scale is used the median is called $M\phi$, and is read off where the 50% line crosses the cumulative distribution. In Fig. 162 curve *a* has a $M\phi$ of -1, and *b* has a $M\phi$ of 2·0. *a* is therefore a coarser deposit than *b*.

Degree of sorting

A well-sorted deposit will have all particles close to the median size, whereas a badly sorted one will have a wide spread of particle size about the mean. A measure of the spread of the curve therefore serves as a measure of the degree of sorting of the deposit ('sorting' is the word generally used, though a weathering deposit that appears 'well-sorted' may have never undergone any actual sorting process).

Quartile deviation (QDϕ) is one measure of sorting. Quartiles are read where the 25% and 75% lines cross the plotted curve, and are known as Q1 and Q3. The quartile deviation is calculated from the formula

$$QD\phi = \frac{Q3 - Q1}{2}$$

Trask's sorting coefficient, *So*, is calculated from the formula

$$So = \frac{Q1}{Q3}$$

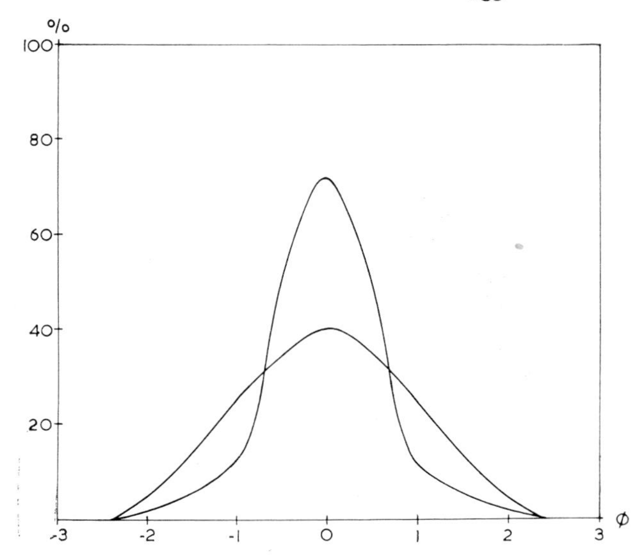

FIG. 163. Two frequency curves of different kurtosis.

A coefficient of 2·5 indicates a well-sorted sediment
3 indicates a normal sediment
4·5 indicates a poorly sorted sediment.

Trask's sorting coefficient is now thought to be rather inefficient, and more elaborate ones have been proposed. Some use the 16 percentile (read off where the 16% line crosses the cumulative curve) and the 84 percentile, and even the 5 and 95 percentiles, though this puts a lot of reliance on the ends of the curve, where accuracy is usually lowest.

Other parameters

Fig. 163 shows how two curves can have the same median and distribution, but differ in peakedness. The property of peakedness is called kurtosis, and there are various formulae for expressing it.

Fig. 164 shows two curves with the same spread but different medians. The symmetrical curve is said to be normal, and the other one skewed. Again there are formulae for determining skewness.

These and a whole host of parameters are now available to research workers, but although many are conscientiously worked out as a routine operation, very few have proved useful, and in weathering studies these more elaborate treatments are not at present justified.

The parameters are mainly described in the journals of sedimentary petrology. A summary of the main ones is given by King (1966).

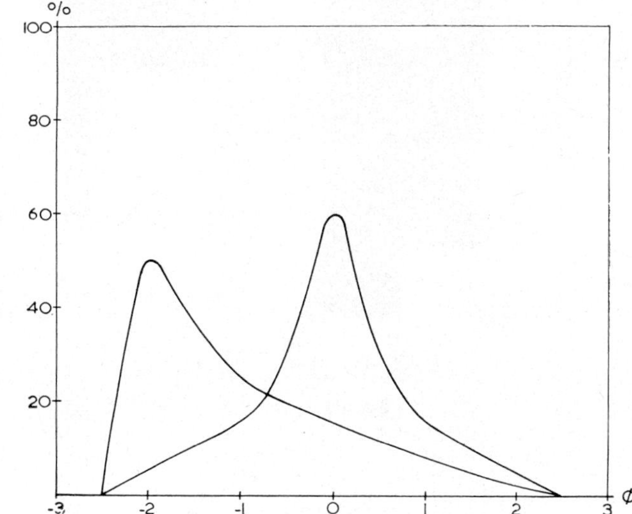

FIG. 164. Frequency curves showing a normal (symmetrical), and a skewed distribution.

MINERAL ANALYSIS

To study the course of weathering one may examine the mineral content of the fresh rock and compare it with the mineral content of the weathered product, possibly through several intermediate stages. Clay minerals are studied by special techniques described later. Sand and silt size minerals are investigated under a polarising microscope, using the techniques of either grain mounting, or thin section. A few other techniques may be used for special reasons.

Thin section investigation

There are routine techniques for the preparation of thin sections of rocks (conventionally 30 microns thick). These are mounted with Canada balsam or other material of the same refractive index (1·54). The slides (normally 3 in × 1 in) are examined under a petrographic microscope, and the minerals recognised by shape, colour, cleavage and by complex optical properties described in textbooks of optical mineralogy.

Thin sections of weathered rocks, sediments and soils are much more difficult to prepare than sections of hard rocks, but numerous techniques are available (see Brewer, 1964). The specimen is first impregnated with a resin, plastic or other material to make it firm, and the section cut or ground in a manner preventing further alteration, such as smearing of clay. Care must be taken to avoid artefacts—features in the section due entirely to the mode of manufacture and not intrinsic properties of the material.

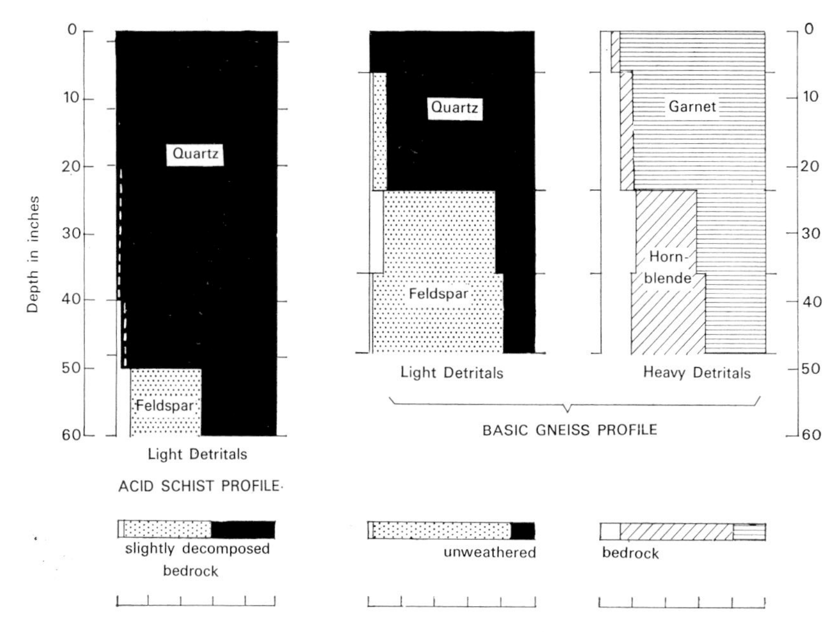

FIG. 165. Distribution of minerals in the fine sand fraction of two tropical soils (from Stephen, 1953).

Minerals are identified as in rock slides, but with soil and weathered material more techniques are required. Clay minerals are not readily identified under the microscope but there are a host of techniques for examining and describing the fabric of clay materials—that is the shape and arrangement of clay fragments, pore spaces, iron oxide distribution and other features which are not part of the original rock but are due at least in part to the nature, degree, and style of weathering. A special science of micropedology is devoted to such studies. Brewer (1964) gives a full account of techniques available.

Thin sections have some advantages over grain studies in investigations of weathering. One can see whether grains are being attacked uniformly, along cleavage planes or only on the interfaces. The formation of new minerals (authigenic minerals) can be seen, and such features as the formation of concretions, the infilling of rock vesicles, and volume changes consequent upon alteration can all be traced. Sometimes colour changes indicate early stages of weathering, as in bleached rims of biotite crystals, or clouding of feldspars.

MINERAL GRAIN ANALYSIS

The specimen is disaggregated, and the various size fractions separated (commonly the different fractions are simply retained after mechanical analysis). Silt, fine sand and coarse sand can all be examined by petrographic microscope. Clays cannot be studied this way, and coarser particles are better studied by a binocular microscope.

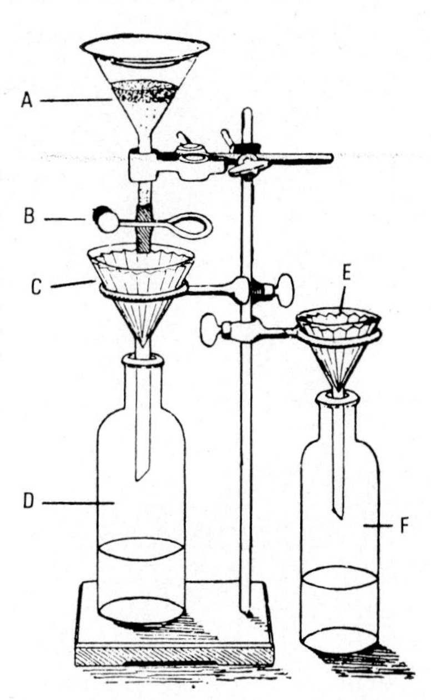

Fig. 166. Simple apparatus for heavy mineral separation (from Milner, 1962).

a Funnel containing bromoform and sample from which heavy minerals are settling. *b* Pinch-cock on rubber tubing. *c* Filter paper for catching heavy minerals. *d* Bromoform bottle. *e* Filter. *f* Benzol (or other solvent) washings bottle.

The aim of the investigation is to determine the minerals present, and also the shape, surface texture and other features of the individual grains. By simply determining the presence or absence of grains in a series of samples of increasing weathering, a weathering sequence can be worked out, as for instance in the Buwekula profile (p. 60).

In soil studies mineral analysis may be used to determine whether a horizon is indeed due to weathering and pedogenesis, or is a layer of different material added by a geological agent, as for instance in the Batcombe profile (p. 156).

The distribution of mineral species in a profile is often best expressed diagrammatically as in Fig. 165.

The appearance of secondary minerals can also be documented, and more advanced techniques, such as the zircon index, may be used to get an estimate of total amount of mineral weathering.

The grains are usually studied by mounting them in a film of liquid of known refractive index. Clove oil, with a refractive index of 1·54 (the same as Canada balsam) is commonly used for routine examination. Sets of refractive index liquids are available commercially or can be made up, and by transferring grains from slide to slide, from one liquid to another, the refractive indices of the mineral are determined, which is a great help in identification. However, many of the common minerals can be identified by colour, form, or a few other simple properties, and for many elementary studies it is possible to work out the major features of weathering without a thorough knowledge of optical mineralogy.

The bulk of sands are usually quartz grains. These so outnumber other grains that it is difficult to get estimates of the amounts of other grains if the whole sample is used. It so happens that most of the less common minerals, known as accessory minerals, are

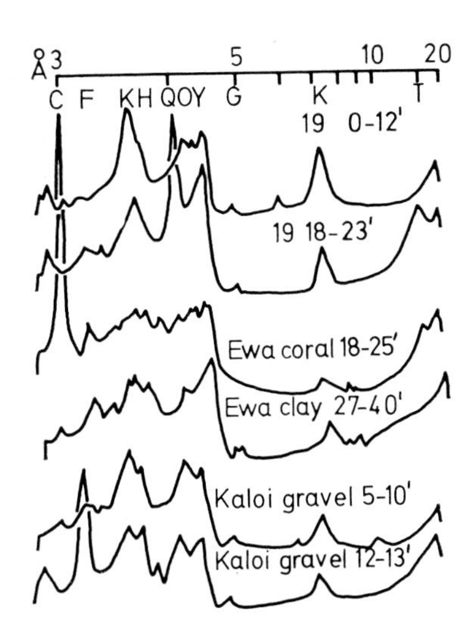

FIG. 167. X-ray diffraction patterns of samples of weathered material from Hawaii (from Ruhe, 1965).

c Calcite. *f* Feldspar. *k* Kaolinite. *h* Hematite. *q* Quartz. *o* Geothite. *y* 110 clay. *g* Gibbsite. *t* 2 : 1 lattice clay.

considerably denser than quartz, and it is common practice to separate the 'heavy minerals' and study them separately. Numerous elaborate methods are available for separating heavies, but the apparatus shown in Fig. 166 is quite serviceable. Bromoform, specific gravity of about 2·8, is placed in the funnel. Sand grains are sprinkled on the surface and stirred, and the heavy grains sink to the bottom. These are tapped off and collected on a filter paper, washed with alcohol or other suitable solvent, dried, and are then ready for investigation. The light minerals can be collected and studied separately. A more detailed account of separation techniques can be found in textbooks of sedimentary petrology, such as Milner (1962) and Pettijohn (1957).

CLAY MINERALS

The commonest techniques used in the study of clay minerals are X-ray diffraction and differential thermal analysis (D.T.A.).

X-ray diffraction is concerned with determining the spacings between the different planes in the crystal lattice, and certain spacings are characteristic of certain minerals. In practice a curve is usually obtained showing the spacings and intensity of lines, as in Fig. 167. Peaks on the curve reveal the various minerals present.

Differential thermal analysis reveals exothermic and endothermic peaks that characterise different minerals. Fig. 168 shows some typical mineral curves, and Fig. 169 shows some D.T.A. results from soils.

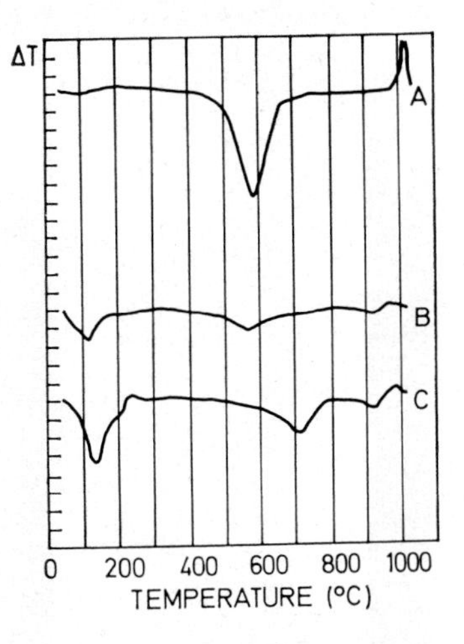

FIG. 168. Differential thermal curves for some clay minerals.

A Kaolinite. *B* Illite. *C* Montmorillonite.

Clay mineralogy in weathering studies is largely concerned with identification of clays, though more advanced studies trace the course of clay genesis or alteration. Quantitative determination of clays is still very difficult, but semi-quantitative determinations of mineral species can be made.

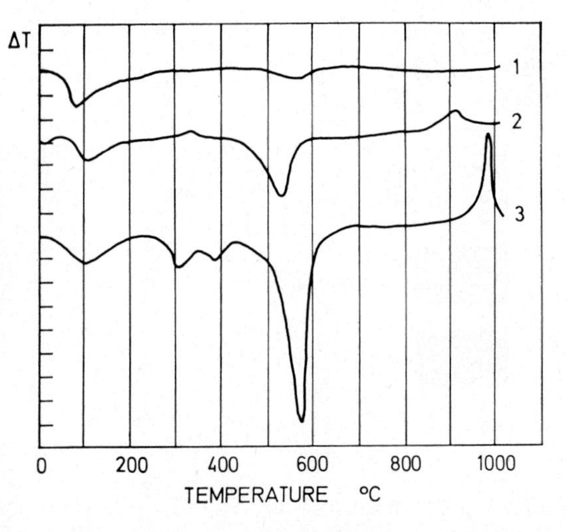

FIG. 169. Typical differential thermal curves for soil clays from:

1 Sweden. 2 Scotland. 3 West Africa. Tropical soils give better defined curves than Arctic soils (from Rich and Kunze, 1964).

The proportion of different minerals may be expressed in tables (as in Table 27), in bar charts as in Fig. 170, or by other means. The bar chart is perhaps the best for comparative studies as it enables easy visualisation of changing proportions.

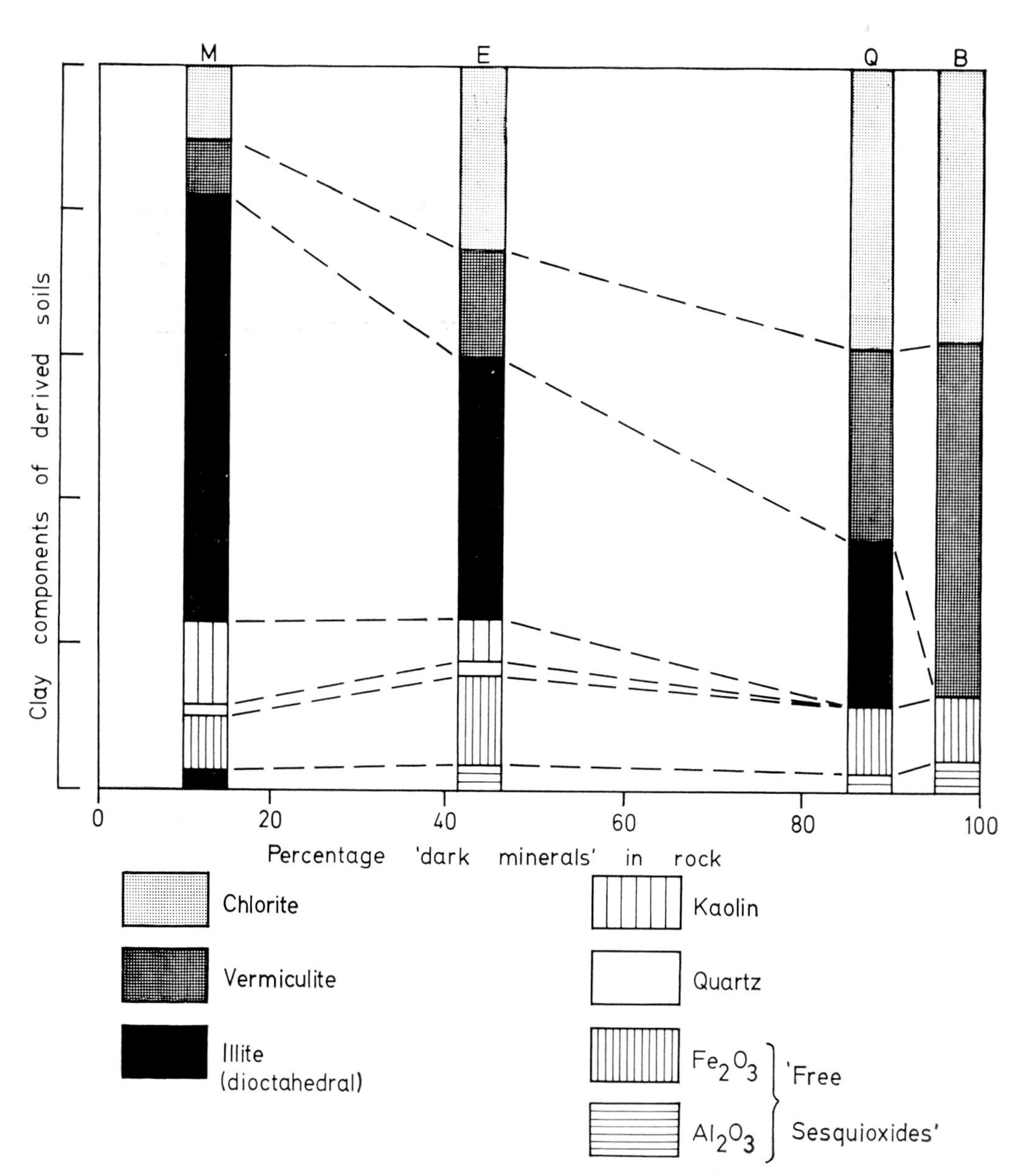

FIG. 170. Bar charts showing variation in the clays of soils derived from different parent rock. M—granite, E—'Ivy-Scar rock', Q—appinite, B—biotitite (from Stephen, 1952).

Table 27. Clay mineral analyses of Chiltern soils, from Avery, Stephen, Brown and Yaalon, 1959 (proportionate amounts of layer-lattice silicates expressed as parts of 10).

Profile and sample no.		Depth (in)	Mo	Vm	M	Ka	Cl
Batcombe (flinty) silt loam	Bu 14/1	1–2	—	5	2	3	Tr.
	14/2	2–8	—	6	1	3	Tr.
	14/3	8–16	—	6	1	3	Tr.
	14/4	16–29	—	5	2	3	—
	14/6	36–60	—	3	3	4	—
	14/7	78–90	2	1	3	4	—
Winchester flinty loam	Bu 54/1	0–2½	Tr.	7	Tr.	3	—
	54/2	2½–7	Tr.	7	Tr.	3	—
	54/3	7–12	6	Tr.	1	3	—
	54/4	12–26	6	—	2	2	—
	54/4a*	—	3	3	1	3	—
	54/5	26–37	6	—	1	3	—
	54/6	37–43	6	—	2	2	—
Charity flinty silt loam	Bu 55/1	0–3	—	5	3	2	Tr.
	55/2	3–11	—	6	2	2	Tr.
	55/3	11–17	—	7	1	2	—
	55/4	17–27	—	3	5	2	—
	55/5	27–33	2	1	5	2	—
	55/6	33–39	4	—	4	2	—
Chalk residue	C 1	—	8	—	2	—	—
	C 2	—	8	—	2	—	—
Reading Beds	RB 1	—	6	1	2	1	—
	RB 2	—	4	Tr.	4	2	—
	RB 3	—	3	Tr.	5	2	—

Mo = montmorillonite; Vm = vermiculite; M = mica; Ka = kaolin; Cl = chlorite.
* Sandy material at 12–26 in.

Many other techniques are available for the study of clays, but have not as yet become as standard as those described above. Infra-red analysis, X-ray spectographic analysis, electron microscopy and other methods all have their particular uses. A summary of these techniques and their application is in Rich and Kunze (1964).

THE EXPRESSION OF CHEMICAL WEATHERING

The methods of chemical analysis are clearly beyond the scope of this book, but the methods by which chemical data are presented are our concern, and must be understood if full benefits are to be derived from inspection of chemical analyses.

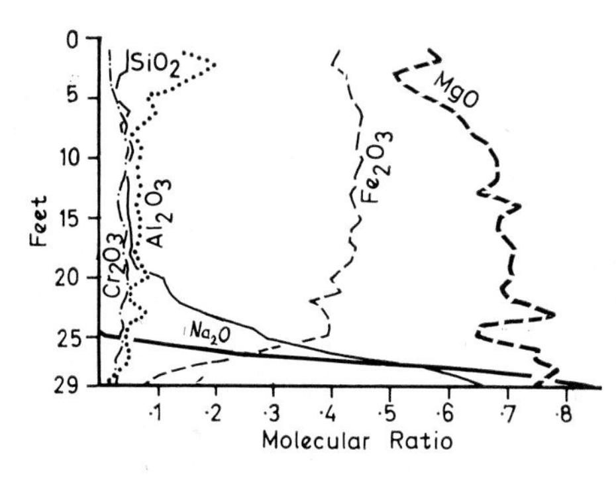

FIG. 171. Variation of chemical composition with depth of an iron ore in Cuba (from Reiche, 1943).

In general, chemical studies of weathering consist of making total chemical analysis of samples from different depths in a weathering profile, and comparing fresh rock with rock in various stages of decomposition. Since weathering consists largely of differential losses of elements, the amount of different elements in fresh and weathered rock are an expression of the type and extent of weathering.

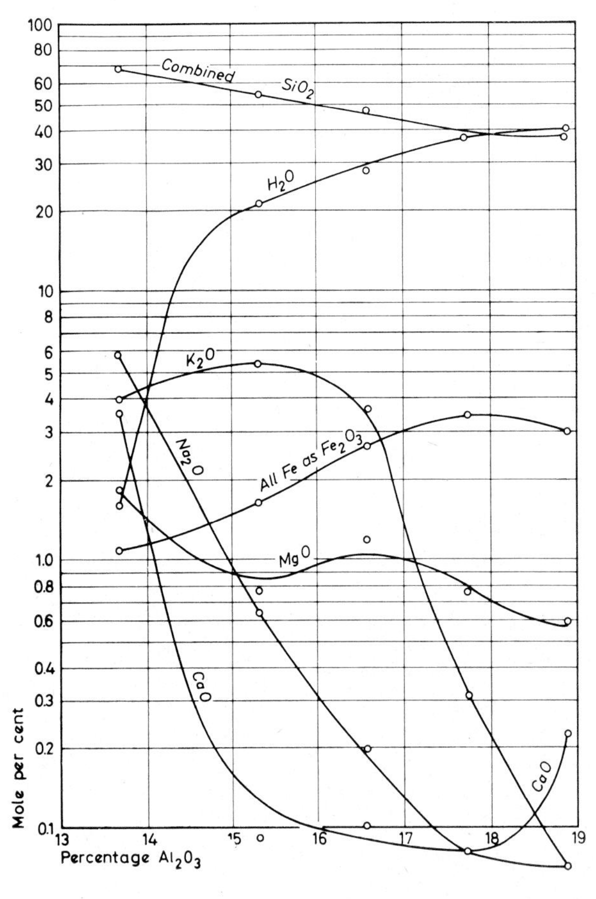

FIG. 172. Variation in chemical composition of specimens plotted against mole percent of alumina in the specimens. The composition of unaltered rock (a granite gneiss) is shown by the points on the vertical at the left (from Reiche, 1950).

Analyses are usually given in weight percentage of the oxides of the elements. For comparative purposes it is often better to use molecular ratios, which are obtained by dividing the weight percentage by the sum of the atomic weights that they represent. If molecular ratios are not used the variations of lighter substances are obscured.

In the simplest methods the quantities of individual substances in fresh and altered rock are compared directly, plotting amount of substance against sample. The actual depth of sample is often used for one axis. Weight percentage or molecular ratios may be used, and log scales can be incorporated to cover a wide range of values with little loss of detail. An example of a simple plot is shown in Fig. 171.

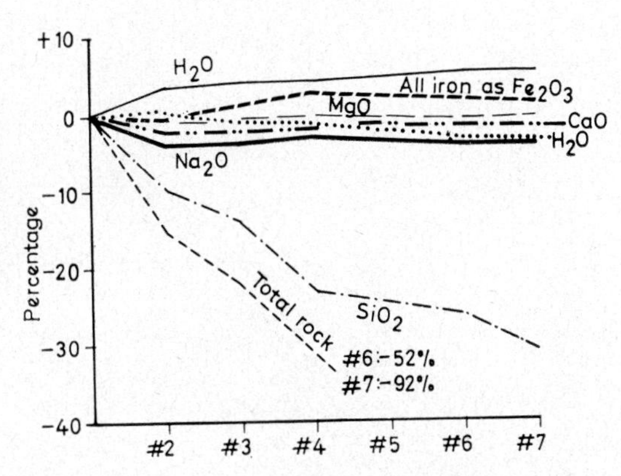

FIG. 173. Loss in terms of weight percentage of unaltered rock assuming Al_2O_3 constant. Weathering Morton gneiss. Sample 5 omitted as definitely aberrant (from Reiche, 1943).

In the example of Fig. 171 the variations of the elements were plotted against depth. If it is believed that one component, say Al, has remained immobile through the weathering process, then the percentage of the stable element may be used as one axis in a variation diagram (the chosen stable substance must vary consistently). Again a log scale is convenient (Fig. 172).

A slight elaboration of the above method provides a diagram in which trends may be seen more easily. If a substance is thought to be immobile, then the losses of other components in a weathered rock can be calculated. The losses (or gains) so derived can be plotted against depth, as in Fig. 173.

The course of weathering can also be represented by plotting specimens on a triangular diagram. The chemical analysis must be reduced to 3 components for this, such as silica, sesquioxides and alkalis plus alkaline earths, as shown in Fig. 174.

Reiche devised a rather complicated method of depicting the course of weathering, shown in Fig. 175, which is not quite as difficult as it looks at first sight. On one axis is the ratio of silica to silica and sesquioxides (the 'product index'). On the other axis is what Reiche calls the 'weathering potential index', which is the mole-percentage ratio of the alkalis and alkaline earths to the total moles present (excluding water in both cases).

This method, and others, are discussed in greater detail by Reiche (1943).

A quite different basis for comparison of rocks and their weathered products (or even their metamorphic equivalent) was proposed by Barth (1948), based on the 'standard cell of a rock'. This is a rock volume containing 160 oxygen atoms. The sum of the cations (silica and metals) associated with this unit is very nearly 100 in all ordinary rocks. This method is especially valuable in constant volume weathering, as no element has to be assumed to be stable.

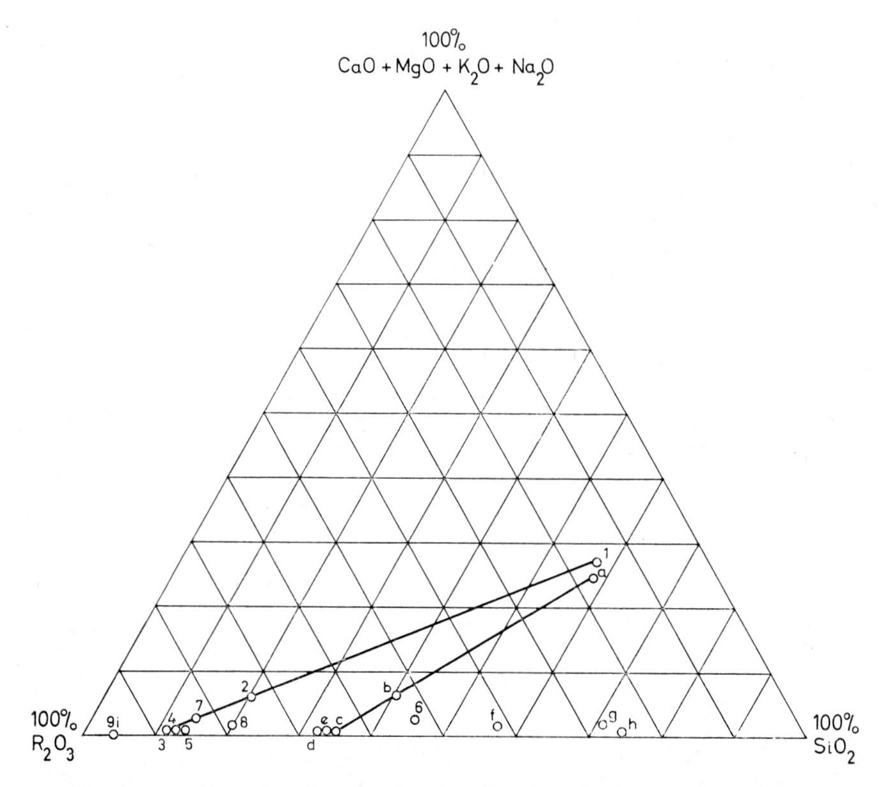

FIG. 174. Triangular diagram showing chemical changes during progressive alterations of a dolerite under humid tropical conditions (from Reiche, 1943).

To consider some examples, fresh Medford diabase has a weight percentage of CaO of 6·31, and weathered Medford diabase has a weight percentage of CaO of 3·52. It can be calculated that a standard cell of fresh Medford diabase would contain 6·48 ions of Ca, and the weathered rock 3·48, so in weathering there has been a loss of 3·00 ions of Ca.

The calculations are prepared for all elements in both fresh and weathered rock, from which losses and gains during weathering are apparent.

Barth gives an example of a rock decomposed (without volume change) by solfataric action. The gains and losses may be simply expressed in a table (as in Table 28), or can be shown diagrammatically (Fig. 176).

Table 28

	K	Na	Ca	Mg	Fe	Al	P	Ti	Si	O	H
Parent Rock	0·5	3·7	10·6	8·4	7·0	17·0	0·2	1·8	48·8	160·0	8·4
Decomposed	0·0	0·0	0·1	0·0	0·2	0·1	0·0	6·4	55·9	160·0	69·6
Gain or Loss	−0·5	−3·7	−10·5	−8·4	−6·8	−16·9	−0·2	+4·6	+7·1	0·0	+61·2

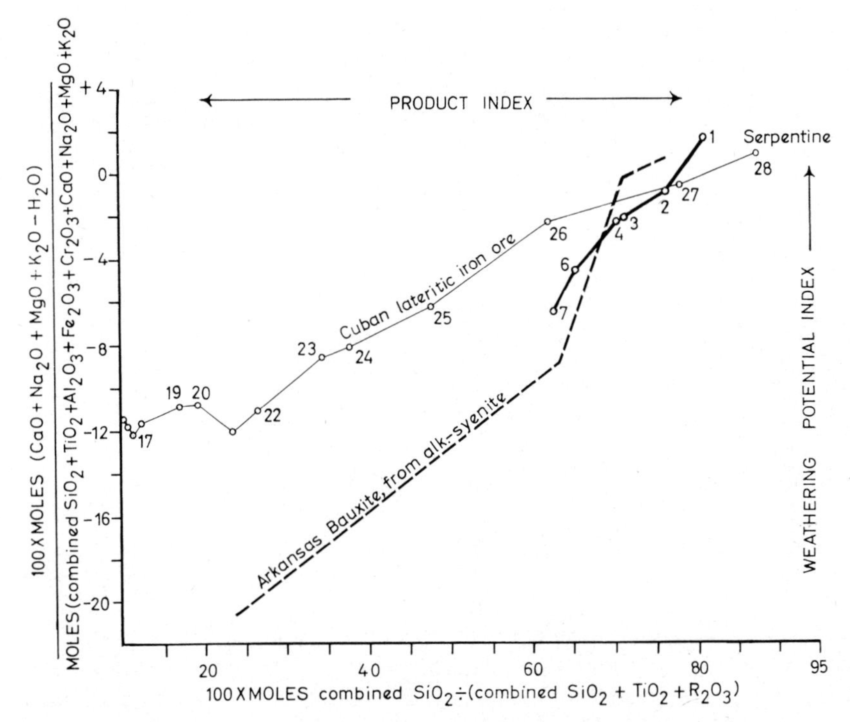

Fig. 175. Reiche's method of plotting the field of chemical weathering, showing progressive weathering on a syenite, a granite gneiss and a serpentine.

Fig. 176A shows how early investigators, arbitrarily using SiO_2 as a constant, interpreted the change from fresh to altered rock, and B shows how the 'standard cell' interpretation, assuming constant volume, gives what is thought to be a truer picture. It seems that the acid hot spring has had the effect of removing all metal ions, with simultaneous introduction of hydrogen, titanium and silica—therefore silica *does not remain constant*.

Table 29 shows the analysis of the Medford diabase as calculated by Keller (1957).

Table 29

	(a)	(b)		(c)	(d)	(e)
SiO_2	51·44	51·38	Si	49·59	48·02	−1·57
Al_2O_3	15·67	15·92	Al	17·82	17·39	—
Fe_2O_3	2·25	10·92	Fe	8·33	9·65	1·32
FeO	8·37	2·60	—	—	—	—
MgO	2·09	1·26	Mg	2·89	1·74	−1·15
CaO	6·31	3·52	Ca	6·48	3·48	−3·00
Na_2O	4·42	4·16	Na	8·21	7·52	−0·69
K_2O	1·91	1·57	K	2·20	1·91	−0·29
H_2O	1·67	2·94	H	10·76	18·29	7·53
H_2O-	0·41	1·70	—	—	—	—
TiO_2	1·95	2·52	Ti	1·39	1·68	0·29
P_2O_5	0·97	1·03	P	0·81	0·79	−0·02
MnO	0·18	0·22	Mn	0·17	0·18	0·01
CO_2	2·14	0·07	C	2·84	—	−2·84
S	0·12	0·01	—	—	—	—
	99·90	99·82	O	160·00	160·00	—

(*a*) Analysis of fresh Medford diabase (Goldich, 1938, No. 9).
(*b*) Analysis of weathered Medford diabase (Goldich, 1938, No. 10).
(*c*) Ions in 160-oxygen rock cell of (*a*).
(*d*) Ions in 160-oxygen rock cell of (*b*).
(*e*) Loss and gain of ions from (*c*) to (*d*).

The formula for weathering may be expressed as:

$$\text{Fresh diabase} + 7{\cdot}53\text{H} + 1{\cdot}32\text{Fe} \rightarrow \text{weathered diabase} + 1{\cdot}57\text{Si} + 0{\cdot}69\text{Na} + 0{\cdot}29\text{K} + 3{\cdot}0\text{Ca} + 1{\cdot}15\text{Mg} + 0{\cdot}2\text{P}$$

As has been said already, this 'standard cell' method is ideally suited for tracing weathering trends when there is no volume change. Whether it can be still useful when considerable volume change occurs is not so clear. Keller (1957) has presented a number of weathering analyses, using the standard cell *recalculated* on the basis of an assumed constant aluminium, which seems to be just the kind of assumption that Barth was trying to avoid.

This compendium of techniques by no means exhausts all the possibilities, and indeed a number of techniques have been mentioned in the body of the book that are not repeated in this chapter (for instance stone orientation measurements; geophysical investigation of regolith thickness; rate of weathering measurements).

The study of weathering is still in the stage where new techniques can play a large part in research, and there is a stimulating need for ingenuity and novelty in weathering investigations.

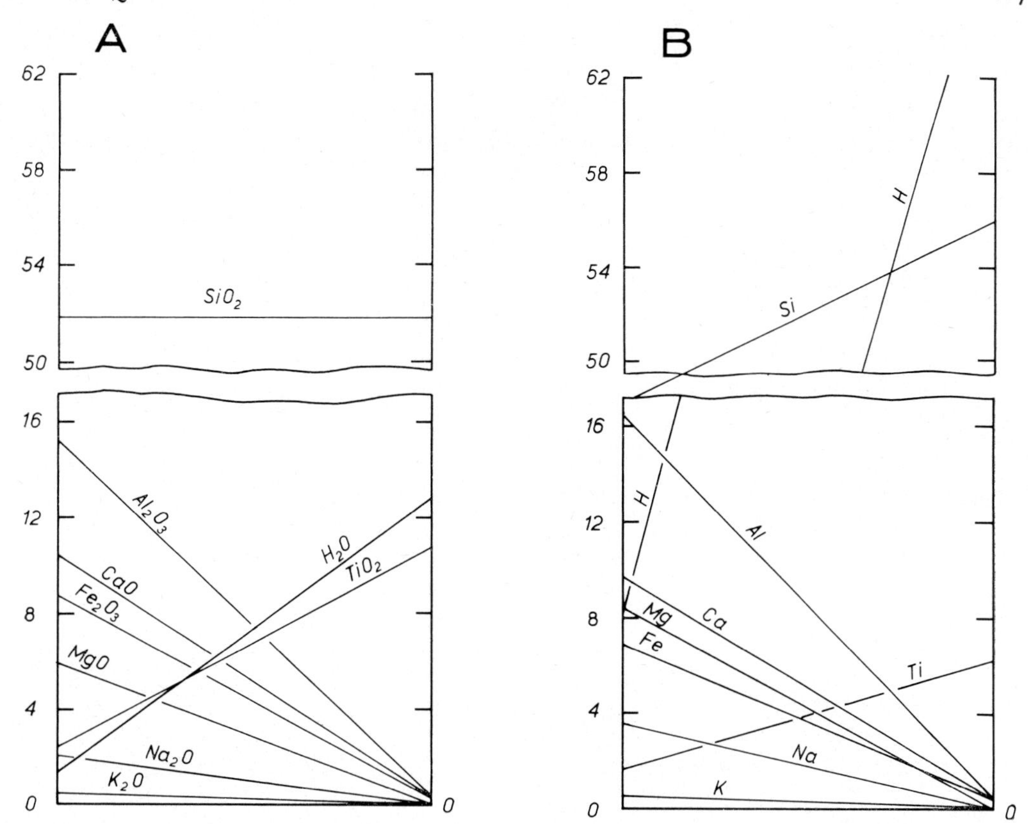

FIG. 176. *A* Alteration diagram according to earlier investigators who assumed SiO_2 to remain constant.

B Alteration diagram in which volumes are kept constant by keeping oxygen constant by the 'standard cell' method (from Barth, 1948).

In general laboratory studies have far outstripped field research in investigations of weathering, largely because routine techniques are available. There is now need for a better balance between field and laboratory research, which can only come from progress in field investigations. I believe that it is the development of field techniques, rather than the assimilation of established laboratory methods, which offers the most promising and exciting challenge to geomorphologists.

BIBLIOGRAPHY

ACKERMANN, E. 1962. Butserstein—Zeugen vorzeitlicher Grundwasserschwankungen. *Zeit. f. Geomorph.*, **6**, 148-82.

ADAMS, F. D. 1910. An experimental investigation into the action of differential pressure on certain minerals and rocks. *J. Geol.*, **18**, 489-524.

AKIMTZEV, V. V. 1932. Historical soils of the Kamenetz-Podolsk fortress. *Proc. 2nd. Intern. Congr. Soil Sci.*, **5**, 132-40.

ALEKSANDROV, V. G., and ZAK, G. A. 1950. Bacteria which decompose aluminium silicates (silicate bacteria). *Mikrobiologiya*, **19**, 99-104.

ANON. 1965. Glossary of soil science terms. *Soil Sci. Soc. Amer. Proc.*, **29**, 330-51.

ANON. 1966. 'Ordered water' molecular pressures offer new clues to rock failure. *Eng. Mining J.*, **167**, 92-94.

AOMINE, S., and KOJI WADA. 1962. Differential weathering of volcanic ash and pumice resulting in the formation of hydrated halloysite. *Amer. Mineral.*, **47**, 1024-48.

ASHTON, K. 1966. Mixture corrosion: a brief review. *Cave Research Group Newsletter*, **100**, 8-11.

AUBERT, G. 1963. La classification des sols. *Cahiers Pedol. ORSTOM*, **3**.

AUGUSTITHIS, S. S., and OTTEMANN, J. 1966. On diffusion rings and sphaeroidal weathering. *Chem. Geol.*, **1**, 201-09.

AVERY, B. W. 1956. A classification of British soils. *Trans. 6th Int. Cong. Soil Sci.*, 279-85.

AVERY, B. W., STEPHEN, I., BROWN, G., and YALLON, D. 1959. The origin and development of brown earths on clay-with-flints and coombe deposits. *J. Soil Sci.*, **10**, 177-95.

AXELROD, D. I. 1966. Origin of deciduous and evergreen habits in temperate forests. *Evolution*, **20**, 1-15.

BAIN, G. W. 1931. Spontaneous rock expansion. *J. Geol.*, **39**, 717-35.

BAKER, A. A. 1959. Imprisoned rocks: a process of rock abrasion. *Vict. Nat.*, **76**, 206-07.

BAKKER, J. P. 1960. Some observations in connection with recent Dutch investigations about granite weathering and slope development in different climates and climate changes. *Zeit. f. Geomorph. Supp.*, **1**, 169-92.

BAKKER, J. P., and LE HEUX, J. W. N. 1952. A remarkable new geomorphological law. *Proc. Koninkl. Akad. Wetenschap. Amsterdam B*, **55**, 399-570.

BALK, R. 1939. Disintegration of glaciated cliffs. *J. Geomorph.*, **2**, 306-13.

BALL, D. F. 1966. Late-glacial scree in Wales. *Biul. Peryglac.*, **15**, 151-63.

BANDAT, H. F. VON. 1962. *Aerogeology.* Gulf Pub. Co., Houston.

BAREN, J. VAN. 1931. Properties and constitution of a volcanic soil built in 50 years in the East-Indian Archipelago. *Comm. Geol. Inst. Agr. Univ. Wageningen, Holland*, **17**.

BARNES, H. L. 1956. Cavitation as a geological agent. *Amer. J. Sci.*, **254**, 493-505.

BARTH, T. F. W. 1948. Oxygen in rocks: a basis for petrographic calculations. *J. Geol.*, **56**, 50-61.

BARTON, D. C. 1916. Notes on the disintegration of granite in Egypt. *J. Geol.*, **24**, 382-93.

BATEMAN, A. M. 1950. *Economic Mineral Deposits.* Wiley, New York.

BAUER, F. 1962. Karstformen in den Osterreichischen Kalkhochalpen. *Actes 2nd Int. Cong. Spel.*, 1958, **1**, 299-329.

BAULIG, H. 1940. Le profil d'équilibre des versants. *Anns. Géog.*, **49**, 81-97.

BEAVIS, F. C. 1959. Pleistocene glaciation on the Bogong High Plains. *Aust. J. Sci.*, **21**, 192.

BERG, N. D. 1964. *Loess as a product of weathering and soil formation.* Israel Program for Scientific Translations, Jerusalem.

BIRD, E. C. F. 1964. *Coastal Landforms.* Australian National University Press, Canberra.

BIRKELAND, P. W. 1964. Pleistocene glaciation of the northern Sierra Nevada, north of Lake Tahoe, California. *J. Geol.*, **72**, 810-25.

BIROT, P. 1949. *Essais sur quelques problèmes de morpholgie génerale.* Lisbon.

BLACKWELDER, E. 1925. Exfoliation as a phase of rock weathering. *J. Geol.*, **33**, 793-806.

BLACKWELDER, E. 1926. Fire as an agent in rock weathering. *J. Geol.*, **35**, 134-40.

BLACKWELDER, E. 1929. Cavernous rock surfaces of the desert. *Amer. J. Sci.*, **17**, 393-99.

BLACKWELDER, E. 1933. The insolation hypothesis of rock weathering. *Amer. J. Sci.*, **26**, 97-113.

BLACKWELDER, E. 1948. Historical significance of desert lacqer (abstract). *Bull. Geol. Soc. Amer*, **59**, 1367.
BLANK, H. R. 1951. 'Rock doughnuts', a product of granite weathering. *Amer. J. Sci.*, **249**, 822-29.
BLOOMFIELD, C. 1964. Mobilization and immobilization phenomena in soils. In *Problems in Palaeoclimatology*, Ed. A. E. M. Nairn. Interscience. New York.
BLUMENSTOCK, D. I., and THORNTHWAITE, C. W. 1941. Climate and the world pattern. *Yearbook of Agriculture for* 1941, 79-92.
BÖGLI, A. 1960. 'Kalklösung und Karrenbildung.' *Zeit. f. Geomorph. Supp.*, **2**. Internat. Beitrage zur Karstmorphologie, 4-21.
BÖGLI, A. 1961. Karrentische, ein Beitrage zur Karstmorphologie *Zeit. f. Geomorph.*, **5**, 185-93.
BOLYSHEV, N. N. 1952. Origin and evolution of Takyr soils. *Pochvoevedenie*, 403-17.
BOSWELL, P. G. H. 1933. *Mineralogy of Sedimentary Rocks*. Murby, London.
BRADLEY, W. C. 1963. Large scale exfoliation in massive sandstones of the Colorado Plateau. *Bull. Geol. Soc. Amer.*, **74**, 519-28.
BREMER, H. 1965. Ayers Rock. *Zeit. f. Geomorph.*, **9**, 249-84.
BRETZ, H. 1960. Origin of Bermuda caves. *Bull. Nat. Spel. Soc.*, **22**, 19-22.
BREWER, R. 1950. Mineralogical examination of soils developed on the Prospect Hill intrusion, New South Wales. *J. Proc. Roy. Soc. N.S.W.* (for 1948), **82**, 272-85.
BREWER, R. and SLEEMAN, J. R. 1960. Soil structure and fabric; their definition and description. *J. Soil Sci.*, **11**, 172-85.
BRIDGES, E. M. 1961. Aspect and time in soil formation. *Agriculture, Lond.*, **68**, 358-63.
BROECKER, W. A., OLSEN, E. Z., and ORR, P. C. 1960. Radiocarbon measurements and annual rings in cave formations. *Nature, Lond.*, **185**, 93-94.
BROOKS, J. H., and WOLFF, K. 1959. Government drilling—Golden Gate Area, Croyden. *Queensland Govt. Mining J.*, **60**, 263-371.
BROWN, C. B. 1924. On some effects of wind and sun in the desert of Tumbez, Peru. *Geol. Mag.*, **61**, 337-39.
BROWN, G. 1961. *X-ray identification and crystal structures of clay minerals*. Mineral. Soc. London.
BROWNE, W. R. 1928. On some aspects of differential erosion. *J. Proc. Roy. Soc. N.S.W.*, **62**, 273-88.
BROWNE, W. R. 1964. Grey billy and the age of tor topography in Monaro, N.S.W. *Proc. Linn. Soc. N.S.W.*, **89**, 322-25.
BRUNSDEN, D. 1964. The origin of the decomposed granite. In *Dartmoor Essays*, Ed. I. G. Simmons.
BRYAN, K. 1922. Erosion and sedimentation in the Papago country. *U.S.G.S. Bull.*, **730**, 19-90.
BRYAN, K. 1925. The Papago Country, Arizona. *U.S.G.S. Water Supply Paper*, **499**, 90-93.
BÜDEL, J. 1951. Die Klimazonen des Eiszeitalters. Trans H. W. Wright in *Intern. Geol. Rev.*, **1**, 72.
BÜDEL, J. 1957. Die "Doppelten Einebnungsflachen" in den feuchten Tropen. *Zeit. f. Geomorph.*, **1**, 201-28.
BÜDEL, J. 1963. Geomorphology based upon climatic causation. *Geog. Rundschau*, **15**, 269-85.
BUNTING, B. T. 1961. The role of seepage moisture in soil formation, slope development and stream initiation. *Amer. J. Sci.*, **259**, 503-18.
BUTLER, B. E. 1959. Periodic phenomena in landscapes as a basis for soil studies. *CSIRO Aust. Soil. Publ. No.*, **14**.
BUTLER, B. E. 1967. Soil periodicity in relation to landform development in southeastern Australia. Ch. 11 in *Landform studies from Australia and New Guinea*, Ed. J. N. Jennings and J. A. Mabbutt. A.N.U., Canberra.
BUTUZOVA, O. V. 1962. Role of the root system of trees in the formation of micro-relief. *Soviet Soil Sci.*, **4**, 364-72.

CAILLEUX, A. 1960. From Tertiary warmth to Quaternary cold; percentages of quartz pebbles. *Biul. Peryglac.*, **9**, 41-6.
CALKIN, P., and CAILLEUX, A. 1962. A quantitative study of cavernous weathering (tafonis) and its application to glacial chronology in Victoria valley, Antarctica. *Zeit. f. Geomorph.*, **6**, 317-24.
CARL, J. D., and AMSTUTZ, G. C. 1958. Three-dimensional Liesegang rings by diffusion in a colloidal matrix, and their significance for the interpretation of geological phenomena. *Bull. Geol. Soc. Amer.*, **69**, 1467-68.
CARROL, D. 1951. Mineralogy of laterites. *Aust. J. Sci.*, **14**, 41.
CARTER, G. F. and PENDLETON, R. L. 1956. The humid soil: process and time. *Geog. Rev.*, **46**, 488-507.

ČERNOHOUZ, J., and ŠOLC, I. 1966. Use of sandstone wanes and weathered basaltic crust in absolute chronology. *Nature, Lond.*, **212**, 806-07.

CHAPMAN, R. W., and GREENFIELD, M. A. 1949. Spheroidal weathering of igneous rocks. *Amer. J. Sci.*, **247**, 407-29.

CHAPMAN, C. A., and RIOUX, R. L. 1958. Statistical study of topography, sheeting, and jointing in granite, Acadia National Park, Maine. *Amer. J. Sci.*, **256**, 111-27.

CHENERY, E. M. 1951. Some aspects of the aluminium cycle. *J. Soil Sci.*, **2**, 97-109.

CHENERY, E. M. 1960. Soils of Uganda. *Uganda Dept. Agric. Mem. Res. Div. Ser.*, 1, No. **1**.

CHURCHWARD, H. M. 1963. Soil studies at Swan Hill, Victoria. IV. Groundsurface history and its expression in the array of soils. *Aust. J. Soil Res.*, **1**, 242-55.

CIGNA, A. *et al.* 1963. Quelques considérations sur l'éffet sel dans la solubilité des calcaires. *Ann. Spéléol.*, **18**, 185-91.

CLAYTON, R. W. 1956. Linear depressions (Bergfutsniederungen) in savanna landscapes. *Geographical studies*, **3**, 102-26.

CLOUD, P. E. 1965. Significance of the Gunflint (Precambrian) microflora. *Science*, **148**, 27-35.

COLEMAN, J. M., GAGLIANO, S. M., and SMITH, W. G. 1966. Chemical and physical weathering on saline high tidal flats, northern Queensland, Australia. *Bull. Geol. Soc. Amer.*, **77**, 205-6.

COLLINI, B., and WALLNER, O. 1961. Ein Bespiel schneller Bausteinzerstorung in Uppsala. *Schweden Uppsala Univ. Geol. Inst. B*, **40**, 211-19.

COOPER, A. W. 1960. An example of the role of microclimate in soil genesis. *Soil Sci.*, **90**, 109-20.

CORBEL, J. 1959. Érosion en terrains calcaires. *Ann. Géogr.*, **68**, 97-120.

COSTIN, A. B., HALLSWORTH, E. G., and WOOF, M. Studies in pedogenesis in New South Wales. III. The Alpine Humus Soils. *J. Soil Sci.*, **3**, 190-218.

COTTON, C. A. 1942. *Geomorphology.* Whitcombe and Tombs, Christchurch.

COTTON, C. A. 1952a. *Volcanoes as landscape forms.* Whitcombe and Tombs, Christchurch.

COTTON, C. A. 1952b. The erosional grading of convex and concave slopes. *Geog. J.*, **118**, 197-204.

COTTON, C. A. 1963. Levels of planation of marine beaches. *Zeit. f. Geomorph.*, **7**, 97-111.

CRAMER, L. W. 1963. Pedestal rocks in the Laramie Range, Albana County. *Wyoming Univ. Dept. Geol. Contr. Geol.*, **2**, 55-7.

CUNNINGHAM, F. 1964. A detail of process on scarp edges of Millstone Grit. *East Midland Geog.*, **3**, 322-6.

CURREY, D. T. 1968. Bellfield Dam, Victoria, Part 1. Site geology. *Inst. Engineers Aust. Ann. Conf. Papers*, 1968, 33-6.

CZEPPE, Z. 1964. Exfoliation in a periglacial climate. *Geog. Polonica*, **2**, 5-10.

DAVIDSON, C. F. 1964. Uniformitarianism and ore genesis. *Mining Mag.* (*London*), **109**, 176-85, 244-53.

DAVIS, S. N. 1964. Silica in streams and groundwater. *Amer. J. Sci.*, **262**, 870-91.

DAVIS, W. M. 1930. Origin of limestone caverns. *Bull. Geol. Soc. Amer.*, **41**, 475-628.

DE SWART, A. M. 1964. Lateritization and landscape development in parts of equatorial Africa. *Zeit. f. Geomorph.*, **8**, 313-33.

DE VILLIERS, J. M. 1965. Present soil forming factors and processes in tropical and subtropical regions. *Soil Sci.*, **99**, 50-7.

DEMEK, J. 1965. Slope development in granite areas of Bohemian massif (Czechoslovakia). *Zeit. f. Geomorph. Supp.*, **5**, 82-106.

DIJK, D. C. VAN. 1959. Soil features in relation to erosional history in the vicinity of Canberra. CSIRO *Aust. Soil Publ. No.* **13**.

DODGE, I. A. 1947. An example of exfoliation caused by chemical weathering. *J. Geol.*, **55**, 38-42.

DONN, W. L., DONN, D. D., and VALENTINE, W. G. 1965. On the early history of the earth. *Bull. Geol. Soc. Amer.*, **76**, 287-306.

DORF, E. 1964. The use of fossil plants in palaeoclimatic interpretation. Ch. 2 in *Problems of Palaeoclimatology*, Ed. A. E. M. Nairn. Interscience, New York.

DOW, W. G. 1946. The effect of salinity on the formation of mudcracks. *Compass*, **41**, 162-6.

DUCHAUFOUR, P. 1960. *Principles de pédologie.* Masson, Paris.

DUPUIS, J., DUTREUIL, J. P., and JAMBU, P. 1965. Historical rendzinas developed on the shell mounds of Saint-Michel-en-l'Herm (Vendée). *C. R. Acad. Sci. Paris*, **260**, 940-3.

DURY, G. H. 1966. *Aspects of the contents of geography in the fifth and sixth forms.* University of Sydney.

EDELMAN, C. H. 1931. Diagenetische Umwandlungserscheinungen an detritischen Pyroxenen und Amphibolen. *Fortschr. Min. Kryst. Pet.*, **16**, 323-4.

ELLIS, J. H. 1938. *The Soils of Manitoba.* Manitoba Econ. Survey Board, Winnipeg.

EMERY, K. O. 1941. Rate of surface retreat of sea cliffs based on dated inscriptions. *Science*, **93**, 617-8.

EMERY, K. O. 1944. Brush fires and rock exfoliation. *Amer. J. Sci.*, **242**, 506-8.

EMERY, K. O. 1960. Weathering of the Great Pyramid. *J. Sed. Pet.*, **30**, 140-3.

EMMONS, W. H. 1940. *Principles of economic geology.* McGraw Hill, New York.

ENGLE, C. G., and SHARP, R. P. 1958. Chemical data on desert varnish. *Bull. Geol. Soc. Amer.*, **69**, 487-518.

ENSLIN, J. F. 1961. Secondary aquifers in South Africa and the scientific selection of boring sites in them. *Inter-African Conference on Hydrology. C.C.T.A. Publ. No.* **66**, 379-89. Nairobi.

EVERETT, K. R. 1966 in: *Environment of the Cape Thompson region*, Alaska, ed. Norman J. Wilimovsky & John N. Wolfe, United States Atomic Energy Commission (Div. Technical Information), 175-220.

EYLES, V. A. 1952. The composition and origin of the Antrim laterites and bauxites. *Mem. Geol. Surv. Northern Ireland.*

FALCONER, J. D. 1911. *The geography and geology of northern Nigeria*, Macmillan, London.

FARMIN, R. 1937. Hypogene exfoliation in rock masses. *J. Geol.*, **45**, 625-35.

FEDER, G. L. 1964. Cause of hatchured weathering patterns produced on dolomite. *J. Sed. Pet.*, **34**, 197-8.

FINCH, L. 1955. *The durability of building stones of Victoria.* Unpublished Ph.D. thesis. Melbourne University.

FITZPATRICK, E. A. 1963. Deeply weathered rock in Scotland, its occurrence, age and contribution to soils. *J. Soil Sci.*, **14**, 33-43.

FOX, C. S. 1935. *Engineering Geology.* Technical Press, London.

FREISE, F. W. 1931. Untersuchung von Mineralen auf Abnutzbarkeit bei Verfrachtung im Wasser. *Tschermaks Min. Petr. Mitt.*, **41**, 1-7.

FREISE, F. W. 1938. Inselberge und Inselberglandschaften in Granit und Gneitsgebieten Brasiliens. *Zeit. f. Geomorph.*, **10**, 137-68.

GAGE, M. 1966. Franz Josef Glacier. *Ice*, **20**, 26-27.

GARRELS, R. M., and CHRIST, C. L. 1965. *Solutions, minerals and equilibria.* Harper and Row, New York.

GEIGER, R. 1965. *The Climate near the ground.* Cambridge, Mass.

GEIKIE, A. 1880. Rockweathering, as illustrated in Edinburgh churchyards. *Proc. Roy. Soc. Edinburgh*, **10**, 518-32.

GÈZE, B. 1965. *La Spéléologie Scientifique.* Éditions du Seuil, Paris.

GLAESSNER, M. F. 1966. Precambrian palaeontology. *Earth Science Reviews*, **1**, 29-50.

GLAZOVSKAYA, M. A. 1950. The weathering of mountain rock in the snow zone of central Tyan-Shan. *Trudy. Pchv. Inst. Dokuchaeva.*, **34**, 28-48.

GOLDICH, S. S. 1938. A study in rock weathering. *J. Geol.*, **46**, 17-58.

GOODCHILD, J. G. 1890. Notes on some observed rates of weathering of limestones. *Geol. Mag.*, **27**, 463-6.

GRAVENOR, C. P. 1954. Mineralogical and size analysis of weathering zones on Illinoian till in Indiana. *Amer. J. Sci.*, **252**, 159-71.

GRAY, W. M. 1965. Surface spalling by thermal stresses in rocks. *Proc. Rock Mechanics Symposium, Toronto.* 85-106. Dept. Mines and Tech. Surveys, Ottawa.

GREEN, K. D., and MAVER, J. L. 1959. The Tarago River Aqueduct. *J. Inst. Engs, Australia*, **31**, 1-19.

GRIGGS, D. T. 1936. The factor of fatigue in rock exfoliation. *J. Geol.*, **44**, 781-96.

GROOM, G. E., and WILLIAMS, H. 1965. The solution of limestone in South Wales. *Geog. J.*, **131**, 37-41.

HACK, J. T. 1966. Circular patterns and exfoliation in crystalline terrane, Grandfather Mountain area, North Carolina. *Bull. Geol. Soc. Amer.*, **77**, 975-86.

HADLEY, R. F. 1965. Erosion rates and process. *U.S. Geological Survey, Prof. Paper* **525-A**, 177.

HALLSWORTH, E. G., and CRAWFORD, D. V. 1965. *Experimental pedology.* Butterworth, London.

HANDLEY, J. R. F. 1952. The geomorphology of the Nzega area of Tanganyika with special reference to the formation of granite tors. *Cong. Geol. Int. Algiers C.R.*, **21**, 201-10.

HARDER, E. C. 1949. Stratigraphy and origin of bauxite deposits. *Bull. Geol. Soc. Amer.*, **60**, 887-908.

HARLAND, W. B. 1957. Exfoliation joints and ice action. *J. Glaciol.*, **3**, 8-12.

HARMS, J. E., and MORGAN, B. D. 1964. Pisolitic limonite deposits in North-West Australia. *Proc. Aust. Inst. Min. Metal.*, **212**, 91-124.

HARPUM, J. R. 1963. Evolution of granite scenery in Tanganyika. *Rec. Geol. Surv. Tanganyika*, **10**, 39-46.

HARRISON, J. B. 1934. *The katamorphism of igneous rocks under humid tropical conditions.* Imp. Bur. Soil Sci., Harpenden.

HARRISON, V. F., GOW, W. A., and IVARSON, K. C. 1966. Leaching of uranium from Eliot Lake ore in the presence of bacteria. *Can. Mining J.*, **87**, 64-7.

HAY, R. L. 1963. Zeolite weathering in Olduvai Gorge, Tanganyika. *Bull. Geol. Soc. Amer.*, **74**, 1281-6.

HAYS, J. 1967. Land surfaces and laterites in the north of the Northern Territory. Ch. 9 in *Landform studies from Australia and New Guinea*, Ed. J. N. Jennings and J. A. Mabbutt, A.N.U., Canberra.

HAYWOOD, B. H. J. 1961. Studies in frost-heave cycles at Schefferville. *McGill Sub-Arctic Research Papers*, No. **11**, 6-10.

HEY, R. W. 1963. Pleistocene screes in Cyrenaica (Libya). *Eiszeitalter und Gegenwart*, **14**, 77-84.

HILGER, A. 1897. Über Verwitterungsvorgange bei Krystallinischen und Sedimentärgesteinen. *Landw. Jahrb.*, **8**, 1-11.

HILLS, E. S. 1949. Shore platforms. *Geol. Mag.*, **86**, 137-52.

HILLS, E. S. 1963. *Outlines of structural geology.* Methuen, London.

HILLS, E. S. 1967. *The Physiography of Victoria.* Whitcombe & Tombs, Melbourne.

HILLS, E. S. 1968. A study of cliffy coastal profiles based on examples in Victoria, Australia. *Zeit. f. Geomorph.* in press.

HILTON, T. E. 1963. The geomorphology of north-eastern Ghana. *Zeit. f. Geomorph.*, **7**, 308-25.

HIRSCHWALD, J. 1908. *Die Prüfung der Naturlichen Bausteine auf ihre Wetterbeständigkeit.* Berlin.

HISSINK, D. J. 1938. The reclamation of the Dutch saline soils and their further weathering under humid climatic conditions of Holland. *Soil Sci.*, **45**, 83-94.

HJULSTRÖM, F. 1935. Studies of the morphological activities of rivers as illustrated by the River Fyris. *Uppsala Univ. Geol. Inst. Bull.*, **25**, 221-527.

HODGKIN, E. P. 1964. Rate of erosion of intertidal limestone. *Zeit. f. Geomorph.*, **8**, 385-92.

HOLLAND, H. D. 1964. On some aspects of the chemical evolution of cave waters. *J. Geol.*, **72**, 36-67.

HOLMES, A. 1923. *Petrographic methods.* Murby, London.

HOLMES, A. 1947. The construction of a geological time-scale. *Trans. Geol. Soc. Glasgow*, **21**, 117-52.

HOLMES, A. 1965. *Principles of Physical Geology.* Nelson, London.

HOVERMANN, J. 1949. Morphologische Untersuchungen im Mittelharz. *Göttingen Geogr. Abh. H.*, **2**.

HUDSON, J. D. 1964. Sedimentation rates in relation to the Phanerozoic time-scale. In *The Phanerozoic time-scale* Geol. Soc. London, 37-42.

HUNT, C. B. 1961. Stratigraphy of desert varnish. *U.S. Geol. Surv., Prof. Paper* **424-B**, 195-96.

HUSSEY, K. M., and TATOR, B. A. 1950. Sandstone spindles. *Amer. J. Sci.*, **248**, 734-40.

HUTCHESON, T. B., and BAILEY, H. H. 1965. Effect of underlying residua on chemical and mineralogical properties of soils developed in a uniform loess overlay. *Soil Sci. Soc. Amer. Proc.*, **29**, 427-32.

IRANI, R. R., and CLAYTON, F. C. 1963. *Particle size: measurement, interpretation and application.* Wiley, New York.

IRVING, E. 1958. Rock magnetism: a new approach to the problems of polar wandering and continental drift. In *Continental Drift.* Univ. of Tasmania, Hobart.

JACKS, G. V. 1953. Organic weathering. *Soils and Fertilizers*, **16**, 165.

JAEGER, J. C. 1962. *Elasticity, fracture and flow.* Methuen, London.

JACKSON, M. L. *et al.* 1948. Weathering sequence of clay size minerals in soils and sediments. *J. Phys. Coll. Chem.*, **52**, 1237-60.

JAHNS, R. H. 1943. Sheet structure in granites: its origin and use as a measure of glacial erosion in New England. *J. Geol.*, **51**, 71-93.

JENNINGS, J. N. 1964. Geomorphology of Punchbowl and Signature caves, Wee Jasper, New South Wales. *Helictite*, **2**, 57-71.

JENNINGS, J. N., and BIK, M. J. 1962. Karst morphology in Australian New Guinea. *Nature, Lond.*, **194**, 1036-8.

JENNY, H. 1941. *Factors of soil formation.* McGraw Hill, New York.

JONES, R. J. 1965. Aspects of the biological weathering of limestone pavement. *Proc. Geol. Assn. Lond.*, **76**, 421-34.

JUDD, J. W. 1886. Report on a series of specimens of the deposits of the Nile delta. *Proc. Roy. Soc.*, 213-27.

JUTSON, J. T. 1934. The physiography of Western Australia. *Geol. Surv. W. Aust. Bull.*, **95**.

KAYE, C. A. 1959. Shoreline features and Quaternary shoreline changes, Puerto Rico. *U.S. Geol. Surv. Prof. Paper* **317-B**, 49-140.

KELLER, W. D. 1957. *The principles of chemical weathering.* Lucas, Columbia, Miss.

KIERSCH, G. A., and Asce, F. 1964. Vaiont reservoir disaster. *Civil Engineering*, **34**, 32-39.

KIESLINGER, A. 1932. *Geologische Diffusionen.* Steinkpf, Dresden.

KING, L. C. 1948. A theory of bornhardts. *Geog. J.*, **112**, 83-6.

KING, L. C. 1957. The uniformitarian nature of hillslopes. *Trans. Edin. Geol. Soc.*, **17**, 81-102.

KING, L. C. 1962. *The morphology of the earth.* Oliver and Boyd, Edinburgh.

KITTLEMAN, L. R. 1964. Application of Rosin's distribution in size-frequency analysis of clastic rocks. *J. Sed. Pet.*, **34**, 483-502.

KLINGE, H. 1965. Podzol soils in the Amazon basin. *J. Soil Sci.*, **16**, 95-103.

KUBIENA, W. L. 1953. *The Soils of Europe.* Murby, London.

KUENEN, Ph.H. 1960. Experimental Abrasion 4: Aeolian Action, *Journal of Geology*, **68**, 427-49.

KUZNETSOV, S. I., IVANOV, M. V., and LYALIKOVA, N. N. 1963. *Introduction to geological microbiology.* McGraw Hill, New York.

LABASSE, H. 1965. Ground stress in longwall and room-and-pillar mining. *Proc. Rock Mechanics Symposium, Toronto.* 47-64. Dept. Mines and Tech. Surveys, Ottawa.

LANGFORD-SMITH, T., DURY, G. H., and McDOUGALL, I. 1966. Dating the duricrust in southern Queensland. *Aust. J. Sci.*, **29**, 79-80.

LARSEN, E. S. 1948. Batholith and associated rocks of Corona, Elsinore and San Luis Rey quadrangles, Southern California. *Geol. Soc. Amer. Mem.*, **29**, 114-9.

LEEPER, G. W. 1964. *Introduction to soil science.* Melbourne University Press.

LEGRAND, H. E. 1952. Solution depressions in diorite in North Carolina. *Amer. J. Sci.*, **250**, 566-85.

LEHMANN, O. 1933. Morphologische Theorie der Verwitterung von Steinschlagwänden. *Viertelj. schr. Natf. Ges. Zurich*, **78**, 83-126.

LENEUF, N., and AUBERT, G. 1960. Attempt to measure the rate of ferrallitisation. *Trans. 7th Congress Soil Science*, **4**, 225-8.

LEOPOLD, L. B., WOLMAN, M. G., and MILLER, J. P. 1964. *Fluvial processes in geomorphology.* Freeman, San Francisco.

LEWIS, C. C., and ENSMINGER, W. S. 1948. Relationship of plant development to the capacity to utilise potassium in orthoclase feldspar. *Soil Sci.*, **65**, 495-500.

LEWIS, W. V. 1954. Pressure release and glacial erosion. *J. Glaciol.*, **2**, 417-22.

LINDGREN, W. 1933. *Mineral deposits.* McGraw Hill, New York.

LINTON, D. L. 1955. The problem of tors. *Geog. J.*, **121**, 470-86.

LIVINGSTONE, D. A. 1963. Chemical composition of rivers and lakes. *U.S. Geol. Surv. Prof. Paper* **440-G**.

LOUGHNAN, F. C. 1962. Some considerations in the weathering of the silicate minerals. *J. Sed. Pet.*, **32**, 284-90.

LOUGHMAN, F. C. *et al.* 1962. Weathering of some Triassic shales in the Sydney area. *J. Geol. Soc. Australia*, **8**, 245-57.

LUMB, P. 1962. The properties of decomposed granite. *Geotechnique.*, **12**, 226-43.

MABBUTT, J. A. 1952. A study of granite relief from South West Africa. *Geol. Mag.*, **89**, 87-96.

MABUTT, J. A. 1961a. A stripped landsurface in Western Australia. *Trans. Inst. Brit. Geogr.*, **29**, 101-14.

MABBUTT, J. A. 1961b. 'Basal surface' or 'weathering front'. *Proc. Geol. Assn. Lond.*, **72**, 357-8.

MABBUTT, J. A. 1966. Mantle-controlled planation of pediments. *Amer. J. Sci.*, **264**, 78-91.

MCCRAW, J. D. 1965. Landscapes of central Otago. *N.Z. Geogr. Soc. Misc. Ser. No.* **5**, 30-45.

McGregor, D. R. *et al.* 1963. Solution caves in gypsum, north central Texas. *J. Geol.*, **71**, 108-115.

Mackenzie, F. T., and Garrels, R. M. 1966. Chemical mass balance between rivers and oceans. *Amer. J. Sci.*, **264**, 507-25.

Mackie, W. 1899. The feldspars present in sedimentary rocks as indicators of the conditions of contemporaneous climates. *Trans. Edin. Geol. Soc.*, **7**, 443-68.

Mackney, D. 1961. A podzol development sequence in oakwoods and heath in Central England. *J. Soil Sci.*, **12**, 23-40.

Marbut, C. 1928. A scheme for soil classification. *Proc. 1st. Int. Cong. Soil Sci.*, **4**, 1-31.

Marshall, C. E. 1964. *The physical chemistry and mineralogy of soils. Vol. 1. Soil materials.* Wiley, New York.

Mason, B. 1966. *Principles of Geochemistry.* 3rd ed. Wiley, New York.

Matthes, F. E. 1930. Geological history of the Yosemite valley. *U.S. Geol. Surv. Prof. Paper* **160**.

Matsumoto, S. 1964. Landforms of accumulated boulders in the Abukuma and Kitikami mountainlands. Science Reports, *Tohoku Univ. 7 Ser. (Geog.)*, **13**, 201-14.

Melton, M. A. 1965. Debris-covered hillslopes of the southern Arizona desert—consideration of their stability and sediment contribution. *J. Geol.*, **73**, 715-29.

Merrill, G. P. 1900. Sandstone disintegration through the formation of interstitial gypsum. *Science*, **11**, 850-1.

Middleton, G. V. 1965. Primary sedimentary structures and their hydrodynamic interpretation. *Soc. Econ. Pal. Mineral. Special Publ.* No. **12**.

Millar, C. E., Turk, L. M., and Foth, H. D. 1966. *Fundamentals of soil science.* Wiley, New York.

Milner, H. B. 1962. *Sedimentary petrography.* Allen and Unwin, London.

Mitchell, B. A. 1959. The ecology of tin mine spoil heaps. *Malay Forester*, **22**, 111-32.

Mohr, E. C., and VanBaren, F. A. 1954. *Tropical soils.* Interscience, New York.

Moore, G. W. 1960. Geology of Carlsbad Caverns, New Mexico. Guide Book to Carlsbad Caverns National Park. *Nat. Spel. Soc. Guide Book Series*, No. 1.

Moore, G. W., and Nicholas G. 1964. *Speleology: the study of caves.* D. C. Heath & Co.

Moss, R. P. 1965. Slope development and soil morphology in a part of south-west Nigeria. *J. Soil Sci.*, **16**, 192-209.

Mulcahy, M. J. 1967. Landscapes, laterites and soils in southwestern Australia. Ch. 10 in *Landform studies from Australia and New Guinea.* Ed. J. N. Jennings and J. A. Mabbutt. A.N.U. Canberra.

Mulcahy, M. J., and Hingston, F. J. 1961. The development and distribution of the soils of the York-Quairading area, Western Australia, in relation to landscape evolution. *CSIRO Soil Publ.* **17**.

Muller, S. W. 1945. Permafrost or permanently frozen ground and related engineering problems. *Spec. Rept. Strategic Engineering Study* **62**, Office of Chief of Engineers, U.S. Army.

Nansen, F. 1922. The strandflat and isostasy. *Skrifter utgit av Videnskapsselskapet Kritsiania*, 1, *Math. Naturvidensk.*, 28-32.

Neiheisel, J. 1963. Heavy mineral investigation of Recent and Pleistocene sands of lower coastal plain of Georgia. *Bull. Geol. Soc. Amer.*, **73**, 365-74.

Nepper-Christensen, P. 1965. Shrinkage and swelling of rocks due to moisture movement. *Medd. Dansk Geol. Forening.*, **15**, 548-55.

Ng, S. K., and Bloomfield, C. 1961. The solution of some minor element oxides by decomposing plant material. *Geochim. Cosmochim. Acta* **24**, 206-25.

Nikiforoff, C. C. 1942. Fundamental formula of soil formation. *Amer. J. Sci.*, **240**, 847-66.

Nikitin, K. K. 1965. Principles of mapping ore deposits in the weathering crust on ultrabasic rocks. *The weathering crust on serpentine massifs; The Weathering Crust, Book 9.* U.S.S.R. Academy of Science, Moscow.

Nishioka, S., and Harada, T. 1958. Elongation of stones due to absorption of water. *Japan Cement Engineering Association. Review of 12th meeting. Tokyo.*

Northcote, K. H. 1962. A factual key for the recognition of Australian Soils. *CSIRO Aust. Div. Soils Rep.* No. **4**, 60.

Norton, E. A., and Smith, R. S. 1930. The influence of topography on soil profile character. *J. Amer. Soc. Agron.*, **22**, 251-62.

Ollier, C. D. 1959. A two cycle theory of tropical pedology. *J. Soil Sci.*, **10**, 137-48.

OLLIER, C. D. 1960. The inselbergs of Uganda. *Zeit. f. Geomorph.*, **4**, 43-52.

OLLIER, C. D. 1963. Insolation weathering: examples from Central Australia. *Amer. J. Sci.*, **261**, 376-87.

OLLIER, C. D. 1965. Some features of granite weathering in Australia. *Zeit. f. Geomorph.*, **9**, 285-304.

OLLIER, C. D. 1966. Desert gilgai. *Nature, Lond.*, **212**, 581-3.

OLLIER, C. D. 1967. Spheroidal weathering, exfoliation and constant volume alteration. *Zeit. f. Geomorph.*, **11**, 103-8.

OLLIER, C. D., and HARROP, J. F. 1958. The caves of Mount Elgon. *Uganda J.*, **22**, 158-63.

OLLIER, C. D., and HARROP, J. F. 1964. The caves of Mont Hoyo, Eastern Congo Republic. *Bull. Nat. Spel. Soc.*, **25**, 73-8.

OLLIER, C. D., and THOMASSON, A. J. 1957. Asymmetrical valleys of the Chiltern Hills. *Geog. J.*, **123**, 71-80.

OLLIER, C. D., and TRATMAN, E. K. 1956. The geomorphology of the caves of north west County Clare, Ireland. *Proc. Univ. Bristol Spel. Soc.*, **7**, 138-57.

OLLIER, C. D., and TUDDENHAM, W. G. 1962. Inselbergs of central Australia. *Zeit. f. Geomorph.*, **5**, 257-76.

OUTCALT, S., and BENEDICT, J. 1965. Photo-interpretation of two types of rock-glacier in the Colorado Front Range, U.S.A. *J. Glaciol.*, **5**, 849-56.

PARFENOVA, E. I., and YARILOVA, E. A. 1965. *Mineralogical investigations in soil science.* Israel Program for Scientific Translations. Jerusalem.

PELTIER, L. 1950. The geographic cycle in periglacial regions as it is related to climatic geomorphology. *Ann. Assoc. Amer. Geog.*, **40**, 214-36.

PETTIJOHN, F. J. 1941. Persistence of heavy minerals and geologic age. *J. Geol.*, **49**, 610-25.

PETTIJOHN, F. J. 1957. *Sedimentary rocks.* Harper, New York.

PETTIJOHN, F. J., and POTTER, P. E. 1964. *Atlas and glossary of primary sedimentary structures.* Springer, Berlin.

PICKNETT, R. G. 1964. A study of calcite solutions at 10°C. *Trans. Cave Research Group of Great Britain*, **7**, 39-62.

POLYNOV, B. B. 1937. *Cycle of weathering.* (Trans. A. Muir). Murby, London.

PONS, L. J., and ZONNEVELD, I. S. 1965. Soil ripening and soil classification. *Int. Inst. Land Reclamation and Improvement.* Publ. No. **13**. Wageningen.

RADWANSKI, S. A., and OLLIER, C. D. 1959. A study of an East African catena. *J. Soil Sci.*, **10** 149-68.

RAESIDE, J. D. 1949. The origin of schist tors in central Otago. *N. Z. Geographer*, **5**, 72-6.

RAGGATT, H. G., OWEN, H. B., and HILLS, E. S. 1945. The bauxite deposits of the Boolara-Mirboo North area, South Gippsland, Victoria. *Comm. Aust. Dist. Supply Shipping. Min. Res. Bull.*, **14**.

RAISTRICK, A., and GILBERT, O. L. 1963. Malham Tarn House: its building materials, their weathering and colonization by plants. *Field Studies*, **1**, 89-115.

RAPP, A. 1960. Recent development of mountain slopes in Kärkevagge and surroundings, North Scandinavia. *Geogr. Ann.*, **42**, 65-200.

RASOOL, S. I., and MCGOVERN, W. E. 1966. Primitive atmosphere of the earth. *Nature, Lond.*, **212**, 1225-6.

RAYNER, J. H. 1966. Classification of soils by numerical methods. *J. Soil Sci.*, **17**, 79-92.

RAZUMOVA, V. N., and KHERASKOV, N. P. 1963. Geologic types of weathering crusts. *Doklady Akad. Nauk. SSSR.*, **148**, 87-9.

REDDEN, J. A. 1963. Exfoliation cave in pegmatites. *Min. Indust. J.*, **10**, 4-5.

REED, J. C. 1963. Origin of some intermittent ponds on quartzite ridges in western North Carolina. *Bull. Geol. Soc. Amer.*, **74**, 1183-7.

REED, R. D. 1930. Recent sands of California. *J. Geol.*, **38**, 223-45.

REICHE, P. 1943. Graphic representation of chemical weathering. *J. Sed. Pet.*, **13**, 58-68.

REICHE, P. 1950. A survey of weathering processes and products. *New Mexico Univ. Publ. Geology*, **3**.

RENAULT, P. 1960. The role of erosion and corrosion in the deepening of a drainage system. *Rev. Géomorph. Dynamique*, **11**, 1-4.

RENWICK, K. 1962. The age of caves by solution. *Cave Science*, **4**, 338-50.

REUTER, G. 1957. Contribution to the nomenclature of soil horizons. *Wiss. Z. Univ. Rostock*, **6**, 207-12.

RICH, C. I., and KUNZE, G. W. 1964. *Soil Clay Mineralogy.* Chapel Hill.

ROGERS, J. J. W., KREUGER, W. C., and KROG, M. J. 1963. Sizes of naturally abraded materials. *J. Sed. Pet.*, **33**, 628-32.

ROTH, E. S. 1965. Temperature and water content as factors in desert weathering. *J. Geol.*, **73**, 454-68.

RUDBERG, S. 1963. Geomorphological processes in a cold semi-arid region. *Axel Heiberg Island Research Reports* (McGill University), 139-50.

RUBEY, W. W. 1933. The size distribution of heavy minerals within a water laid sandstone. *J. Sed. Pet.*, **3**, 3-29.

RUBEY, W. W. 1951. Geological history of sea water. *Bull. Geol. Soc. Amer.*, **62**, 1111-48.

RUHE, R. V. 1965. Relation of fluctuations of sea level to soil genesis in the Quaternary. *Soil Sci.*, **99**, 23-9.

RUXTON, B. P. 1966. The measurement of denudation rates. *Inst. Aust. Geog. 5th meeting.* Sydney.

RUXTON, B. P., and BERRY, L. 1957. The weathering of granite and associated erosional features in Hong Kong. *Bull. Geol. Soc. Amer.*, **68**, 1263-92.

RUXTON, B. P., and BERRY L. 1961a. Notes on faceted slopes, rock fans and domes on granite in east central Sudan. *Amer. J. Sci.*, **259**, 194-205.

RUXTON, B. P., and BERRY, L. 1961b. Weathering profiles and geomorphic position on granite in two tropical regions. *Rev. Géomorph. Dynamique*, **12**, 16-31.

ST ARNAUD, R. J., and WHITESIDE, E. P. 1963. Physical breakdown in relation to soil development. *J. Soil Sci.*, **14**, 267-81.

SALISBURY, E. J. 1925. Note on the edaphic succession in some dune soils with special reference to the time factor. *J. Ecol.*, **13**, 322-8.

SAXENA, S. K. 1966. Evolution of zircons in sedimentary and metamorphic rocks. *Sedimentology*, **6**, 1-33.

SCHEFFER, F., MEYER, B., and KALK, E. 1963. Biologische Ursachen der Wüstenlackbildung. *Zeit. f. Geomorph.*, **7**, 112-9.

SCHEIDEGGER, A. E. 1961. *Theoretical geomorphology.* Prentice-Hall, New Jersey.

SCHNATTNER, I. 1961. Weathering phenomena in the crystalline of the Sinai in the light of current notions. *Bull. Res. Council Israel*, **10**, **G**, 247-66.

SCHRECKENTHAL-SCHIMITSCHEK, G. 1935. Der Einfluss des Bodens auf die Vegetation im Moränengelände des Mittelbergferners (Pitzal, Tirol) *Z. Gletscherkunde*, **23**, 57-66.

SCHUMM, S. A., and CHORLEY, R. J. 1966. Talus weathering and scarp recession in the Colorado Plateaus. *Zeit. f. Geomorph.*, **10**, 11-36.

SCHWARZBACH, M. 1963. *Climates of the past: an introduction to paleoclimatology.* Van Nostrand, London.

SCHWARZBACH, M. 1966. Bemerkenswerte Konglomerat-Verwitterung. *Zeit. f. Geomorph.*, **10**, 169-82.

SHARP, K. R., Boucaut, W. R. P., and SVENSON, D. 1963. *Report on the geology of the Khancoban project and associated works.* Snowy Mountains Hydro-Electric Authority.

SHAW, C. F. 1928. Profile development and the relationship of soils in California. *Proc. 1st. Int. Cong. Soil Sci.*, **4**, 291-317.

SIMONETT, D. 1959. The role of rainfall in soil formation on the basalts of north Queensland. *Ann. Assoc. Amer. Geog.*, **49**, 211.

SIMONSON, R. W. 1957. What soils are. *Yearbook of Agriculture*, 1957, 17-31. U.S. Dept. Ag.

SIMPSON, D. R. 1964. Exfoliation of the Upper Pocohontas sandstone Mercer County, West Virginia. *Amer. J. Sci.*, **242**, 545-51.

SIVARAJASINGHAM, S., ALEXANDER, L. T., CADY, J. G., and CLINE, M. G. 1962. Laterite. *Advances in Agronomy.* Academic Press, New York.

SMALLEY, I. J. 1966. Formation of quartz sand. *Nature, Lond.*, **211**, 476-9.

SMITH, D. INGLE 1965. Some aspects of limestone solution in the Bristol region. *Geog. J.*, **131**, 44-9.

SMITH, L. L. 1941. Weather pits in granite of the southern piedmont. *J. Geomorph.*, **4**, 117-27.

SMITHSON, F. 1953. The micro-mineralogy of North Wales soils. *J. Soil Sci.*, **4**, 194-210.

SMYTH, C. H. 1913. The relative solubilities of the chemical constituents of rocks. *J. Geol.*, **21**, 105-20.

SNEATH, P. H., and SOKAL, R. R. 1963. *Principles of numerical taxonomy.* Freeman, San Francisco.

SPARKS, B. W. 1960. *Geomorphology*, Longmans, Green, London.

STARKEY, R. L. 1966. Oxidation and reduction of sulfur compounds. *Soil Sci.*, **101**, 297-306.

STEPANOV, I. N. 1965. Weathering processes in glacial type lithogenesis. *Int. Geol. Rev.*, **7**, 2182-3.

STEPHEN, I. 1952. A study of rock weathering with reference to the soils of the Malvern Hills. *J. Soil Sci.*, **3**, 20-33, 219-37.

STEPHEN, I. 1953. A petrographic study of a tropical black earth and a grey earth from the Gold Coast. *J. Soil Sci.*, **4**, 211-9.

STEPHENS, C. G. 1966. Origin of silcretes of Central Australia. *Nature, Lond.*, **209**, 496.

STEWART, A. J. 1966. The petrography, structure, and mode of emplacement of the Cobaw granite, Victoria. *Proc. Roy. Soc. Victoria*, **79**, 275-317.

STICHER, H., and BACH, R. 1966. Fundamentals in the chemical weathering of silicates. *Soils and Fertilizers*, **29**, 321-5.

STRAKHOV, N. M. 1967. *Principles of lithogenesis.* Vol. 1. Trans. J. P. Fitzimmons. Oliver and Boyd, Edinburgh.

SWEETING, M. M. 1950. Erosion cycles and limestone caverns in the Ingleborough district. *Geog. J.*, **115**, 63-78.

SWEETING, M. M. 1960. The caves of the Buchan area, Victoria. *Zeit. f. Geomorph. Supp.*, **2**, 81-91.

SWEETING, M. M. 1966. The weathering of limestones. In *Essays in Geomorphology*, Ed. G. H. Dury. Heinemann, London.

TABER, S. 1930. The mechanics of frost heaving. *J. Geol.*, **38**, 303-17.

TALSMA, T. 1963. The control of saline groundwater. *Meddelingen van de Landbouwhogeschool te Wageningen*, **63**, 1-68.

TAMM, O. 1920. Bodenstudien in der Nordschwedischen Nadelwaldregion. *Medd. Statens Skogsförsöksandstalt*, **17**, 49-300.

TAMM, O. 1932. Der braune Waldboden in Schweden. *Proc. 2nd. Int. Cong. Soil. Sci.*, **5**, 178-89.

TANNER, W. F. 1958. The zig-zag nature of type I and type IV curves. *J. Sed. Pet.*, **28**, 372-5.

TARR, R. S. 1915. A study of some heating tests, and the light they throw on the cause of disaggregation of granite. *Econ. Geol.*, **10**, 348-67.

TEODOROVICH, G. I. 1961. *Authigenic minerals in sedimentary rocks.* Consultants Bureau, New York.

TERJESEN, S. G. ERGA, O., THORSEN, G., and VE, A. 1961. On effects of trace elements in limestone solution. *Chem. Eng. Sci.* **14**, 227.

THIEL, G. A. 1940. The relative resistance to abrasion of mineral grains of sand size. *J. Sed. Pet.*, **10**, 103-24.

THOMAS, H. H. 1909. Detrital andalusite in Tertiary and Post-Tertiary sands. *Mineral. Mag.*, **15**, 241-4.

THOMAS, M. F. 1965. Some aspects of the geomorphology of tors and domes in Nigeria. *Zeit. f. Geomorph.*, **9**, 63-81.

THOMAS, M. F. 1966. Some geomorphological implications of deep weathering patterns in crystalline rocks in Nigeria. *Trans. Inst. Brit. Geog.*, **40**, 173-93.

THOMAS, T. M. 1954. Solutional subsidence outliers of Millstone Grit on the Carboniferous Limestone of the North Crop of the South Wales Coalfield. *Geol. Mag.*, **91**, 220-6.

THOMAS, T. M. 1963. Solution subsidence in south-east Carmarthenshire and south-west Breconshire. *Trans. Inst. Brit. Geog.*, **33**, 45-60.

THOMASSON, A. J. 1965. Review of books by Berg and Bespalov. *Geog. J.*, **131**, 414-5.

THOMASSON, A. J., and AVERY, B. W. 1963. The soils of Hertfordshire. *Trans. Hertfordshire Nat. Hist. Soc.*, **25**, 247-63.

THOMPSON, J. G. 1957. Granite soils in Southern Rhodesia. *Rhod. Agric. J.*, **54**, 121-8.

THORNTHWAITE, C. W. 1931. The climates of North America. *Geog. Rev.*, **21**, 633-54.

THORP, J. 1931. The effects of vegetation and climate upon soil profiles in northern and northwestern Wyoming. *Soil Sci.*, **32**, 283-301.

THOULET, J. 1913. Notes de lithologie sousmarine. *Ann. Inst. Oceanogr.*, **5**, fasc. 9.

TILLER, K. G. 1963. Weathering and soil formation on dolerite in Tasmania. *Aust. J. Soil Sci.*, **1**, 74-90.

TRENDALL, A. F. 1962. The formation of 'apparent peneplains' by a process of combined lateritization and surface wash. *Zeit. f. Geomorph.*, **6**, 183-97.

TRICART, J. 1960. Experiences de desagregation de roches granitiques par la cristallisation du sel marin. *Zeit. f. Geomorph. Supp.* **1**, 239-40.

TRICART, J., and CAILLEUX, A. 1962. *Le modêle glaciaire et nival.* Centre de Documentation Universitaire, Paris.

TWIDALE, C. R. 1962. Steepened margins of inselbergs from north-western Eyre Peninsula, South Australia. *Zeit. f. Geomorph.*, **6**, 52-69.

TWIDALE, C. R. 1964. A contribution to the study of domed inselbergs. *Trans. Inst. Brit. Geog.*, **34**, 91-113.

TWIDALE, C. R., and CORBIN, E. M. 1963. Gnammas. *Rev. de Géomorph. Dynamique*, **14**, 1-20.

TYRREL, G. W. 1926. *The principles of petrology.* Methuen, London.

U.S. DEPT. AGRIC. 1960. *Soil Classification. A comprehensive system. 7th Approximation.*

UZIELLI, G. 1875. Sopra la zircone della Costa Terrena. *Atti R. Accad. Lincei.*, **3**, 862-877.

VAGELER, P. 1930. Grundriss der tropischen und subtropischen Bodenkunde. *Verlag. f. Ackerbau. Berlin.*

VAN DER MERWE, C. R., and WEBER, H. W. 1963. The clay minerals of South African soils developed from granite under different climatic conditions. *S. Afr. J. Agric. Sci.*, **6**, 411-54.

VERSTAPPEN, H. T. 1960. Some observations on karst development in the Malay Archipelago. *J. Tropical Geog.*, **14**, 1-10.

VISHER, S. S. 1945. Climatic maps of geological interest. *Bull. Geol. Soc. Amer.*, **56**, 713-36.

VOUTE, C. 1963. Some geological aspects of the conservation project for the Philae temples in the Aswan area. *Geol. Rundschau*, **52**, 665-75.

WAGNER, P. A. 1913. Negative spheroidal weathering and jointing in a granite of southern Rhodesia. *Trans. Geol. Soc. S. Africa*, **15**, 155-64.

WALKER, E. H. 1963. Relative rates of erosion under grass and forest in a valley of Western Wyoming. *Northwest Sci.*, **37**, 104-11.

WARD, W. T. 1952. The tors of central Otago. *N.Z. Journ. Sci. Tech.*, B**33**, 191-200.

WARTH, H. 1895. The quarrying of granite in India. *Nature, Lond.*, **51**, 272.

WATANABE, T., YAMASAKI, M., KOJIMA, G., NAGAOKA, S., and HIRAYAMA, K. 1954. Geological study of damages caused by atomic bombs in Hiroshima and Nagasaki. *Japanese J. Geol. Geog.*, **24**, 161-70.

WATSON, E. 1966. Two nivation cirques near Aberystwyth, Wales. *Biul. Peryglac.*, **15**, 79-101.

WATSON, E., and WATSON, S. 1967. The periglacial origin of the drifts at Morfa-Bychan, near Aberystwyth. *Geol. J.*, **5**, 419-40.

WATSON, J. P. 1965. A soil catena on granite. *J. Soil Sci.*, **15**, 238-57; **16**, 158-70.

WAYLAND, E. J. 1933. Peneplains and some other erosional platforms. *Ann. Rep. Bull. Uganda Geol. Surv. Dept.*, 77-9.

WEBLEY, D. M., HENDERSON, M. E. K., and TAYLOR, I. F. 1963. The microbiology of rocks and weathered stones. *J. Soil Sci.*, **14**, 102-12.

WEBSTER, R. 1960. Soil genesis and classification in Central Africa. *Soils and Fertilizers*, **23**, 77-9.

WEINERT, H. H. 1961. Climate and weathered Karroo dolerites. *Nature, Lond.*, **191**, 325-29.

WEINERT, H. H. 1965. Climatic factors affecting the weathering of igneous rocks. *Agric. Meteorol.*, **2**, 27-42.

WELLMAN, H. W., and WILSON, A. T. 1965. Salt weathering, a neglected erosive agent in coastal and arid environments. *Nature, Lond.*, **205**, 1097-1098.

WELLS, N. 1959. Soil sequences from basalt and greywacke. *DSIR (N.Z.) Soil Bur. Pub.*, **197**.

WENTWORTH, C. K. 1922. A scale of grade and class terms for clastic sediments. *J. Geol.*, **30**, 377-92.

WENTWORTH, C. K. 1938. Marine bench-forming processes: water level weathering. *J. Geomorph.*, **1**, 6-32.

WILFORD, C. E. 1966. 'Bell holes' in Sarawak Caves. *Bull. Nat. Spel. Soc.*, **28**, 179-82.

WILHELMY, H. 1958. *Klimamorphologie der Massengesteine.* Westermann, Braunschweig.

WILLDEN, R., and MABEY, D. R. 1961. Giant desiccation features on the Black Rock and Smoke Creek Deserts, Nevada. *Science*, **133**, 1359-60.

WILLIAMS, H., TURNER, F. J., and GILBERT, C. M. 1954. *Petrography.* Freeman, San Francisco.

WILLIS, B. 1936. East African plateaus and rift valleys. *Carnegie Inst. Wash. Publ.*, **470**.

WOOD, A. 1942. The development of hillside slopes. *Proc. Geol. Assoc. London.*, **53**, 128-39.

WOOLNOUGH, W. G. 1927. The duricrust of Australia. *J. Proc. Roy. Soc. N.S.W.*, **61**, 25-53.

WRIGHT, R. L. 1963. Deep weathering and erosion surfaces in the Daly River basin, Northern Territory. *J. Geol. Soc. Aust.*, **10**, 151-64.

YARDLEY, D. H. 1951. Frost-thrusting in the Northwest Territories. *J. Geol.*, **59**, 65-9.

YARILOVA, E. A. 1950. The transformation of the minerals of syenite in the first stages of soil formation. *Trudy. Pochv. Inst. Dokuchaeva*, **34**, 110-42.

YEHLE, L. A. 1954. Soil tongues and their confusion with certain indicators of periglacial climate. *Amer. J. Sci.*, **252**, 532-46.

YOUNG, A. 1963. Some field observations of slope form and regolith, and their relation to slope development. *Trans. Inst. Brit. Geog.*, **32**, 1-29.

INDEX

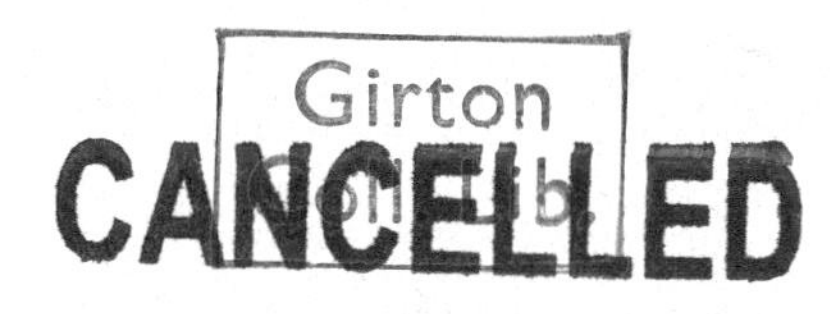
Girton
CANCELLED